AF650246

Ecologically Based Integrated Pest Management

PART-II

www.nipaers.com
Browse, Search, Read & Buy...
Print Books eChapters
eBooks Publishers
Forthcoming Subject Catalogues
New Titles
Online Resources on:
✓ Current Affairs
✓ Reasoning, Logic & Aptitude
✓ Competitive Examinations
✓ English Language Lab
✓ Writing & Pronunciation Tools
✓ Personality Development
✓ Online Programmes for Professionals
✓ Interview Preparation
Pay Using
PhonePe Google Pay Paytm PayPal UPI CCAvenue

Ecologically Based Integrated Pest Management

PART-II

Edited by :

D.P. ABROL

Professor and Head

Division of Entomology, Faculty of Agriculture, Main Campus Chatha

Sher-e-Kashmir University of Agril. Sciences & Technology

Jammu -180 009 (J&K)

and

UMA SHANKAR

Division of Entomology, Faculty of Agriculture, Main Campus Chatha

Sher-e-Kashmir University of Agril. Sciences & Technology

Jammu -180 009 (J&K)

New India Publishing Agency

Pitam Pura, New Delhi-110 088

Published by
Sumit Pal Jain *for*
New India Publishing Agency
101, Vikas Surya Plaza, CU Block, L.S.C. Mkt.,
Pitam Pura, New Delhi- 110 088, (India)
Phone: 011-27341717, Fax: 011-27341616
E-mail: info@nipabooks.com
Web: www.nipabooks.com

© Editors, **2012**

ISBN : 978-93-80235-95-0

All rights reserved, no part of this publication may be reproduced, stored in a retrieval system or transmitted in any form or by any means, electronic, mechanical, photocopying, recording or otherwise without the prior written permission of the publisher/authors.

This book contains information obtained from authentic and highly regarded sources. Reasonable efforts have been made to publish reliable data and information, but the authors/editor(s)/contributors and publisher cannot assume responsibility for the validity of all materials or the consequences of their use. The authors/editor(s)/contributors and publisher have attempted to trace and acknowledge the copyright holders of all materials reproduced in this publication and apologize to copyright holders if permission and acknowledgements to publish in this form have not been given. If any copyright material has not been acknowledged please write and let us know so we may rectify it.

Contents

Part-II

Section
IV

BREEDING FOR RESISTANCE

Ecologically Based Integrated Pest Management, 2012
© D.P. Abrol and Uma Shankar (eds.), pp. 493-501
New India Publishing Agency, New Delhi (India)
e-mail : info@nipabooks.com; website : www.nipabooks.com

25

Breeding for Resistance to Major Insect-Pests of Cereals

B.B. GUPTA AND R.K. SALGOTRA

Introduction

Worldwide damage due to insect pest attack accounts for 15 per cent crop losses, despite the use of insecticides which represents over US $100 billion (Krattiger, 1997). The annual cost of insect control itself amounts to US $8 billion, thus warranting immediate and economical control measures. The crop varieties developed in the past 30 years were high yielders, but unfortunately had poor storage characteristics and were highly susceptible to pest damage. Insect-pests are capable of evolving to biotypes that can adapt to new situations, for instance they overcome the effect of toxic materials or bypass natural or artificial plant resistance, which further confounds the problem (Roush and McKenzie, 1987). Under these circumstances, provision of food to the rapidly expanding population has always been a challenge facing mankind. This problem is more acute in the tropics and sub-tropics, where the climate provides a highly conducive environment for a wide range of insects and necessitates massive efforts to suppress the population densities of different pests in order to achieve an adequate supply of food. In developing countries, the problem of competition from insect-pests is further complicated with a rapid annual increase in human population (2.5-3.0 percentage) against a meager 1.0 per cent increase in food production. In order to feed the ever expanding population, crop protection plays a vital and integral role in the present day agricultural production to minimize yield losses. Currently, the crop protection practice in agricultural systems relies exclusively on the use of agrochemicals, although a few specific cases do exist, where inherent varietal resistance and biological control have been successfully employed. The exclusive use of chemical pesticides not only results in rapid build-up of resistance to such compounds, but their non-selectivity affects the balance between pests and natural predators, and is generally in favour of pests (Metcalf, 1986). Therefore, an integrated pest management (IPM) programme, comprising a combination of practices including the judicious use of pesticides, crop rotation, field sanitation and above all exploitation of inherently resistant plant varieties would provide the best option (Meiners and Elden, 1978). The last option includes the use of transgenic crops, expressing foreign insecticidal genes which could make a significant contribution to sustainable agriculture and thus could be an integral component of IPM (Boulter, 1993). The production of transgenic crops has seen rapid advances during the last decade with the commercial introduction of *Bt* transgenics, but the major concern with these crops has been the development of resistance by pest and public acceptability. Hence, there has been a need to discover new effective plant genes which would offer resistance/protection against these pests.

Main Insects that Cause Damage to the Crop

Yield of cereal crops lost because of direct insect damage is estimated to cost at least several billion dollars. It is also known that several devastating diseases in different crop plants are caused by virsuses and transmitted by insects. The main insects that cause the damage are lepidopteran, homopteran, dipteran, thysanopteran, coleopteran and hemipteran. They cause varying degrees of damage depending on the stage of growth of the crop and the environmental conditions relevant for that growth.

The presence of genes with resistance against the insect-pests of above said orders has been found only among the germplasm of different cereal crops. Some progress in developing resistance cultivars against these insects has also been attained, following classical breeding methods.

Classical Breeding for Insect Resistance

The pest management in modern agriculture is heavily dependent upon pesticides, which has become an unbearable drain on the economy of farmers especially in developing countries. Plants and insects have an evolutionary history of co-existence and many insects are expected to have become pests only under practices of intensive agriculture. The development of resistant varieties provides a sound alternative not only to reduce the input cost but to produce food and fibre that is safe for human and animal consumption. During the process of co-evolution of plants and insects the crop plants acquired genetic resistance against commonly encountered insects. But under intensive breeding and cultivation, genetic resistance has usually been lost or bred-out while eliminating some uneconomic and undesirable features of crops. Presence of gossypol in cotton provides protection against many insects but gossypol is poisonous to make cotton-seed oil unsuitable for humans and animals. The glandless cottons are free from this toxic substance but that makes these highly susceptible to boll worms and grey weevils. The inherent potential of plants to survive in the presence of insect populations has been systematically exploited only during recent times. A systematic research on exploitation of insect resistance in crop plants started after the pioneering work by Painter (1951). In spite of considerable similarities in the evolution of pathogens and insects especially in relation to plants, relatively less efforts have been directed to develop pest resistant than disease resistant varieties. A relatively slow rate of multiplication and spread of insects than pathogens makes use of chemicals more effective to control pests than diseases. But management of host-plant resistance needs to be used singly or in combination with other approaches to reduce the use of pesticides in agricultural production. The genetic resistance provides a cheap method which is environment friendly. Furthermore unlike the chemical and biological control methods the protection of resistant varieties is not influenced and dependent upon vagaries of weather like rains which affects the efficacy of insecticides. Chemical control of insects is not economical and justifiable particularly in cases of low incidence of pests where resistant varieties provide the best solution. Breeding of insect resistant varieties requires knowledge about the genetic control of resistance in host and virulence of pest to adopt appropriate breeding strategies and procedures.

Systematic Breeding Programme for Development of Resistant Varieties

Breeding strategies for insect resistance are similar to those of disease resistance where major as well as minor genes can be exploited in the form of simple or multiple resistance. The polygenic resistance is relatively more stable but it is difficult to incorporate in high yielding varieties. Screening techniques to evaluate plants for resistance usually favour selection of plants with high level of resistance which makes the selection for horizontal resistance ineffective. Such a type of resistance can be incorporated through any form of recurrent selection that is applicable to the crop. Use of molecular markers is expected to aid in the identification of genetic factors underlying horizontal resistance. The marker-aided selection can be very effective to transfer this type of resistance even from the unadapted wild and weedy crop species. The strategies to use major gene resistance include: sequential release of varieties with major genes; pyramiding of major genes; rotation of varieties with major genes; development of varietal mixtures and multilines. Major gene resistance has mostly been exploited for insect resistance but it is known to be prone to quick breakdown of resistance due to evolution of new biotypes of insects. A single gene *'Bph-1'* conferred resistance to brown plant hopper in variety IR 26 of rice but broke down within two years. Another recessive gene for resistance, *'bph-2'*, was deployed in the form of variety IR 36 and still another *'Bph-3'* in IR 56 in quick succession to match new biotypes of pest. But major gene resistance has also been reported to be effective for long time e.g. for resistance to jassids in cotton in Africa (Gallun and Khush, 1980). Identification of diverse sources of resistance on the basis of reaction to a specific biotype is essential to prolong the life of insect resistant varieties. The search for new genes conferring resistance has to be continued since resistance genes become ineffective after some time. Spatial deployment of resistance genes in different zones is not common practice in insect management on account of restricted migration of insect population. Biotechnological approaches hold great promise to produce insect resistant transgenic plants through insecticidal genes especially with tissue specific and temporal effects to minimize selection pressure on insect population.

Mechanisms of Insect Resistance

Painter (1951) classified resistance to insects into three categories:

i) Non-Preference

A plant may have some inherent features that make it undesirable for feeding, colonization, oviposition or shelter by the insect. Such characters may include colour, type of pubescence, leaf angle, odour, taste etc. which act as a deterent for the insect to harbour it for feeding and multiplication. The cotton plant with thick hair on plant body is not preferred by jassids, and soybean without pubescence can be extensively damaged by potato leafhopper. Some authors prefer non-acceptance over non-preference because the latter stipulates choice for the insect to prefer a particular type whereas under forced feeding, as on monocultures, it may even attack other types also.

ii) Antibiosis

The host plant may produce certain chemicals that adversely affect the biology of pest such as reduced size and weight, shortened life span, decreased fecundity and rate of reproduction once the insect invades the host. It is the true form of resistance but exerts maximum selection pressure on the pest population.

iii) Tolerance

It carries the same meaning as for disease reaction i.e. host plant gives a normal performance even in the presence of pest population of a size that would severely damage the susceptible host. Tolerant varieties may minimize damage from pest through high regeneration properties to grow more after the damage. Russel (1978) cited a number of instances like tolerance of sugar beet lines to aphids, tolerance of cereals to green bugs where this phenomenon is operative.

Resistance of plant to insects represents their ability to restrict, retard or overcome pest infestation. The level of resistance depends upon relative success or failure of insect to survive and reproduce on host plant and is expressed as relative damage to plant in qualitative or quantitative terms. Painter (1951) defined plant resistance as "the relative amount of heritable qualities possessed by the plant which influence the ultimate degree of damage done by the insect. In practical agriculture, it represents the ability of a certain variety to produce a larger crop of good quality than do ordinary varieties at the same level of insect population". Panda and Khush (1995) have described a number of tests to determine the probable mechanism of resistance for different crops and pest situations.

Existence of classical 'gene-for-gene' system of genetic control of resistance has been established for brown plant hopper in rice. The genetical control of insect resistance can be ascertained like those of pathogens for the breeding of resistant varieties but requires special attention on the following aspects:

1. A reliable procedure of screening for resistance to an insect pest is essential to manipulate host plant resistance. Screening under natural insect population pressure increases unreliability due to seasonal fluctuations in insect populations. Mass collection of insects from field populations for release on test plants is simplest and ensures adequate insect population. Some trap crops can also be interplanted with the test plants to attract the target insects. But mass rearing of insects under laboratory conditions is most dependable method to get adequate and regular supply of desired type of insect population.
2. Artificial rearing of insects of known biotype is essential to screen for resistance but it is time consuming, costly and has the risk of reducing genetic diversity in pest population. Insect population reared in laboratory for several generations may genetically change from the natural population. Such populations are not only inadequate for the

purpose of screening for resistance but also lead to incorrect genetical information. So field collected insects should be regularly introduced into rearing populations.

3. Proper conditions for artificial pest infestation and multiplication on susceptible plants. Stage of the insect, number of insects, growth stage of plant and environmental conditions must be adequate for uniform infestation of the whole of the host plant materials.
4. Efficient screening techniques and methods to evaluate the level of resistance must be developed and standardized for each crop and pest under consideration.

Genetics of Resistance to Insects

Khush and Brar (1991) reviewed the information on genetics of resistance to insects in most of the important food, feed and fibre crops. A wide range of genetic systems ranging from monogenic control to highly complex polygenic control is operative under different situations. Genetical analysis must be performed for a specific crop and pest situation prevailing in the target environmental conditions. Panda and Khush (1995) have described various aspects of the genetic analysis of resistance which include:

i) Plant Material

The plant material to be used as parents should be purified for reaction to insect and for morphological characters. The lines to be used should be uniform for resistance or susceptibility and the plants to be evaluated for reaction to insects should be free from diseases and nutritional deficiencies.

ii) Insect Population

The insect population should be of known genetic constitution multiplied on a standard susceptible variety preferably from a single-pair mating of insects of known virulence.

iii) Scoring for Resistance

The screening of host populations for resistance can be in the form of damage to host plant or effect on insect population. The reaction of the plant to the insect is preferred over the reaction of insect to plant in scoring for resistance. The individual plants or progenies are evaluated on a suitable scale as dictated by requirements of host damage.

iv) Inheritance of Resistance

The segregating populations are developed from crosses of resistant and susceptible varieties. Use of highly susceptible variety as a parent minimizes the chances of modifier genes that may influence the segregation ratios. Observations on plant response in F_1 and F_2 reveals the dominance/ recessive nature of resistance and the probable number of genes involved in resistance. Segregation ratios observed in F_2 should be verified through behaviour of F_3 progenies. The resistance to a particular insect in different crosses may be caused by same or different genes which requires a test of allelism. Panda and Khush (1995) elaborated very schematic procedure, along with specific illustrative examples, to test for new alleles particularly with respect to reaction to specific biotypes of the insect.

Breeding Methods for Insect Resistance

The choice of breeding method for insect resistance depends upon the reproductive system of the crop and nature of genetical control of resistance to insects. The most commonly used breeding methods for self-pollinated crops are pure-line selection, mass selection or hybridization followed by pedigree method, bulk method, single seed descent and backcross method.

Recurrent selection methods and hybrid breeding are used in cross pollinated crops. As an instance, a part of the seed of selected plants in pure line selection is used to test for insect resistance in laboratory or greenhouse while the remaining seed is grown in progeny rows for selection as in standard pure line selection for improvement of yield.

Hybridization is the most commonly used method where an otherwise desirable but susceptible lines are crossed with an insect resistant variety. Segregating generations are evaluated mostly by the pedigree method. A part of the seed of F_2 plants is used to screen for resistance to insects under artificial conditions whereas remaining seed of each plant is grown to evaluate F_3 progeny rows for yield and other desirable characters. The selection among F_3 progenies is based on insect resistance along with other characters considered in routine pedigree method for yield improvement.

Single plant selections are made within selected F_3 progenies to repeat the process in F_4 and F_5 generations. In most of the situations the reaction of plants to insect is available before maturity and thus can be used for final selection at maturity. Selection for insect resistance can also be performed through its association, if any, with morphological markers e.g. high hair density on plant body in cotton is a strong indication of resistance to jassids. Molecular markers can also be used provided the resistance gene is tagged and sufficient facilities are available to screen large number of plants usually evaluated in such populations.

The bulk method can be used where a high population of target insect can be released under natural conditions. In contrast to a random bulk of whole seed lot, the seed of only resistant plants is harvested and bulked to raise the next generation. Bulk method does not find wider application not only for insect and disease resistance but even for yield improvement. Similarly single seed descent can be used for rapid advancement of generations with minimum space. Single plant selections in F_5 or F_6 are evaluated both for insect resistance in screen house and for yield performance under field conditions to identify the uniform, high yielding and insect resistant progenies to be bulked. Since the resistant plants have to be identified and crossed in the same season so screening for resistance has to be conducted in field or preferably in screening-cum-crossing block under green house where selected plants can further be backcrossed to recurrent parent in the same crop season. Insect resistance controlled by major genes in cross-pollinated crops can easily be incorporated into hybrids provided one of the parents carries dominant gene for resistance. Several hybrids of rice with resistance to green leafhopper have been developed in China where susceptible male sterile lines were crossed with the resistant parents like IR 26, IR 54 and IR 8761-19-1 (Panda and Khush, 1995). The recurrent selection is the commonly used approach to incorporate polygenes for horizontal resistance. Open-pollinated populations are challenged with high insect population and resistant plants are identified to be intercrossed. The resistant plants can be selfed to lock-up the genes for resistance. The selfed seed of all resistant plants is bulked and grown next year for intercrossing. Such an intermating population should be exposed to high insect population to identify and remove highly susceptible plants so that only resistant plants intercross with each. But it would require two years to complete one cycle of recurrent selection. Such a type of cyclic selection can be repeated to raise level of insect resistance in the population of cross-pollinated crops.

A Transgenic Approach for Developing Insect Resistant Varieties

It is satisfying to realize that we have advanced into the expansion phase of crop biotechnology, in which genetically modified (GM) plants are playing an important role in crop improvement. The future security of the food supply will depend on science providing the tools to allow efficient agricultural production that is sustainable in every sense. Amazing technical innovations have resulted in plant transformations coming into the market place. Transgenic insect resistant plants have a track record of success that will become progressively more difficult for opponents of genetic engineering to ignore. Field and greenhouse tests of *Bt* crops produced by public institutions have demonstrated the effectiveness of *Bt* technology in controlling pests in China (High *et al.*, 2004), in India (Khanna and Raina, 2002), and Pakistan (Bashir *et al.*, 2004). *Bt* rice effectively controls lepidopteran pests without inducing the emergence of other rice pests. As a component of integrated pest management (IPM), *Bt* rice has the potential to increase yields and greatly reduce insecticide use in Asia (Toenniessen *et al.*, 2003).

Although there has been no evidence of detrimental effects, the potential of *Bt* crops to provide pest resistance and bring about indirect damage to non target species calls for caution in how we deploy transgenic plants expressing insecticidal genes. Thus, rational and innovative strategies need to be explored to test different hypothesis and arrive at strategies need to be explored to test different hypothesis and arrive at strategies that provide an overall balance of cost versus benefit.

References

Bashir, K., Husnain, T., Fatima, T., Latif, Z., Mehdi, S. A. and Riazuddin, S. 2004. Field evaluation and risk assessment of transgenic indica basmati rice. *Mol. Breed.* 13: 3001-3012.

Boulter, D. 1993. Insect-pest control by copying nature using genetically engineered crops. *Biochemistry,* 34: 1453-1466.

Gallun, R.L. and Khush, G.S. 1980. Genetic factors affecting expression and stability of resistance. In: F.G. Maxwell and R.P. Iennings. *Breeding Plants Resistant to Insects (eds).* John Wiley and Sons, New York.

High, S.M., Chhen, M.B., Shu, Q.Y. and Altosaar, I. 2004. Achieving successful deployment of *Bt* rice. *Trends Plant Sci.* 9: 286-292.

Khanna, H. and Raina, S. 2002. Elite indica transgenic rice plants expressing modified *Cry1Ac* endotoxin of *Bacillus thuringiensis* show enhanced resistance to yellow stem borer (*Scripophaga incertulas*). *Transgenic Res.* 11:411-423.

Khush, G.S. and Brar, D.S. 1991. Genetics of resistance to insects in crop plants. *Adv. Agron.* 45: 223-274.

Krattiger, A.F. 1997. Insect resistance in crops: a case study of *Bacillus thuringiensis* (*B*t) and its transfer to developing countries. *ISAAA Briefs,* 2: 42.

Meiners, J.P. and Elden, T.C. 1978. Resistance to insects and diseases in *Phaseolus.* In: *Advances in legume science.* International Legume Conference, (1978, Kew Surrey, UK). Summerfield, R.S and Bunting, A.H. eds., pp. 359-364.

Metcalf, R.L. 1986. The ecology of insecticides and the chemical control of insects. In: Kogan, M. ed. *Ecological Theory and Integrated Pest Management.* New York, John Wiley and Sons, pp. 251-297.

Painter, R.H. 1951. *Insect Resistance in Crop Plants.* MacMillan Co., New York.

Panda, N. and Khush, G.S. 1995. *Host Plant Resistance to Insects.* CAB International, Wallingford, UK.

Roush, R.T. and Mckenzie, J.A. 1987. Ecological and genetics of insecticide and araricide resistance. *Annual Review of Entomology,* 32: 361-381.

Russell, R.R. 1978. *Plant Breeding for Pest and Disease Resistance.* Butterworths, Lond. Boston.

Toenniessen, G.H., O'Toole, T.C. and De Vries, J. 2003. Advances in Plant Biotechnology and its adoption in developing countries. *Curr. Opin. Plant Biol.* 6: 191-198.

□□□

Ecologically Based Integrated Pest Management, 2012
© D.P. Abrol and Uma Shankar (eds.), pp. 503-518
New India Publishing Agency, New Delhi (India)
e-mail : info@nipabooks.com; website : www.nipabooks.com

26

Vegetable Improvement through Transgenics

SATESH KUMAR AND SHEETAL DOGRA

Introduction

Biotechnology, in the recent years, has created unprecedented opportunities not only for the manipulation of biological systems for the benefit of mankind, but also to understand studies for better understanding of the fundamental life processes. Consequently, it has become the fastest and most rapidly growing technology in the world. As per US National Science Foundation, *"Biotechnology is defined as any technique that uses living organisms or parts of organisms to make or modify products, to improve plants, amend animals or to develop microorganisms for specific uses"*. Biotechnology offers new strategies that can be used to develop transgenic crop plants with improved growth, yield and quality and tolerance to biotic and abiotic stresses. By transgenic crop, we mean any plant containing a gene(s) which have been artificially inserted instead of the plant acquiring them through pollination. A plant in which a foreign gene has been transferred through genetic engineering is called transgenic plant and the gene so transferred is called transgenic. The inserted gene sequence known as transgene may come from another unrelated plant or from completely different species. Plants containing transgene are often called genetically modified or GM crops. Transgenic technologies have opened up many exciting possibilities to improve products with added value having application in food, agriculture, animal husbandry, environment, medicine and industry. It also offers uncommon opportunities for improvement in genetic potential of plants and animals by introduction or removal of gene(s) that regulate a specific trait.

Importance of Transgenics

Transgenic technology enables plant breeders to bring together, in one plant useful genes not just from within the crop species or from closely related plants but from a wide range of living sources. This technology provides the means for identifying and isolating genes controlling specific characteristics in one kind of organism, and for moving copies of those genes into another quite different organism, which will then also have those characteristics. This powerful tool enables plant breeders to do what they have always done not only to generate more useful and productive crop varieties containing new combinations of genes but also expands the possibilities beyond the limitations imposed by traditional cross-pollination and selection techniques. Transgenic plants have been developed for a variety of reasons like longer shelf life, disease and pest resistance etc. The first transgenic crop approved for sale in the US in 1994 was the FlavrSavr tomato, which was intended to have a longer shelf life. An emerging class of transgenic plant, sometimes known as pharmacrops, aims to use plants to manufacture other products such as pharmaceuticals and industrial chemicals.

Present Status

China was the first country to commercialize transgenics in the early nineties with the introduction of virus resistant tobacco which was later followed by virus resistant tomato (James, 1998). Subsequently, the global area planted with transgenics crops has risen from 1.7 million hectares in 1996 to 90 million hectares in 2005 (James, 2005). In 2006, this figure touched 100 million hectare mark. Out of the total area of the world under transgenic crops, three countries namely USA, Argentina and Canada cover more than 90 per cent. The major crops occupying about more than 80 per cent area are herbicide-tolerant soybean, Bt corn, herbicide tolerant canola and herbicide tolerant cotton. Several transgenic cultivars of major food crops, such as soybean, maize, canola, potato and papaya have been commercially released incorporating genes for resistance to herbicides, insects and viruses. In India, three varieties in Bt cotton namely Mach 12, Mach 162 and Mach 184 and Bt brinjal have been released for commercial cultivation.

Transgenic Crops under Commercial Production Worldwide

According to Manjunath 2005, 04 field crops, 03 vegetables, 01 fruit crop and 01 tobacco crop are presently produced commercially across the globe as depicted in Table 1.

Table 1. Transgenic crops under commercial production

Type of the crop	Name of the crop	Name of the country
Vegetables	Tomato & Pepper	China
	Squash	USA
Field crops	Soybean	Canada, USA, and Argentina South Africa, Brazil, East Europe, Uruguay, Paraguay and Chile
	Canola	Canada and USA
	Cotton	Argentina, Mexico, China, South Africa, India and Columbia
	Maize	Canada, USA, Western Europe, Argentina, South Africa, Uruguay, Philippines and Chile
Fruits	Papaya	Hawaii (USA)
Other crops	Tobacco	USA

Till date, no transgenic crop variety tolerant to biotic stress has so far been reported to be released for cultivation. However, 7 transgenic varieties exhibiting tolerance to a range of biotic stresses underwent field testing in Bolivia (frost tolerant potato variety), China (cold tolerant tomato), Egypt (salt tolerant wheat variety) and India (moisture stress tolerant variety) and Thailand (salt tolerant and drought tolerant rice varieties). Most of the R&D activities on development of abiotic stress tolerant crops are being carried out in six countries of the Asian region namely Bangladesh, China, India, Indonesia, Pakistan and Thailand. In India, genetically altered seed has not been commercialized as the transgenic varieties developed are undergoing field trials and evaluation. The private sector is emerging fast in this regard and most of the Indian companies have now collaborations with multinationals like Proagro-PGS, Rallis India Ltd etc. (Table 2)

Major Components for the Development of Transgenic Plants

Despite significant advances over the past decade, development of efficient transformation methods can take many years. Effective regeneration and transformation systems are the prerequisite for successful genetic transformation (Sharma *et al.*, 2003). Genetic transformation technology relies on the conceptual framework and the technical approaches of plant tissue culture and molecular biology to develop commercial process and products which comprises mainly the nine major components:

Table 2 : Vegetable transgenic developmental activities in India

Institution	Vegetable crop	Transgenic	Target
Public sector			
CPRI Shimla	Potato	Bt cry 1Ab	Develop plant resistant to potato tuber moth
		Osmotin	Develop water stress to lerant plants
		Coat protein	Develop plants resistant to potato virus Y (PVY)
Delhi University, Delhi	Tomato	Ctx-B and Tcp antigens of *Vibrio chlorea*	Edible vaccine development
	Brinjal	Chitinase, glucanase and thaumatin encoding genes	Create disease resistant plants
	Brinjal	Cry 1Ab	Generate lapidopteran resistant plants
IARI, New Delhi	Tomato	Cry 1Ab	Generate lapidopteran resistant plants
	Cauliflower	Cry 1Ab	Generate diamond back moth resistant plants
	Cabbage	Cry 1Ab	Generate diamond back moth resistant plants
	Tomato	ACC synthase	Control fruit ripening
		Replicase gene	Develop leaf curl virus resistant plants
IIHR, Bangalore	Muskmelon	Rabies glycoprotein	Double edible vaccine
	Tomato	Leaf curl virus	Develop leaf curl virus resistant plants
JNU, ND	Potato	Ama 1	Develop protein rich potatoes.
Private Sector			
IAHS, Bangalore	Tomato	Alfafa glucanase and tomato leaf curl virus genes	Develop plants resistant to viral and fungal diseases
Proagro-PGS (India) Ltd., Gurgaon	Tomato	Cry 1Ab	Generate lepidopteron resistant plants
	Brinjal	Cry 1Ab	Generate lepidopteron resistant plants
	Cauliflower	Cry 1H/Cry 9C	Generate lepidopteron resistant plants
		Bar, Barnase, Barstar	Superior hybrid culture
	Cabbage	Cry 1H/Cry 9C	Generate lepidopteron resistant plants

Source : Kalia, 2004

- Development of reliable tissue culture regeneration systems.
- Preparation of gene construct and transformation with suitable vectors.
- Efficient techniques of transformation for the introduction of genes into crop plants.
- Recovery and multiplication of transgenic plants.
- Molecular and genetic characterization of transgenic plants for stable and efficient gene expression.
- Transfer of genes to elite cultivars by conventional breeding methods, if required.
- Evaluation of transgenic plants for their effectiveness in alleviating the biotic or abiotic stresses and field performance.
- Bio-safety assessment including food and environmental safety.
- Commercialization of genetically engineered crop.

Methodology of Transgenic Plant

Introduction to DNA

DNA contains genes that encode necessary information for assembly of specific proteins via mRNA.These proteins then function as enzymes to catalyze biochemical reactions, or as structural or storage units of a cell, to contribute to expression of a plant trait. (The general sequence of events by which the information encoded in DNA is expressed in the form of proteins via an mRNA intermediary is shown in the diagram below).

Transcription Translation

DNA -------------------> mRNA -----------------> Protein ------------> Trait

This mechanism is same in every species either similar or dissimilar e.g., a DNA segment from bacteria can be interpreted and translated into a functional protein when inserted into a plant. Enzymes which recognizes and cuts the DNA at a specific region is restriction enzyme (EcoR1) and other enzymes known as ligases, join the ends of two DNA fragments.

Identification and Location of Genes for Plant Traits

This is the most crucial part in the transgenic process and usually identifying a single gene involved with a trait is not sufficient. How the gene is regulated, how it interacts with other genes active in the same biochemical pathway and what other effects it might have on the plant are to be understood.

Designing Genes for Insertion

Once a gene has been isolated and cloned (amplified in a bacterial vector), it must undergo several modifications before it can be effectively inserted into a plant e.g., it should be fitted with a constitutive promoter (causing gene expression throughout the life cycle of the plant in most tissues) like CaMV35S, from the cauliflower mosaic virus, for high degree of expression in plants and also a selected marker for identification purpose.

Transforming Plants through Gene Transfer

Transformation is the heritable change in a cell or organism brought about by the uptake and establishment of introduced DNA. There are four main methods of transforming plant cells and tissues:

- *The "Gene Gun" method* (also known as micro projectile bombardment or biolistics): This technique employs a system which accelerates micro projectiles coated with DNA for transformation of plants. In this procedure, small metal (gold or tungsten) particles (<4 milli micron diameter) coated with DNA are accelerated and allowed to hit the target plant tissue at a high velocity. In some plant cells, the DNA thus introduced gets incorporated into the nuclear or organelle genome.
- *The Agro bacterium mediated gene transfer*: In general, the agro-bacterium method is considered preferable to the gene gun, because of the greater frequency of single-site insertions of the foreign DNA, making it easier to monitor. *Agrobacterium tumefaciens* is a remarkable species of soil-dwelling bacteria that has the ability to infect plant cells with a piece of its DNA through wounds. The exudates produced upon wounding induce vir genes of A. *tumefaciens* which in turn stimulates the transfer of T-DNA into the plant cell where it ultimately becomes integrated into the plant chromosome.
- *PEG mediated gene transfer* : This method involves treatment of plants with chemicals like PEG (Polyethylene Glycol) that create reversible channels in the plasma membrane. The DNA passively diffuses through these channels into the cytoplasm, where upon a small portion enters the nucleus and integrates into the plant genome.
- *Micro-injection* : This method delivers DNA directly into the cytoplasm or the nucleus of a cell using finely drawn glass capillaries (~1 micron in diameter).

Selection

Following the gene insertion process, plant tissues are transferred to a selective medium containing an antibiotic or herbicide, depending on which selectable marker was used. Only plants expressing the selectable marker gene will survive and possess the transgene of interest and in subsequent steps in the process only these surviving plants will be used.

Regeneration of Whole Plants

To obtain whole plants from transgenic tissues such as immature embryos, they are grown under controlled environmental conditions in a series of media containing nutrients and hormones, a process known as tissue culture. Once whole plants are generated and produce seed, evaluation of the progeny begins.

Risks and Concerns

The introduction of transgenic crops and foods into the existing food production system has generated a number of questions about possible negative consequences. The issues surrounding objections to transgenic crops can be broadly grouped into concerns like :

Concerns about Human Health

- Allergenicity

There is possibility that transgenic food may cause allergic reactions, but no such report has been published yet.

- Antibiotic resistance

The use of antibiotic resistance markers in the development of transgenic crops has raised concerns about whether transgenic foods will play a part in our loss of ability to treat illnesses with antibiotic drugs. At several stages of the laboratory process, developers of transgenic crops use DNA that codes for resistance to certain antibiotics, and this DNA becomes a permanent feature of the final product although it serves no purpose beyond the laboratory stage. Transfer of resistance gene from transgenic food to microorganisms that normally inhabit our stomach and intestine, or to bacteria that we ingest along with food, could help those micro-organisms to survive an oral dose of antibiotic medicine.

- Changed nutrient levels

It is still controversial point to say that there is any qualitywise difference in GM and non GM crops, because there have been only a few studies to date comparing the nutritional quality of genetically modified foods to their unmodified counterparts.

Concerns about Damage to the Environment

- Effect on insects

The suggestion that Bt corn pollen might kill Monarch butterfly larvae galvanized public interest in the effect of transgenic crops on the environment.

- Crop-to-weed gene flow

Hybridization of crops with nearby weeds may enable weeds to acquire traits such as resistance to herbicides. Research results indicate that crop traits may escape from cultivation and persist for many years in wild populations. Genes that provide a competitive edge, such as resistance to viral disease, could benefit weed populations around a crop field. Many cultivated crops have sexually compatible wild relatives with which they hybridize under favorable circumstances. The likelihood that transgenes will spread can be different for each crop in each area of the world.

- Antibiotic resistance

There is also concern that transgenic plants growing in the field will transfer their antibiotic resistance genes to soil micro-organisms, thus causing a general increase in the level of antibiotic resistance in the environment. However, many soil organisms have naturally occurring resistance as a defense against other organisms that generate antibiotics, so genes contributed occasionally by transgenic plants are unlikely to cause a change in the existing level of antibiotic resistance in the environment.

- Leakage of GM proteins into soil

There are concerns that transgenic plants may leak chemical compounds different than conventional plants into the soil through their roots because of their changed DNA. Speculation that this may be happening leads to concern about whether the communities of micro-organisms living near transgenic plants may be affected.

- Low / Poor reduction in pesticide spraying

Pest-resistant crops such as Bt corn and Bt cotton have been promoted as a means to reduce the spraying of pesticides, while herbicide-tolerant crops such as Roundup Ready soybeans are said to reduce the application of herbicides. Only Bt cotton is the crop for which claims of reduced spraying are clear. In other crops, reductions in the spraying of chemicals is not clear.

Concerns about Damage to Current Farming Practices

❐ Crop- to-crop gene flow

Hybridization of transgenic crops with nearby conventional crops raises concerns about isolation distances to ensure purity of crops.

Some achievements in vegetables

Improved quality

Parthenocarpy has been achieved in brinjal and tomato using def H9-γ laaM heterologous gene. Line transformed with this gene give normal sized parthenocarpic fruits even from the emasculated flowers, Table 3 and 4.

Table 3 : Engineered vegetables for parthenocarpic fruits.

Transgenic against	Transgene	Origin of transgene	Transformed plant	Reference
Tomato	Rol B gene	*A. rhizogenes*	Tomato	Carmi *et al.*, 1997
Brinjal	Iaah gene	*P. syringae pv.savas*	Brinjal	Rotino *et al.*, 1997

Table 4 : Commercial release of transgenic vegetable varieties

Crop	Name	Company	Trait	Product status
Tomato	Flavr Savr	Calgene	Quality (Shelf life)	Released 1994
Tomato	Endless summer	DNA Plant Technology	Quality (Shelf life)	Blocked by patent claims
Tomato	-	Zenca	Quality (Paste consistency)	Released in 1995
Tomato, capsicum	-	China	Virus resistant	Released in 1993-94
Squash	Freedom-II	Asgrow	Virus resistant	Released in 1995

Insect Resistance

Bacillus thuringiensis is a bacterium that is pathogenic for a number of insect pests. Its lethal effect is mediated by a toxin protein it produces. Through recombinant DNA methods, the toxin gene can be introduced directly into the genome of the plant where it is expressed and provides protection against insect pests of the plant. The insecticidal protein gene from Bt conferred for reaching resistance to classes of certain lepidopteron species when transgenic tomato, potato, maize and cotton plants have been produced. There are several Bt genes encoding proteins that are largely insecticidal. These proteins are not harmful to mammals but kill insects by interfering with the ion transport in the midgut. Different types of proteins are available that selectively kill specific groups of insects raising the possibility of producing plants resistance to particular pests (Table 5).

Table 5 : Engineered vegetables expressing insect resistance gene.

Transgenic product	Origin of transgene	Target insect	Transformed plants	Reference
Cry 1 AB	*Bacillus thuringiensis*	Lepidoptera	Tomato	Delannay *et al.*, 1989
Cry 1 AB	*Bacillus thuringiensis*	Lepidoptera	Tomato	Kumar *et al.*, 1998
Cry 1 H	*Bacillus thuringiensis*	Lepidoptera	Corn	Duck and Evola, 1997
CpT1 (Cowpea trypsin inhibitor)	Cowpea	Lepidoptera	Tomato	Gatehouse *et al.*, 1992
Tomato proteinase inhibitor I	Tomato	Lepidoptera	Tomato	McGurl *et al.*, 1994
Tomato proteinase inhibitor II	Tomato	Lepidoptera	Tomato	McGurl *et al.*, 1994
Amylase inhibitor of the common bean	Common bean	Lepidoptera	Common bean	Schroeder *et al.*, 1995
Snow drop lectin (GNA)	Common bean	Lepidoptera	Tomato	Gatehouse *et al.*, 1992
Bovine pancreatic trypsin inhibitor	Common bean	Lepidoptera	Lettuce	Schuler *et al.*, 1998

Viral Resistance

Genes that provide resistance against plant viruses have been successfully introduced into such crop plants as tobacco, tomatoes and potatoes (Table 6).

Bacterial Resistance

A novel approach for engineering plants resistant to bacterial diseases has been undertaken to isolate tabtoxin resistance genes (ttr) from bacteria

Pseudomonas syringae pv.*tabaci* and subsequently to introduce into susceptible plants for transgenic resistance (Yoneyama and Anazi, 1991). In a similar way, Phaseolotoxin insensitive OCTase (Ornithine Carbamyl Transferase) gene has been isolated from *Pseudomonas syringae pv. phaseolicola* and introduced to beans to observe total resistance (Herreara Estrella and Simpson, 1995). This simple and useful technology for molecular breeding of disease resistant plants by detoxification of pathogenic toxins has a great to be exploited against bacterial diseases in vegetable crops (Table 7).

Table 6 : Transgenic plants male against various important viruses

Purpose of genetic manipulation	Transgenic product	Transformed plant	Origin of transgene	Reference
Potato virus X (PVX)	Viral coat protein	Potato	PVX	Homenway *et al.*, 1988
Potato virus Y (PVY)	Viral coat protein	Potato	PVY	Stark and Beachy, 1989
Tomato yellow leaf curl virus	Coat protein	Tomato	TYLCV	Kunik *et al.*, 1994
Tomato spotted wilt virus	N-gene	Tomato	TSWV	Pang *et al.*,
Tomato yellow leaf curl virus	Truncated version of CI gene of TYLCV	Tomato	TYLCV	Brunetti *et al.*, 1997
Tomato mosaic virus	Antisens RNA	Tomato	TOMV	Beachy *et al.*, 1990
Cucumber mosaic virus	Satellite RNA	Tomato	CMV	Stommel *et al.*, 1998
Cucumber mosaic virus	Viral coat protein	Cucumber	CMV	Gonsalves *et al.*, 1992

Table 7 : Engineered Vegetables for bacterial resistance.

Resistance against	Transgene product	Origin of transgene	Transformed plant
Erwinia carotovora	Lysozyme	T4 bacteriophage	Potato
Erwinia carotovora	Tachyplesin	Horseshoe crab	Potato
Pseudomonas syringae pv phaseolicola	Phaseolotoxin insensitive OCTase	*Pseudomonas syringae pv phaseolicola*	Bean
Erwinia carotovora	Pestate lyase	*Erwinia carotovora*	Potato
Erwinia carotovora	Glucose oxidase	*Aspergillus niger*	Potato

Source : Singh *et al.*, 2001

Fungal Resistance

One major approach for engineering plants resistant to fungal diseases include use of chitinase genes of plant or bacterial origin. Chitinase enzyme degrades chitin rich walls of fungus and delay symptom development. Other antifungal protein genes include osmotin,glucanase thionin, peroxidase etc.Recently, an effective antifungal protein from barley seeds called ribosome inactivating protein (RIP) has been isolated.RIP acts by inhibiting protein synthesis via glycosidase modification of fungal 28SrDNA. Transgenic crop plants expressing RIP proteins are now being tested against many fungi (Sharma and Kumar, 1995).

Herbicide Resistance

Genes for resistance to some of the herbicides have been introduced into some crop plants and enable them to thrive even when exposed to the weed killer (Table 8).

Table 8 : Mode of action of herbicides and basis of achieving resistance.

Active principle of herbicide	Inhibited pathway	Target product	Use	Basis of resistance
Glyphosate (Round up)	Aromatic amino acid biosynthesis	EPSPS	Broad spectrum	Over expression of EPSPS gene: bacterial ar A gene
Sulphonylurea and Imidazolinones	Branched chain amino acid	ALS	Selected crops	Mutant ALS gene
Phosphinothricin	Glutamine biosynthesis	GS	Broad spectrum	Gene amplification bar gene: detoxification
Photosynthesis inhibitors				
Atrazine (Lasso)	Photosystem II	QB (32K Da protein)	Selected crops	Mutant *psbA* gene:GST gene:detoxifi cation
Bromoxynill (buctril)	Photosynthesis	-	Selected crops	bxn gene:detoxifi cation

Source : Gupta, 2001

Salt Tolerance

Salt tolerant tomato was developed in US which showed highly-expressed sodium/proton anti port pump that sequestered excess sodium in the vacuole of leaf cells. There was no sodium buildup in the fruit. *L.peruvianum* and *L.pennelli* have been found tolerant to sodium chloride. Similarly, salt tolerant lines have also been isolated in capsicum using liquid media 1/2w/v sodium chloride.

"Terminator" Genes

This term is used for transgenes introduced into crop plants to make them produce sterile seeds (and thus force the farmer to buy fresh seeds for the following season rather than saving seeds from the current crop). The process involves introducing three transgenes into the plant. A gene encoding a toxin which is lethal to developing seeds but not to mature seeds or the plant. This gene is normally inactive because of a stretch of DNA inserted between it and its promoter. A gene encoding a recombinase -an enzyme that can remove the spacer in the toxin gene thus allowing to be expressed. A repressor gene whose protein product binds to the promoter of the recombinase thus keeping it inactive. When the seeds are soaked (before their sale) in a solution of tetracycline, synthesis of the repressor is blocked, the recombinase gene becomes active, the spacer is removed from the toxin gene and it can now be turned on. Because the toxin does not harm the growing plant, the crop can be grown normally except that its seeds are sterile which can not be sown in the next year.

Delayed Fruit Ripening

Polygalacturonase is responsible for the degradation of pectin fraction of the fruit cell wall thus plays a role in fruit softening. It has been shown that a single antisense gene could inhibit expression of tomato Pgu gene during ripening. Using the antisense technology, antisense RNA from polygalacturonase has been inserted. When normal PGU mRNA is synthesized the antisense RNA binds to it and prevents synthesis of Pgu. Antisense RNA against ACC synthase (ACC oxidase). It has been observed that when it expressed in tomato plants and this inhibited by ethylene biosynthesis by 99.5% production of which appears to be involved in the initiation, modulation and coordination of expression of many of the genes required for the ripening process (Table 9).

Socio-Economic Concerns

The introduction of transgenic plants into agriculture has been vigorously opposed by some. There are a number of issues that worry the opponents. One of them is the potential risk of transgenes in commercial crops endangering native or non-target species.

Table 9 : Engineered vegetables for post harvest traits.

Origin of transgene	Transgene product	Purpose of genetic manipulation	Transformed plant	Reference
Tomato	Antisense polygalactouranase	Improved shelf life and storage	Tomato	Sheehy *et al.*, 1988
Tomato	Antisense ACC oxidase	Ripening	Tomato	Hamilton *et al.*, 1990
Tomato	Antisense ACC synthase	Ripening	Tomato	Oellar *et al.*, 1991
Tomato	Protein synthase gene	Fruit pigmentation	Tomato	Bird *et al.*, 1991

Examples

- A gene for herbicide resistance e.g. corn, escaping into a weed species could make control of the weed far more difficult.
- The gene for Bt toxin expressed in pollen might endanger pollinators like honeybees.
- To date, field studies on Bt cotton and maize (corn) show that the numbers of some non-target insects are reduced somewhat but not as much as in fields treated with insecticides.
- Another worry is the inadvertent mixing of transgenic crops with no transgenic food crops. Although this has occurred periodically, there is absolutely no evidence of a threat to human health.

Future Developments

- New techniques for producing transgenic plants will improve the efficiency of the process and will help resolve some of the environmental and health concerns. Among the expected changes are the following:
- More efficient transformation
- Better marker genes to replace the use of antibiotic resistance genes.
- Better control of gene expression through more specific promoters so that the inserted gene will be active only when and where needed.
- Transfer of multi-gene DNA fragments to modify more complex traits.

References

Beachy, R.N., Loesh Fries, S. and Tumer, N.E. 1990.Coat protein mediated resistance against virus infection. *Annual Review of Phytopathology* 28:451-74.

Bird, C.R., Ray, J.A., Fletcher, J.D, Boniwell, J.M., Bird, A.S., Teulieres, C., Blain, I., Bramley, P.M. and Schuch, W. 1991. Using antisense RNA to study gene function: inhibition of carotenoid biosynthesis in transgenic tomatoes. *Biotechnology* 14:875-79.

Brunetti, A., Tavazza, M., Noris, E., Tavazza, R., Caciagli, P., Ancora, G., Crespi, S. and Accotto, G.P. 1997. High expression of truncated viral REP protein confers resistance to tomato yellow leaf curl virus in transgenic tomato plants. *Molecular Plant-Microbe Interactions* 10:571-79.

Carmi, Y., Salts, Y., Shabtai, S., Dedicovz, B., Pilowsky, M. and Barg, R. 1997. Transgenic parthenocarpy due to specific over sensitization of the ovary to auxins. Acta *Horticulturae* 447:579-81.

Delannay, X., LaVallee, B.J., Proksh, R.K., Fuchs, R.L., Sims, S.R., Greenplate, J.T., Marrone, P.G., Doson, R.B., Augustine, J.J., Layton, J.G. and Fischoff, D.A. 1989. Field performance of transgenic tomato plants expressing the Baccillus thuringinensis var. Kurstaki insect control protein. *Biotechnology* 7:1265.

Duck, N., and Evola, S. 1997. In: Advances in Insect Control: The role of transgenic plants. Carozzi, N and Kooziel, M (Eds), Taylor and Francis. pp1-20.

Gatehouse, A.M.R, Boulter, D. and Hilder, V.A. 1992. In: Plant Genetic Manipulation for Crop Protection. Gatehouse A.M.R, Boulter, D. and Hilder, V.A.(Eds).CAB International, pp, 155-81.

Golsalves, D., Chee, P., Providenti, R. and Slightom, J.L.1992. Comparison of coat protein mediated and genetically derived resistance in cucumbers to infection by cucumber mosaic under field conditions with natural challenge inoculations by vectors. *Biotechnology* 10:1562-70.

Gupta, P.K. 2001. Elements of biotechnology. Rastogi Publications, Meerut. India.

Hamilton, A.J., Lycett, G.W and Grierson, D. 1990. Antisense gene that inhibits synthesis of the hormone ethylene in transgenic plants. *Nature* 336:284-87.

Herrera Estrella, L. and Simpson, J. 1995. Genetically engineered resistance to bacterial and fungal pathogens. *World Journal of Microbiology and Botany* 11:383-92.

Hemenway, C., Fang, R.X., Kaniewsk, W.K., Chua, N.H. and Turner, N.E. 1988. Analysis of the mechanism of protection in transgenic plants expressing the potato virus X coat protein or its antisense RNA. *EMBO Journal* 7:1273.

James, C. 1998. Movement towards food security accelerates. *Asian Seeds* 5:18-21.

James, C. 2005. Global review of commercialized transgenic crops. *Current Science* 84: 303-309.

Kalia, P. 2004. Fruits of biotechnology in vegetables : a boon to farmers, consumers, processors and environment. *Indian Horticulture* 49:26-34.

Kumar, P.A., Ajin Mandaokar, Sreenivasan Karra, Chakabarti, S.K., Suman, B., Sharma, S.R. Sarvjeet, K. and Sharma, R.P. 1998. Insect resistant transgenic tomato plants. *Molecular Breeding* 4:33-37.

Kunik, T., Salomon, R., Zamir, D., Navot, N., Zeidan, M., Michelson, I., Gafni, Y. and Czosnak, H.1994. Transgenic tomato plants expression in tomato yellow leaf curl virus capsid protein are resistant to the virus. *Biotechnology* 12:500-4.

Manjunath, T.M. 2005. A decade of commercialized transgenic crops. Analysis of their global adoption, safety and benefits. The sixth Dr. S. Pradhan Memorial Lecture delivered at IARI, New Delhi on 23rd March, 2005.

McGurl, B., Orozo-Cardenas, M., Pearce, G. and Ryan, C.A. 1994. Over expression of the pro-eyetemin gene in transgenic tomato plants generates a systemic signal that constitutively induces proteinase inhibitor synthesis. *Proceedings of National Academy Sciences, USA*, 91:9799-9802.

Oeller, P.W., Ming-Wong, L., Teylor, L.P.., Pike, D.A and Theologis, A. 1991. Reversible inhibition of tomato fruit senescence by antisense RNA. *Current Science* 254:437.

Pang, S.Z., Nagpal, P., Wang, M., Slightem, J.L. and Gonsalves, D. 1992. Resistance to heterologous isolates of tomato spotted wilt virus in transgenic tobacco expressing its necleo-capsid protein gene. *Phytopathology* 82:1223-29.

Rotino, G.L., Perri, E., Zottini, M., Sommer and Spena, A.1997. Genetic engineering of parthenocarpic plants. *Nature Biotechnology* 15:1398-1401.

Schroeder, H.E., Gollash, S., Moore, A., Tame, L.M., Craig, S., Hardie, D.C., Chrispeels, M.J., Spencer, D. and Higgins, T.J.M. 1995. Bean alpha amylase inhibitor confers resistance to pea weevil in transgenic peas. *Plant Physiology* 107:1233-39.

Stark, D.M. and Beachy, R.N. 1989.Protection against polyvirus infection in transgenic plants. Evidence from broad spectrum resistance. *Biotechnology* 7:1257-62.

Singh, M., Chakraborty, S., Srivastava Kirti and Kalloo, G. 2001. Biotechnological advances in vegetable crops: In: *Emerging Scenario in Vegetable Research and Development.* Kalloo,G and Singh, Kirti (Eds). Research Periodical and Book Publishing House, pp 58-84.

Sharma, R.P. and Kumar, P.A. 1995. Revolutionizing plant breeding through biotechnology, sustaining crop and animal productivity. The challenge of the decade, Deb, D.L (Ed). Associated Publishing Co,New Delhi,pp. 147-156.

Sharma, M., Charak, K.S. and Ramanaiah, T.V. 2003. Agricultural biotechnological Research in India: Status and Policies. *Current Science* 84:297-365.

Sheehy, R.E., Kramer, Mand Hiatt, W.R. 1988. Reduction of poly galactouranase activity in tomato fruit by antisense RNA. *Proceedings of National Academy Sciences*, USA 85:8805-09.

Stommel, J.R., Tousignnant, M.E., Wai, T.Pasini, R. and Kaper, J.M. 1998. Viral satellite RNA expression in transgenic tomato confers field tolerance to cucumber mosaic virus. *Plant Diseases* 82: 391-96.

Schular, T.H., Popy, G.M., Kerry, B.R. and Denholm, I. 1998. Insect resistant transgenic plants. *Trends in Biotechnology* 16:168-175.

Yoneyama, K. and Anzai, H. 1991. Gene technological study on disease control by the inactivation of pathogenic toxins in transgenic plants. *Journal of Pest Science* 16:291.

❑❑❑

Ecologically Based Integrated Pest Management, 2012
© D.P. Abrol and Uma Shankar (eds.), pp. 519-535
New India Publishing Agency, New Delhi (India)
e-mail : info@nipabooks.com; website : www.nipabooks.com

27

Conventional Breeding Strategies for Insect Resistance in Vegetable Crops

SANJEEV KUMAR, PUJA RATTAN AND JAG PAUL SHARMA

Introduction

Vegetables are preferable host to many insects because of their succulence and suitable nutritional status. All vegetables are attacked but degree of damage and number of pests varies depending upon the crop, variety and climate. Pest infestation in vegetables has been the main bottleneck in their economical production. The problem has more socio-economic dimensions because of rapid expansion of area under vegetables, intensive cultivation, round the year cultivation, export enhancement and health consciousness among the people.

The diamondback moth (DBM), *Plutella xylostella* (L.), is a major pest of brassica crops worldwide. DBM has been estimated globally to cost US$ 1 billion in direct losses and control costs. (Grzywacz *et al.*, 2010)

Vegetables are receiving heavy doses of insecticides in comparison to field crops. This may be attributed to:

i) Farmers start pest control at later stage
ii) Development of resistance against pesticides against by pests
iii) Development of new biotypes by pests
iv) Poor quality pesticides
v) Large scale cultivation of pest preferred variety
vi) Integrated pest management technique either not developed or not known to farmers
vii) Farmers unaware of proper spraying schedule

As the vegetables are mostly consumed fresh and many rather consumed raw so indiscriminate use of pesticides can lead to serious health hazards through contamination of foodstuff, water, air and soil. Further chemical control measures add, to production cost, accelerates racial evolution in insects. Hence breeding varieties for insect resistance is highly desirable from human health and ecological points of view.

Losses Due to Pest

Losses due to pests are relatively more in vegetables, because they serve as a better host to many pests. It could be attributed to their features offering reproductive advantage and conducive environment for survival. Pests cause damage to the host plant in various ways:

i) Tissue Feeding Insects

They feed on plant parts such as leaf, stem, root, fruit etc. For instance, fruit borer, stem or shoot borer, root borers, leafhoppers, beetles, weevils etc. Host performance is badly affected because of poor quality fruits, reduced photosynthetic area and fewer numbers of fruits.

ii) Sucking Pests

These insects suck the cell sap of the plant and weaken the host physiology for instance, aphids, jassids, mites, thrips etc.

iii) Pathogen Vector Insects

These insects act as vector for disease causing organisms. Thus, help in spreading the disease in addition to loss caused by them. Aphids, white flies, mites and even some leafhoppers serve as vector to many viruses.

The losses due to insect pests in vegetables are around 40%. Borers are most devastating pests where losses vary from 12% to total crop failures. For instance fruit and shoot borer infestation in brinjal has become a national problem. Tomato fruit borer is another problem that invites heavy pesticide applications on tomato. Resistant breeding seems the best option to reduce pesticide poisoning to human fellows as well as environment. Transgenic vegetables with Bt genes have shown promise in successful controlling of the devastating pests with fewer pesticide applications.

Important Insect- Pests of Vegetables

Pests are the key factors in destabilization the vegetable production. Both quality and quantity are adversely affected. Fruit borer, fruit fly and beetles often account for heavy losses, namely fruit borer in tomato 60-70%, fruit and shoot borer in brinjal 60-80%, midge in chilli 30-50%, red pumpkin beetle in cucurbits 20-80% (Puri *et al.,* 2001).

Some pests of economic importance are enlisted below:

Crop	Insect pests
Potato	Potato tuber moth (*Phthorimea operculolla*), Bihar hairy caterpillar (*Spilosoma Oblique*), ragi cutworm (*Spodoptera exigua*), tobacco caterpillar (*Spodoptera litura*), Fruit borer (*Helicoverpa armigera*), Semilooper (*Plusia chalcites*)., Tuber grubs, cut worms
Tomato	Fruit borer (*Helicoverpa armigera),* Epilachua or Hadda beetle (*Epilachna vigintioctopunctata),* White fly (*Bemisia tabaci*)
Brinjal	Shoot and fruit borer (*Leucinodes orbonalis*), **Epilachna or Hadda beetle** (*Epilachna vigintioctopunctata),* **Leaf roller** (*Eublemma olivacea*), **Jassids** (*Amrasca biguttula biguttula, Hishimonas phycitis*), **Aphids** (*Myzus persicae* and *Aphis gossypii*), **Red spider mite** (*Tetranychus neocoledonicus,T.cinnaarinus,T.telarius, Paratetranychus indicus*), **White fly** (Bemesia tabaci).
Chilli and pepper	**Thrips** (*Scirtothrips dorsalis*), **Aphids** (*Aphis gossypii* and *Myzus persicae*), **Mites** (*Polyphagotarsonemus latus, Tarsonemus translucens and Tetranychus cinnabarinus*)
Cole crops	**Cabbage butterfly** (*Pieris brassicae L.*), **Diamond back moth** (*Plutella xylostella*)
Cucurbits	**Fruitfly** (*Dacus cucurbita, Dacus dorsalis*), **Red pumpkin beetle** *(Aulacophora foveicollis)*
Legume vegetables	**Pod borers** (*Etiella Zinckenela; Adisura atkinsoni, Exelastic atomosa* and *Helcoverpa armigera*), **Stem fly** (*Ophiomyia phaseoli),* **Pea leaf miner** (Agromyza flaveola), **Aphid** (*Macrosiphum pisi, Aphis craccivora, Myzus persicae*)
Okra	**Shoot and fruit borer** (*Earius vitella*), **Jassids/ leaf hopper**(*Amrasca biguttula biguttula),* **White fly** (*Bemisia tabaci*)

Mechanisms of Insect Resistance

Manifestation of resistance to an insect-pest by a host species could be due to environment unfavorable to pest. This is known as **pseudo resistance** or **ecological resistance**. However, when pest population is abundant and host variety relatively performs better and restricts pest proliferation, this is due to inherited resistance.

Non-Preference or Antixenosis

The host varieties are not preferred by the insect. This may be attributed to unsuitability of host to insect-pest for colonization and reproduction. It means the mechanism is characterized by unattractive host and related to food, shelter and breeding problem of pest. The degree of non-preference may vary from complete to partial as in some cases insects will not accept the resistance variety even if there is no alternative source of food.

Leafminer (*Liriomyza langei* Frick) is a major pest that causes considerable damage to a wide variety of vegetable crops including lettuce, and resistant cultivars remain the most economic means of insect control. Mou *et al.*, 2002 studied eighty-four lettuce cultivars and introduced lines were grown in an insect cage with eight replications for resistance screening. Leafminer flies were released in the cage to feed on the plants. Significant genetic variation for leafminer stings per unit leaf area was observed among genotypes tested. Resistant lines with fewer leafminer stings were found in *Lactuca sativa, L. saligna, L. serriola, and L. virosa,* and the resistance was confirmed in a field experiment.

Morphological Features Associated with Non-Preference

- Thick layer of cellulose hairs
- Leaf thickness
- Hard midrib
- Trichome length and density
- Stiffness of stem
- Floral nectaries presence
- Glandular trichomes
- Exclusive examples of bio-chemical non-preference are rare.

Antibiosis

The mechanism refers to adverse effects of feeding on resistant host on reproduction and survival of insect pests.

Feeding on resistant host may lead to:

i) Metabolic abnormalities
ii) Failure to pupate
iii) Failure to adult emergence
iv) Abnormal or undersized adults
v) Disturbed hibernation
vi) Decreased fecundity
viii) Abnormal behavior

Fifty cultivars of taro, *Colocasia esculenta* (L.) Schott (Araceae), collected from islands in Micronesia and Polynesia, eight cultivars from the University of Hawaii's taro germplasm collection, and a closely related aroid, *Xanthosoma sagittifolium* (L.) (Araceae), were screened for antibiosis and antixenosis to *Aphis gossypii* Glover. Life history data for *A. gossypii* were collected by assessing survivorship and fecundity of aphids caged on taro leaves in the field. Significant differences in aphid reproductive rate and longevity were observed among the taro cultivars, and cultivars were ranked from most resistant to most susceptible. Antixenosis was assayed in the laboratory in a multiround choice test where *A. gossypii* were offered four leaf discs excised from different taro cultivars. 'Iliuaua', 'Rumung Mary', 'Maria', 'Ketan 36', and 'Agaga' were the most resistant in terms of reducing aphid fecundity and survivorship, whereas the Iliuana, 'Purple', 'TC-83001', and 'Putih 24' were least preferred in aphid choice tests.

Antixenosis and antibiosis in the resistance of common beans (*Phaseolus vulgaris* L.) to the melon thrips, *Thrips palmi* Karny, were investigated under laboratory and field conditions. Experiments were conducted for four moderately resistant genotypes ('Brunca', BH-130, EMP 486, and FEB 115) in comparison with one susceptible genotype (APN 18). Multiple-choice tests recorded most thrips on EMP 486 and least on FEB 115. Dual-choice tests conducted in both laboratory and field confirmed the antixenotic effect of FEB 115 and the attractant effect of EMP 486 on thrips. These results demonstrate the significance of antixenosis in the resistance of common beans to *T. palmi* (Andrea Frei *et al.,* 2005).

Tolerance

It is the ability of a host plant to recover or tolerate pest attack without significant loss in yield. In case of tolerance, insect population is comparable to a susceptible variety.

Avoidance

The host evades the pest attack because of its developmental alterations that don't match with active stages of the insects. For instance insect attacking stage and susceptible stages of the host do not coincide.

Genetics of Insect Resistance

Knowledge of the genetic control of resistance helps in adopting relevant breeding strategies and breeding methodology to develop pest resistant varieties. Genetics of resistance involves interaction of two living entities i.e, host and pest, so elucidation of the phenomenon requires the simultaneous study of resistance in host and virulence of the pest.

Oligogenic Resistance

One or few genes govern this type of insect resistance. In segregating generations there are distinct genotypes falling in either resistant or susceptible class. This is also known as vertical resistance where each gene has large effect. This type of resistance often succumb to insect pest pressure and is less stable barring some example against aphid resistance. Monogenic resistance segregates in ratio of 3: 1 and is easy to handle and also to transfer through backcross method of breeding.

'True French' is an open-pollinated cultivar of the Zucchini (Courgette) Group of *Cucurbita pepo* and is susceptible to *Zucchini yellow mosaic virus* (ZYMV). Using *C. moschata*'Menina' as the source of ZYMV resistance and following six generations of backcrossing, a true-breeding line nearly isogenic to 'True French', designated 381e, was recovered that carried ZYMV resistance, albeit not at as high a level as in 'Menina'. 'True French' and accession 381e were crossed, and their reciprocal F_1, F_2, and backcross progenies were grown in a chamber and inoculated with a highly virulent, non-aphid-transmissible strain of ZYMV. Nearly all F_1 plants and all plants of the backcross to 381e were classified as resistant. Segregation to resistant and susceptible individuals occurred in the backcross to the susceptible parent, in accordance with a 3:5 three-gene ratio of resistant: susceptible. The F_2 segregated in accordance with a ratio of 45 resistant : 19 susceptible, which would be obtained if there was one major gene for resistance, *Zym-1* (*Zym*), and two other genes, herein designated *Zym-2* and *Zym-3*, both of which were complementary to *Zym-1*. The presence of *Zym-1* and either *Zym-2* or *Zym-3* is necessary for resistance to be expressed in young plants, but the presence of all three might be necessary for resistance to continue to be expressed during subsequent development of the plants Paris and Cohen (2005).

Polygenic Resistance

Resistance reaction of different individuals in a population sometimes does not fall in to distinct classes rather it shows a continuous range of expression. This type of host response cannot be analyzed by Mendelian ratio and is known as polygenic resistance. It is also known as horizontal resistance, general resistance, quantitative resistance or biotype non-specific resistance. Several genes control it and each gene has small and cumulative effect. Polygenic resistance is relatively stable.

Cytoplasmic Resistance

This type of insect resistance shows extrachromosomal inheritance. The genes conferring resistance are localized in the cytoplasm, for example resistance to root aphid in lettuce. It exhibits maternal impact and thus can easily be transferred to the required variety through back cross method by using resistant variety as female and non recurrent parent.

Sources of Resistance Gene

There are mainly following sources:

Land Races or Improved Varieties

Resistance may occur in some traditionally grown varieties which otherwise do not have much agronomic superiority. Several improved varieties may also serve the source of resistance where the resistance has been transferred from wild species.

Germplasm Collection

Genetic resources collection and evaluation has become the central activity to provide sustainable base to breeding programs. Moreover, search for resistant genotypes has top priority in evaluation of germplasm collections.

Wild Species

Wild species are searched for resistance when it is not available in cultivated forms. Wild species are usually found in primary and secondary centers of diversity. Wild species have been used for transfer of insect resistance in case of potato, okra, tomato and onion etc.

Gene from Unrelated Organisms

Biotechnology techniques have made the crossibility barriers meaningless as it facilitates the transfer of gene not only across the family but also across the kingdom. Direct and indirect gene transfer techniques are available and have been successfully employed to integrated gene from unrelated species. The varieties so developed are designated as transgenic plants.

Important sources of resistance in vegetables

Crop	Source	Insect/Pest
Tomato	*L. peruvianum, L. hirsutum*	Leaf Curl Virus
	PI 126449	Fruit worm
	IVT 74453	White fly
Brinjal	Anamalai, Pant Samrat, bhagyamati, Pusa Purple Cluster, *Solanum gilo, S. anomalum*	Shoot and fruit borer
Bell pepper and chilli	Pant C1	Leaf curl
	Caleapin Red, Chamatkar,	Thrips
	Kalayanpuri Red Punjab Lal	Mites
Okra	Long Green Smooth	Nematodes
	IIHR21, Crimson Smooth Long, *A. manihot ssp. manihot*	Jassids
	Red I, Red II, Red Wonder I, Pusa Sawani, Long Green	Fruit/ shoot borer
Garden Pea	EC 16704, 21711, 25173	Leaf minor
	Bonneville, Asaugi, Draft Gray Sugar	Stem fly
Muskmelon	PI 161375	CMV
	PI 414723	WMV
	C. callossus	Fruit fly
Summer squash	Punjab Chappan Kaddu	Red pumpkin beetle
Bottle gourd	PI 27153	CMV, WMV

At first, breeders found and used resistance genes from the adapted, local landrace populations that also were the initial gene pool as a source of resistance genes for their new varieties. As years went by, these gene pools began to dry up and breeders looked further a field, turning to exotic (unadapted) landraces, and even to wild relatives of their crop. Sometimes they made extraordinary efforts to hybridize the domestic crop with a very distant wild relative—making a cross that could not succeed under natural conditions. Embryo rescue and even x-ray treatments were used to make "unnatural" crosses and derive breeding progeny from them. The breeders fooled around with Mother Nature; they moved genes farther than natural processes would allow.

But the breeders as a whole preferred not to breed from exotic varieties or distant and often wild relatives. They used exotic material only when there was no other choice. This preference was due not only to the difficulty of wide hybridization, but also to the fact that exotic germplasm exacerbates the problem of undesirable linkages. Few or none of the foreign genes—except the desired resistance genes—were suitable for the needs of high yielding, locally adapted varieties. But often the breeders had no choice; either they got the needed resistance genes from a distant relative, or they got nothing at all.

Resistant Breeding before Mendel

Wild relatives of some crop plants are not uniformly resistant to insect and disease pests. This can be demonstrated in simple fashion—when selections of these wild populations are set out in plant-rows, some of them are highly susceptible, others are resistant, and some are intermediate in resistance to the common pests of the region. The first plant breeders, were those women and men who domesticated crops and could save only those genotypes that had some level of resistance, i.e., those individual plants that did not succumb to pest depredation. In fact, therefore, they selected for pest resistance and thus changed the population structure of their crop species in favor of resistance genes. This change made it possible to grow the crops in monoculture, which was convenient for food production and harvest. It was also convenient for multiplication of disease and insect pests that might not be affected by the limited sample of resistance genes.

Plant breeding thus set the stage for sequential cycles of pest resistance and pest susceptibility of crop plants. We have no direct record of the consequences of this ancient ecological meddling, but myth and historical accounts tell of disastrous disease epidemics and insect outbreaks, so one can assume that from time to time large plantings of crops that were uniformly susceptible to a new kind of insect or disease fostered increased of that pest to epidemic proportions. Resistance genes were essential for crop domestication and monoculture but they did not guarantee perfect safety.

Resistant Breeding after Mendel

Genetics-based plant breeding, launched in the early years of the 20th century, produced new crop varieties with improved resistance to major disease and insect pests. Usually such resistance was developed as a second phase—a rescue operation—after new varieties, selected primarily for high yield, was discovered to be susceptible to a particular insect or disease. Breeders found early on that they could identify single gene (usually dominant) that conferred essentially complete resistance to the pest in question. Varieties containing such excellent resistance were developed and released for large-scale farmer use. But breeders then discovered, that the "perfect" resistance lost its effectiveness after a few seasons. They soon learned, with the aid of entomologists and plant pathologists that insect and disease pests are highly diverse genetically, and that almost without fail a rare pest genotype will turn up (or perhaps be created *de novo* by natural mutation) that is not affected by the newly-deployed resistance gene. The new pest genotype multiplies and the crop variety's resistance "breaks down".

As years went by, breeders found that some kinds of resistance did not fail, and that such resistance often was less than complete; the plants suffered some damage but gave satisfactory performance overall. This longer lasting resistance was dubbed "durable" resistance. Further, the breeders discovered that durable resistance usually (but not always) was governed by several genes rather than by one major gene. The multi factorial kind of resistance has been called "horizontal resistance." The major-gene resistance has been called "vertical resistance."

The good news, then, was that breeders could identify and breed for durable resistance. The bad news was that the breeding was more difficult because several genes had to be transferred at one time, thus requiring larger populations for selection, as well as multiplying the usual problems with "linkage drag" (undesirable genes that are tightly linked to the desired ones). To this day, breeders use both kinds of resistance in varying proportions, according to the crop and where it is grown.

So at the end of the 20th century, plant breeding for pest resistance had laid out the genetic framework of vertical and horizontal resistance, and identified important sources of new resistance genes, i.e., plant germplasm from anywhere in the world. Sources were limited, however, to the crop species itself or its relatives, either wild or cultivated. All of the introduced genes therefore came from plants.

Plant breeders selected not only for tolerance or resistance to disease and insect pests, they also selected for tolerance to abiotic stresses such as heat and drought, cool temperatures, or nutrient imbalance. Much of this selection was involuntary; in selecting varieties with top performance over many seasons and many locations the breeders necessarily selected varieties with tolerance to the prevailing abiotic stresses of the diverse seasons and localities. In selecting for tolerance to environmental stresses, breeders necessarily changed the genetic makeup of the crop species, altering it still further from that of the original wild species, which had been restricted to certain environmental niches.

Breeding Methods for Insect Resistance

Breeding methods for insect resistance are similar to other aspects namely, disease resistance or agronomic improvement but most crucialy part of the programme is availability of simple, efficient and reliable screening techniques.

The following breeding methods are commonly used for development of insects resistant varieties.

Introduction

Many resistant varieties in vegetables have been directly introduced. Some of the examples are ad follows:

Crop	Varieties
Tomato	Sioux
Pea	Arkel and Bonneville
Cabbage	Golden Acre
Capsicum	California Wonder
French bean	Contender
Carrot	Nantes

These varieties have been introduced many years back, but they still dominate the cultivation, which indicates their better resistance to biotic and abiotic stresses, prevalent in the new environment.

Introduction is the easiest and quickest method of developing insect resistant variety, some examples of insect resistant introduction are:

Crop	Variety	Resistance
Tomato	Red Cloud	White fly
Capsicum	California Wonder	Glass house fly
Carrot	Nantes	White fly

Selection

It is the core activity in any variety development program. Selection is effective on genetic variable populations such as local land races, introductions showing genetic diversity or segregating generations of crosses. Pure line and mass selections are effective in self pollinated crops. Selection follows complete evaluation procedure and selected material may qualify for release as a new variety (Fig. 1).

Mutation Breeding

Mutation is sudden heritable change in the genome of an organism. Frequency of useful mutations is very low, so mutation breeding is advised only when resistance is not available in cultivated forms and related species. Nonetheless, mutation breeding so far has yielded more than 600 varieties in field, vegetable, fruit and ornamental crops.

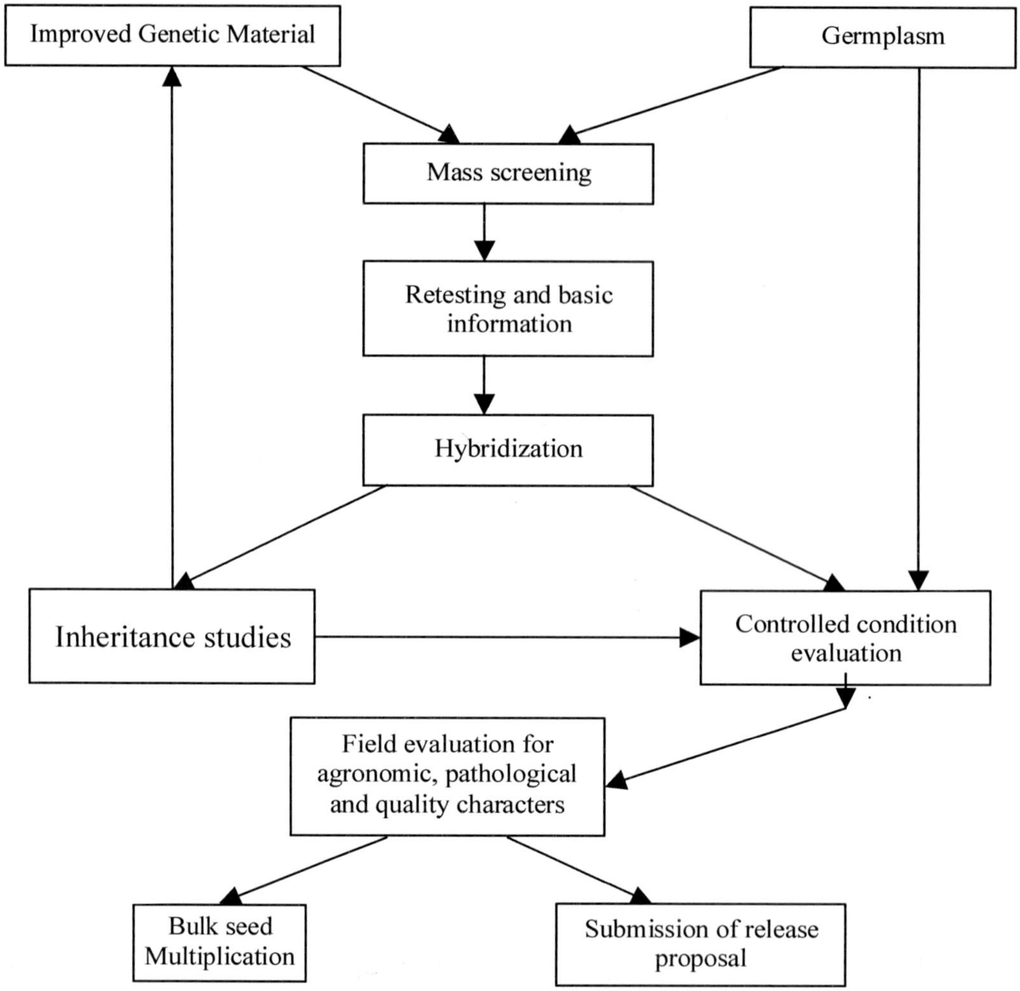

Fig. 1 : Schematic presentation of program for breeding for pest resistance.

Resistant varieties developed through mutation breeding are:

Crop	Variety	Resistance
Chilli	Horgoskaslatka-X-3	CMV
Okra	EMS-8	YVMV and fruit borer

Hybridization

It involves crossing of two genetically different lines. It is an effective means of transfer of insect resistance from one variety to another. Resistance available in poor yielding land races and wild species is incorporated into agronomically potential genotypes through hybridization. Then backcrossing is carried using potential genotype as recurrent parent. 5 to 6 backcrosses are normally attempted to eliminate the undesirable genes of wild parent and recover the cultivated parent. The procedure is simple when resistance is monogenic, while in case of polygenic control of resistance, large selfing generation after each backcross is to be raised to identify the resistant plants before attempting the next backcross. In wide crosses, backcross method is most simple and desirable approach.

Some resistant varieties developed through hybridization followed by selection are:

Crop	Variety	Resistance
Pea	Pant Uphar (IP-3), Bonneville, Asaugi	Stem fly
	Ooty-1	White fly
Brinjal	Anamalai, Pant Samrat, Pusa Purple Cluster	Shoot and fruit borer
Chilli	Pusa Jawala	Thrips, Mites and Aphids
	Pusa Sadabahar	CMV, TMV and Leaf Curl Virus
Okra	IIHR 21, Crimson Smooth Long	Jassids
	Red WonderI, Pusa Sawani, Long Green	Fruit/Shoot Borer
Pumpkin	Arka Suryamukhi	Fruit fly
Summer squash	Pujab Chappan Kaddu	Red pumpkin beetle

Modifications of Backcross Cross Method

The back cross method can be modified in many ways to suit the breeder's requirements. Some important modifications are described here.

Double Backcross Method

In this method F1 hybrid is backcrossed to both the parents separately for three generations. BC3 selected plants from both sides are crossed to produce F1 generation. F1 plants are advanced to F2 generation and selection for desirable genotypes is carried. Back crossing is accompanied by selection for the character received from the non recurrent parent. Diagrammatic presentation of this procedure is given in Fig. 2.

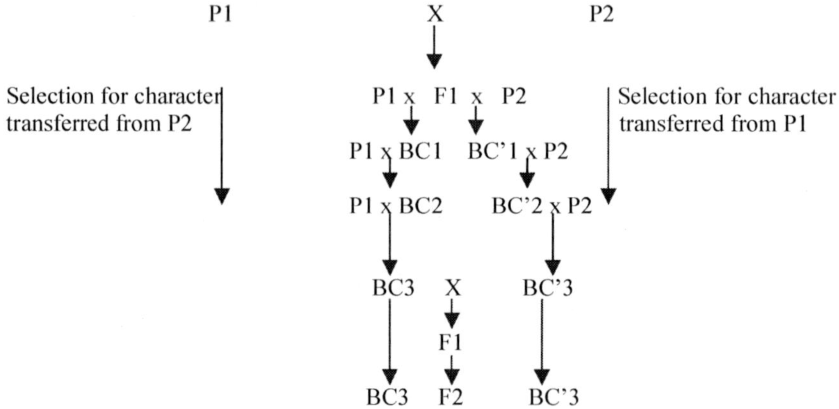

Fig. 2 : Double back cross method (P1: Recurrent parent, P2: Non recurrent parent, BC: backcross Progeny)

Multi-Recurrent Parent Backcrossing

The approach consists of backcrossing of resistant genotype to two or more recurrent parents. The purpose is to develop genotypes with desirable traits having many noble genes from diverse outstanding varieties. One to two backcrosses with each selected variety are undertaken. This may result in development of several desirable varieties.(Fig. 3).

Recurrent parent A X Non recurrent parent
↓
F1 x Recurrent parent A
↓
BC1 x Recurrent parent B
↓
BC2 x Recurrent parent B
↓
BC3 x Recurrent parent B
↓
BC4 x Recurrent parent B
↓
BC5 generation
↓
Selfing
↓
Advancing to F2 generation, single plant selection is practiced

Fig. 3 : Multi recurrent parent backcrossing

Backcross-Pedigree Method

The method is started as backcross method and F1 hybrid is backcrossed one, two or three times depending upon the choice of the breeder. Then the single plants are selected and advanced to F5 and F6 progenies according to pedigree method of selection. The method is effective when non-recurrent parent does not have many undesirable genes and secondly many good genes from recurrent parent have been transferred to the resultant genotype (Fig. 4).

Recurrent parent X Non recurrent parent
↓
F1 x Recurrent parent (BC1)
↓
BC1 x Recurrent parent (BC2)
↓
F1/BC2generation
↓
F2 population (single selection plant practiced)
↓
F3 progenies
↓
F4 progenies
↓
F5 progenies
↓
F6 progenies
↓
Yield trails

Back cross method (Recurrent parent X Non recurrent parent to F2 population)

Pedigree method (F3 progenies to Yield trails)

Fig. 4 : Backcross- pedigree method

Tomato fruitworm (TFW), *Helicoverpa armigera* (Hübner), is the most destructive pest of tomato (*Lycopersicon esculentum* Mill) in South and Southeast Asia. Research was initiated at AVRDC in 1980 to identify sources of resistance to this pest and to breed TFW-resistant tomato lines suitable for the region. Germplasm screening revealed high levels of TFW resistance only in the wild *Lycopersicon* species, particularly *L. hirsutum* and *L. pennellii*. Consequently TFW resistance breeding research focused on introgressing resistance from the wild species into the cultivated tomato. In the first resistance breeding efforts from 1981 to 1988, *L. hirsutum* accessions 76W-PI134417-A and 77W-PI134417-1 were selected as donor parents in a backcrossing program with *L. esculentum*. Backcross lines with significantly lower pest damage were derived from this program but all were invariably small-fruited and not commercially acceptable. From 1988 to 1995, a second attempt was undertaken using *L.*

pennellii accession LA 716 as the resistance donor parent. Selection of resistant progenies was based initially on TFW damage and subsequently on acylsugar content, the compound that confers insect resistance, in leaves. No backcross progenies with acylsugar levels near that of the donor parent were obtained. Furthermore, progenies with higher acylsugar levels also showed low pollen production and other undesirable horticultural traits. The research demonstrated the presence of high levels of TFW-resistance in several wild tomato accessions but efforts to introgress this resistance into the cultivated tomato in order to combine it with acceptable fruit qualities did not succeed. (Talekar *et al.*, 2006)

Insect resistant accessions of wild *Lycopersicon* species were identified long ago, yet insect resistance has not been successfully introgressed into tomato cultivars. The effectiveness of the inbred backcross method for introgressing resistance to beet armyworm (*Spodoptera exigua* Hübner), tomato fruitworm (*Helicoverpa zea* Boddie), and potato aphid (*Macrosiphum euphorbiae* Thomas) from the resistant *Lycopersicon pennellii* (Corr) D'Arcy acc. 'LA 716' into *Lycopersicon esculentum* Mill. 'Peto 95' was examined. Inbred backcross populations at two backcross levels (BC_2 and BC_3) were developed and tested at two levels of selfing over 2 years (i.e., $3 in 1994 and $4 in 1995). Variation among inbred backcross lines (IBL) in the amount of fruit damage from beet armyworm and tomato fruitworm and the proportion of potato aphid infested plants was observed. Significant line x year interactions suggest that genotype x environment effects may be important to beet armyworm and tomato fruitworm resistance. Differences between observed and predicted gain in beet armyworm and tomato fruitworm resistance of post hoc selected IBL were found, which might be attributed to underestimation of genotype x environment effects, or to non-additive genetic effects. Large genetic correlations (r^2_g) between reduced fruit damage and later maturity, lower fruit yield, and smaller fruit mass were estimated. The inbred backcross method was used to successfully introgress insect resistance from *L. esculentum* to cultivated tomato (Hartman and Clair, 1998).

References

Andrea Frei, Hainan Gu, Juan M. Bueno, Cesar Cardona, and Silvia Dorn. 2005. Antixenosis and Antibiosis of Common Beans to Thrips Palmi Karny (Thysanoptera: Thripidae) *Journal of Economic Entomology* 96(5):1577-1584.

Chahal, G.S. 2000. Genetics of insect resistance. In, resistance breeding in crop plants (*Nanda et al.*, ed.) CAS, Deptt. of Plant Breeding, PAU, Ludhiana. pp. 96-98.

Grzywacz, D., Rossbach, A., Rauf, A., Russell, D.A., Srinivasan, R., Shelton, A.M. 2010. Current control methods for diamondback moth and other brassica insect pests and the prospects for improved management with lepidopteran-resistant Bt vegetable brassicas in Asia and Africa. *Crop Protection*, 29: 68-79.

Hartman, J.B. and Clair, D.A. St. 1998. Variation for Insect Resistance and Horticultural Traits in Tomato Inbred Backcross Populations Derived from *Lycopersicon pennellii*. *Crop Science Society of America* 38:1501-1508.

Kalloo, G. 1988. Vegetable breeding Vol. II CRC Press, Florida, Pp. 95-140.

Mou, B., Ryder, E.J., Tanaka, J., Liu, Y.B., Chaney, W.E. 2002. Breeding for resistance to leafminer in lettuce. ISHS *Acta Horticulturae* 637.

Paris, H.S. and Cohen, S. 2005. Oligogenic inheritance for resistance to *Zucchini yellow mosaic virus* in *Cucurbita pepo*. *Annals of Applied Biology*. 136 (3):209-214.

Pink, D.A.C., Kift, N.B., Ellis, P.R., McCliment, S.J., Lynn, J. and Tatchell, G.M. 2008.Genetic control of resistance to the aphid *Brevicoryne brassicae* in the wild species *Brassica fruticulosa*. *Plant Breeding* 122 (1): 24-29.

Sharma, J.P. 2009. Principles of Vegetable Breeding. Kalyani Publishers, Ludhiana, Pp. 263-276.

Talekar, N.S., Opena, R.T. and Hanson, P. 2006. *Helicoverpa armigera* management: A review of AVRDC's research on host plant resistance in tomato 25(5): 461-467.

□□□

Ecologically Based Integrated Pest Management, 2012
© D.P. Abrol and Uma Shankar (eds.), pp. 537-551
New India Publishing Agency, New Delhi (India)
e-mail : info@nipabooks.com; website : www.nipabooks.com

28

Breeding for Resistance to Major Insect-Pests of Pulses

BIMAL SINGH JAMWAL

Introduction

Within the very small portion of the plant kingdom that includes the world's crop plants, the legumes are a large group second to the cereals as sources of human food. Legumes belong to the family leguminosae with a group of bacteria, *Rhizobium,* which is able to utilize atmospheric nitrogen. This symbiosis which occurs in the root hairs may be the key to the success of legumes worldwide distribution. The family leguminosae includes about 18,000 species, characterized by their fruits, which are pods, by their (usually) alternate, pinnate or trifoliate leaves (Cobley and Steele, 1976). The diversity in the family ranges from hardwood trees of the tropical rain forests to the herbs of temperate pastures.

The leguminosae has three sub-families, the Caesalpiniodeae, the Mimosoideae and the Papilionoideae. The first two consist mainly of tropical trees and shrubs with few economically important species and it is the Papilionoideae, that are of agricultural importance. Papilionoideae supply three kinds of human food; edible tubers; leaf, green pods, and unripe seed (vegetable legumes); and ripe dry seeds (grain legumes, pulses). The paipilionoideae are classified in 10 botanical tribes, three of which together contain the major vegetable and grain legumes. They are the *Vicieae* which includes the genera *Vicia, Pisum, Lens. Lathyrus* and *Cicer*, the *Phaseoleae* which includes *Phaseolus, Vigna, Glycine, Dolichos* and *Cajanus*; and the Hedysareae which contains *Arachis* (Smart, 1978).

In the developed countries, grain legumes are an important indirect source of protein, being animal feeds of good biological value, However, for many of the developing countries pulses constitute the cheap and readily available source of dietary protein. Therefore, the only practical means of solving the protein malnutrition problem in the developing countries is to increase greatly the production of pulse crops.

The pulse crops, in general, give lower yields than the cereal crops (Bahl *et al.*, 1979; Borlaug, 1975). This fact has been misinterpreted in relation to the genetic potential of these crops for high grain yield. One school of thought believes that because pulses are rich in protein and it takes more energy weight for weight to synthesize protein thatn carbohydrates, the grain yields of the pulse crops must necessarily be some what lower (Boulter, 1980; Fabson *et al.*, 1978) From comparisons of the known energy requirements of vaiours metabolic pathways, one gram glucose can give rise to 0.8g carbohydrate but on average, only about 0.5 g protein and even less oil. According to Penning de Varies *et al.* (1974), equivalent seed yields would be about 70 units for soybeans, 90 for peas and 100 for cereals.

Such large concentrations of protein (and hence nitrogen) in seeds may require early mobilization of protein of leaves, thus impairing their capacity for prolonged photosynthesis. The maintenance of symbiotic dinitrogen fixation in root nodules requires prolonged use of photosynthate and thus may reduce the energy available for storage in seeds,. These energy requirements are essentially critical for legumes because of their indeterminate growth habit and progressive flowering and seed setting compared with the synchronous flowering of cereals and may be responsible for their reduced yield potential (Evans, 1980). While the argument in terms of bioenergenetics appears to be quite sound, the conclusion in respect to the relation between grain yield and protein content may not (Rao *et al.*, 1975; Bahl *et al.*, 1979; Sharma and Kharkwal, 1982). For those grain legumes that have received serious attention from plant breeders during the past 50 years, grain yields are as high as those of cereals (e.g. soybean and fababean). The reasons for the low yield of pulses may rather be that they have not received enough attention concerning intensive breeding efforts. Even more important may be the fact that pulses (particularly in developing countries) have been traditionally grown under marginal conditions, while better inputs (land, fertilizer, irrigation) reserved for cereals. Under these stress conditions pulses have been selected more for adaptation for low level of management rather than for high yields. The plants have responded under these conditions by developing particular characteristics such as bushy, spreading or indeterminate growth habit, uniform development, and specific photoperiod or thermo reactions often associated with extremely early or late maturity. The aggressive growth of many of the pulse crops means that they must be efficient photosynthesizers; in other words, their biological yields are high (Jaranowski and Micke, 1985). A high biological yield, however, does not automatically lead to a high yield. A major requirement is to construct the plant type so as to improve the grain straw ratio, to generate adoption to higher plant populations per unit area etc.

Grain legumes occupy a unique position in the world agriculture by virtue of their higher protein content and their capacity to fix atmospheric nitrogen. Legumes are two three times richer in grain protein that cereals. The question is whether it is more feasible and economic to improve the protein yield per unit area by increasing protein content in seeds or by increasing yield. From the point of view of breeding, improving grain yield seems easier and higher grain yields would contribute to increased protein production per hectare.

Genetics of Resistance to Insect Pests of Chickpea

Several insect-pests damage the chickpeas. However, only few of which cause considerable losses to the yield are dealt in the ensuing paragraphs.

Pod-borers : *Heliothis armigera* (Hub.), *H. virescens* (FAB) and *H. viriplaca* (Hufnagel) are the most serious pests of chickpea in most areas of the world. The larvae feed voraciously on the crop from the seedling stage to maturity (Lateef, 1985).

Variation due to general combining ability were highly significant. indicating a preponderance of additive genetic variation.

Semi-looper : The larvae of semi-looper *(Autographa nigrisigna* Wlk. syn = *Plusia nigrisigna).* Sometimes causes serious damage to the chickpea crop by feeding on leaf buds. flowers and pod. Rizvi and Singh (1980) reported that it is a common pest in eastern Uttar Pradesh (India). It was predominant in the month of March.

Bruchids : All the three species *(Callosobruchus maculatus* Fabr. C. *chinensis* Linn. and C. *analis* Fabr.) are serious stored pests of gram. Non-preference to egg-laying is observed in resistant cultivars (Saxena and Raina, 1970).

Breeding for Resistance to Insect-Pests of Chickpea

Screening for resistance to insect-pests : The germplasm lines may be grown in 3-5 m row length alongwith infector rows for screening against pod-borer and pod-fly (Chhabra and Kooner, 1980). The lines with lowest damage are tested again in replicated trial surrounded by infector rows. The best entry may be tested in hot spots/artificial conditions. Since 1976, at ICRISA T an open field screening technique, using natural populations of pod-borer *(Heliothis armigera* Hub.), occasionally supplemented by laboratory-reared insects was used to screen> 12000 accessions of chickpea. Some of the lines were found to suffer considerably less pest damage than others. This was due to differences in oviposition and larval preference (Lateef, 1985). A correlation between malic acid content and borer damage was observed. The threshold of low borer damage being 250 mg malate per ml exudate from plants. Moderate damage was correlated with maltate contents between 130 and 250 mg per ml and high damage with less than 130 mg per ml (Rembold, 1981).

It was observed that seed coat of resistant variety was very rough and the coarse surface of the seed might be an important factor responsible for the inhibition of egg-laying by the bruchids (Saxena and Raina, 1970).

For screening the germplasm against root-knot nematode *(Meloidogyne incognita* K. and W.) the seeds are sown in 10 to 30 cm earthware pots filled with Sterilized soil. Sharma and Mathur (1985) maintained three seedlings per pot and inoculated with 4000 freshly hatched larvae of the nematodes. Fifteen plants of each line were uprooted after 60 days and categorized according to the scale given by Mani and Sethi (1984); where, 1 = no galling; 2 =1-25 percent root system- with galls; 3= 26-50 per cent root system with galls; 4= 51-75 per cent root system with galls; and 5= 76-100 per cent system with galls.

Sources or resistance to insect-pests. Limited work has been done to find out the sources of resistance against the insect-pests in chickpea. Only one line against pod-border and one line against bruchids is reported to be resistant. However, several resistant lines are available against root-knot nematodes.

Resistance to Insect - Pests of Pigeonpea

Several insect-pests are reported to cause damage to pigeonpea at the various stages of its growth, (Lateef and Reed, 1981). Of these, pod-borer, pod-fly and bruchids are important. Therefore, these insects alongwith the mechanisms of resistance are described in the following paragraphs.

Pod-borer *(Heliothis armigera* Hub.). It is a very destructive and polyphagous pest of pigeon-pea, particularly in southern India (Lateef and Reed, 1981). Larvae in *Atylosia* spp. grew more slowly, took longer to pupate, and formed smaller pupae. The adult females reared on *A.scarabaeoides* laid the fewest eggs and the viability of these eggs was relatively low, suggesting that antibiosis was operating. Pod walls of *A. scarabaeoides* is tough and the pod borer damage is limited due to scarification of the pod surface and the seeds are left intact. This suggest that this species of *Atylosia* has antibiosis as a mechanism of resistance (Lateef *et al.,* 1981).

Pod-fly *(Melanagromyza obtusa* Mall.). It is an oligophagous pest and its common occurrence in northern and central India (Bindra and Singh, 1972) and causes considerable damage (Lateef and Reed. 1981). It appears during pod formation stage (Singh and Singh, 1978). It is reported to cause 40 to 60 percent damage of grains in northern India (Ahmad, 1938) and damage from 11.4 to 86.4 percent to the grains in central India (Gangrade, 1963; 1964).

Investigations were conducted on the biochemical basis of resistance in pigeonpea pod-walls to pod-fly by Dass and adak (1978). In green pods of pigeonpea, wax, total phenols, crude fibres, proline content and ascorbic acid were negatively correlated while reducing 'and non-reducing sugar contents were positively correlated with pod-fly infestation. Correlation between moisture content, total free soluble sugars, orthodihydric phenols, total amino acids protein/nitrogen content and pod infestation was found to be no significant. Narrow ratio of sugar to protein/nitrogen was recorded both.in the least and a highly susceptible group, however, no consistent trend was noted in different susceptibility groups.

Bruchids (Bruchus sp.) - Bruchids are best known as pests of stored grain and surveys showed considerable infestations in the pods, particularly in southern India (Lateef and Reed, 1981).

Sources of resistance to insect-pests. The sources of resistance to some of the insect-pests are presented in Table 3.8. Of these AS 71-31 was resistant to pod borer and pulse beetle, Basant was resistant to pulse beetle and reniform nematode, and ICP 11295, ICP 11299 and ICP 12757 was resistant to root knot and reniform nematodes. Out of 40 entries screened none of the entries were completely free from the pod-borer attack. However, some of the entries were less susceptible than others and are presented in Table 3.8. Mid-late entries were significantly less damaged than late and early entries. The early entries were more severely damaged than mid-late and late (Bhosale and Nawale, 1983).

Genetics of Resistance to Insect - Pests of Field-Pea

Pea aphid *(Acyrthosiphon pisum).* This aphid suck the cell sap which results in yellowing and subsequent drying of leaves. Hanington (1945) demonstrated occurrence of physiological races of the insect. The susceptible varieties generally contained a higher concentration of free and total amino-acids than the resistant varieties (Auclair *et al.*. 1957).

Pod borers *(Heliothis obsoleta* P. and *Etiella zinc/cenella* T.). The larvae of pod-borers feed on the flowers and bore into the pods eating up the seeds. The pod borers are very serious and causes a heavy loss in yield.

Pea leaf-miner *(Phytomyza atricornis* M.). It is also a key pest of fieldpeas in India (Kooner and Chhabra, 1980). The larvae feed by making tunnels in the leaves and pupate inside the mines which are zig-zag. The flowering and fruiting of the infested plants are reduced and the yield is affected adversely.

The leaf-eating caterpillar *(Laphygma exigua* H.). It feeds on the leaves, generally during morning and evening and in serious cases of attack it defoliates the plants.

Pea thrips This cause damage to the young crop by sucking the cell-sap (Kooner and Chhabra, 1980).

Pea moth *(Laspeyresia nigricana* Steph.). Nolte (1959) showed that short varieties with an early flowering period suffered least from attacks by the insects.

Pea stem fly *(Ophiomyia phaseoli* Tryon.). The fly is very small and larvae are cream coloured, 2-4 mm long. The infested young plants become yellow and may die in the course of time. In older plants vigour is very much reduced, leading to poor growth and low yield. The pest is a serious problem in the early sown crop (Kooner and Chhabra, 1980).

Pea weevil *(Bruch* us *pisorum* L.). The eggs are laid *on* the surface of the seed and the newly hatched grubs bore the seeds.

Pea varieties with green seeds were least susceptible to injury by *B. pisorum* (Sulyndin, 1947). Green seeded varieties of peas such as P. *sativum* var. *glaucospermum* and var. *medullare* could give useful plant material *for* breeding resistant lines (Ufer, 1948).

Breeding for Resistance to Insect - Pests in Field (Pea)

Screening techniques for pod-borer are presented in portion on chickpea and against jassids and bruchids in portion on mungbean. Similar techniques can be followed in peas.

Sources of resistance to insect-pests of pea. Chhabra (1980) reported EC 16704, EC 21711 and EC 25173 as resistant against leaf minor, *Phytomyza atricornis*. Sources of resistance against other important insect-pests may be found out.

Genetics of Resistance to Insect-Pests of Cowpea

Many insect-pests attack all parts of cowpea plants at every stage of growth as well as seeds in storage (Singh and Allen, 1980). In the following paragraphs the important insect-pests, their damage and inheritance of resistance, if discovered, are described.

Leafhoppers : *Empoasca kerri* (Pruthi) breeds throughout the crop growth period. Damage to the crop is caused by the adults as well as nymphs, both of which are very agile and more brisky, forward and side ways. The attacked leaves turn pale and then rust-red (Yadav and Yadav, 1983).

Whiterfly : The whiterfly (*Bemisia tabaci* Genn.) is noticed throughout the crop growth. It is known to transmit a number virus diseases including the leaf curl disease in cowpea..

Aphhids : The nymphs and adults of *Aphis craccivora* (Koch.) suck sap from leaves, stem, inflorescence or developing pods. Leaves acquire a curly appearance, flowers fail to form pods and the developing pods do not produce healthy seeds. It is an important pest of cowpeas in India and the Philippines. In Africa it is an important vector of cowpea virus (Singh, 1977).

The plant bugs : Both nymphs and adults of the green plant bug *(Nezara* viridula Linn). Scuck the sap from tender parts of the plants. The pest usually appears at the time of flower formation.

The pod sucking bug (*Acanthomia horrida* Germ.) is one of the most common pod sucking bugs of cowpea in the West Africa (Singh, 1977). Both nymphus and adults feed on pod.

Surface grasshopper : The surface grasshopper (*Charotogonus trachypterus* blanch) damages the germinating plants of cowpea. The attack is more in the areas of waste lands and low rainfall.

Cowpea leafminor: The damage of *Acrocercops* sp. Is seen throughout the crop growth with whitish tunnels in the leaves.

The caterpillars : The Bihar hairy caterpillar (Diacrisia oblique walker) eat leaves and soft portions of stems and branches. In case of severe infestation, the plants are completely denuded of leaves (Yadav and Yadav, 1983).

The larvae of tobacco caterpillar (*Spodoptera litura* F.) feed on leaves and fresh growth. They are mostly active at night and cause extensive damage to developing pods.

Blue butterfly : The blue butterfly (*Polyomatus bosticus* Linn) is a very active pest and found in all stages from September and onwards in India. The larve start boring into flower buds and green pods after 3-5 hrs of hatching (Yadav and Yadav, 1983).

Flower thrips : The flower thrip (*Taeniothrips sjostedti* Trtb.) feeds on flower buds and flowers and causes distortion of these parts and the flower drop. It is a serious pest of flower buds of the cowpea in Africa (Singh, 1977). The yield reduction towards yield may be as high as 50 percent (IITA, 1973).

Pod borers : The gram pod borer (*Heloithis armigera Hubner)* feeds at early stages on tender leaves, flower and flower buds. As the larva grows and simultancouly pod formation takes place, it is often found feeding on the developing grains after cutting a round shape hole in the pod and thrusting its head therein.

Beetles : The galarucid beetle (*Madurasia obscurella* Jacoby) is an important pest of Vigna species. The damage is caused by adults during night by making holes in the leaf lamina and perforate tham. On account of these injuries the normal growth of the plant is hampered (Yadav ad Yadav, 1983).

Weevils : Grey weevil (*Myllocerus* sp.) is active during the month of July and August in India i.e. at the time of flower buds appearance. They feed on leaves, leaf buds and flower petal, cutting 'U' shaped notches. The damage is visible throughout the crop period.

The cowpea weevil (*Callosobruchus maculates* Fab.) is the most important pest of stored cowpea seeds (Singh, 1977). Often after 3-5 months storage 100 percent infestation by cowpea weevil is recorded in West Africa. It infests pod in the field and is active as a storage pest. The larvae feed on dry cowpea seeds in storage.

Maruca testulalis

Breeding for resistance to insect pests

Screening for resistance to insect-pests : About 5000 accessions of the world gennp1asmcollections as well as several thousand breeding lines were screened *for* resistance to the major pests and techniques have been developed. to produce uniform infestations of pest populations in the field conditions (Singh and Allen., 1980).

A method for screening resistance of cowpea to the pod borer was developed by Jackai (1982). Flower, pod, and seed damage measurements provided the most important assessment of resistance to pod borer, *Maruca testulalis* Geyer. The larval population in the flowers showed marked 'differences in infestation levels between cultivars and was suggested as a means of comparing results.

Genetics and Breeding of Pulse Crops

One hundred and three cultivars were rated in replicated trials in the field and glasshouse for their reactions to the root knot nematode. Forty-one cultivars considerably showed sOme resistance to the nematode in both the field and glasshouse (Ogbuji, 1978).

Sources of resistance to insect-pests : The identification of resistance to insect-pests in cowpea began in the U.S.A. in the beginning of this century by Orton (1902) and Webber and Orton (1902) who noticed that cv. Iron was resistant to root knot nematode. Subsequent studies have led to the identification of other sources of resistance in the U.S.A. (Hare, 1959; Thomason and McKinney, t960), Nigeria (Amosu, 1974; Ogbuji, 1978), India (Sharma and Sethi, 1976; Patel *et al.*, 1977) and Brazil (Lemos and daPonte, 1918).

Both antibiosis and non-preference (Karel and Malinga, 1980) mechanisms of resistance were responsible against the leafhoppers and pea aphids attack. Severe aphid mortality was observed as the cause of resistance (Singh, 1977).

Sources of resistance against some of the insect-pest is are compiled in Table 5.10. of these, TVu 4557 (VITA 5) was resistant to leafhoppers and pod borers, TVu 410 was resistant to aphids and pea aphids and TVu 1509 was resistant to leafhoppers and thrips.

Insect-pests : Lentil crop is attacked by a number of insect-pests. The most depredations (Haw tin and Chancellor, 1979) are due to pod borers *(Etitlla zinckenella)*, aphids *(Aphids* ssp.), weevils *(Stona lineatus* L.), bruchids *(Bruchus* ssp.) and cutworms *(Agrotis* ssp.). Lentils taken from archaeological sites in Egypt have revealed the same degree of infestation and by a member of the same Bruchidae genus as could be expected today (Southgate, 1978). Pea aphids *(Acyrthosiphon pisum* Harris), cowpea aphids *(Aphis craccivora* Koch), thrips *(Megalurothrips* ssp.), seed com maggots *(Hylemya platura* Meigen), and wire worms *(Limonius* and *Ctenicera* ssp.) are the most commonly recorded pests of lentils in the Palouse, U.S.A. (Summerfield *et al.*, 1982). Many of these are common in alfalfa fields from where they migrate when the crop is harvested. According to Kooner and Chhabra (1980) the lentil pod borer *(Etiella zinckenella* T.) is the key pest in India. It damages the floral parts and pods. Black aphid *(Aphis craccivora)* is a secondary pest and causes the leaf damage.

Breeding for resistance to insect-pests in lentil : Sandhu and Malhotra (1980) screened germ plasm of lentil against pod-borer by sowing susceptible material all round and between the experimental materials and selected the resistant lines. They reported 1942 and *P202* as resistant to pod-borer. Chhabra (1980) reported *P202,* P248, *P262, P289,* P772, P803, 1'927 as resistant to pod-borer.

Insect-Pests of Moongbean

No information is available on the genetics of resistance in insect-pests in mungbean. However, the major insect pests or the one which are minor but may cause considerable damage in certain localities/seasons are described as follows:

Whitefly : The *Bemisia tabaci* (Genn.) is the most important species of whitefly reported as a vector of plant diseases considered to be due to virus (Coasta, 1976). It is the vector of more than 25 different diseases (Costa, 1969; Varma, 1963). It transmits MYMV in mungbean and also in certain other grain legumes. There are about 12 overlapping generations of whitefly in a year but it is most active during May to September (Kooner and Chhabra, 1980), In addition to transmitting MYMV, the nymphs and adults suck the sap of the plant and lower its vitality.

Hairy caterpillar : Hairy caterpillars *Diacrisia oblugua* Walker, *Amsacta moorei* Butler and *A. albistriga* Walker) though polyphagous, are sporadic. It feeds on the foliage of nearly 40 cultivated plants. The attacked leaves show a fine network like skeleton of the veins of the leaves. The advanced stage caterpillars nibble at the leaves in the solitary phase and eat away the entire leaf tissue, leaving only the midrib. The pest can damage the crop completely if the population is very high and it is not controlled with insecticides.

Semilooper : The semilooper (*Anomis flava* F.) is considered a minor pest of mungbean. It is sporadic but in certain years, it may cause serious damage. The green larvae feeds voraciously on the leaves and other tender parts and may defoliate the whole field within a few days (Kooner and Chhabra, 1980)

Gacucid beetle (*Madurasia obscurella* J.) is a small but very active dark-brown insect. The adults feed on foliage and stipple the leaves. With small, circular holes. It cause considerable damage to mungbean crop from July to October in India.

Jassids : The nymphs and adults of *Empoasca* sp. suck the juice from the mesophyll of the leaves. Consequently, the leaves turn brown and the leaf surface becomes uneven. In case of servere infestations, the leaves dry up.

Grey weevil : The weevil (*Myllocerus* sp.) is more common on soybean than on mungbean and blackgram. It feeds on leaves and makes hole in them from the early stages of growth to the harvesting.

Stemfly : The stemfly (*Melanogromyza phaseoli*) is a very small fly which cause serious damage to young plants. The maggots after mining the leaves travel towards the stem through the petiole and generally kill the young plants completely.

Breeding for Resistance to Insect-Pests in Mungbean

Mungbean is also a host to number of insect-pests which cause considerable damage to the crop from the seedling stage to reproductive stage in field and also during storage. The information on some of the important insect-pests pertaining to screening and sources of resistance is presented in the following paragraphs.

Screening for resistance to insect-pests in mungbean : The germplasm of mungbean can be screened against whitefly (*Bemisia tabaci* Genn.) and jassids (*Empoasca* sp.) in the field. All the entries may be grown in a single or paired row of 3 to 5 m length, alongwith the infector. When the incidence reaches its peak on the infector it may be cut down and left in situ to facilitate the migration of insect-pests to neighboring test entries. The population of insect-pest(s) may be recorded from 2 to 3 spots in each entry after 10 to 15 days of the infector is cut down. The split wooden cages of 60 cm height and 25 cm diameter (covered with black cloth from all sides except one side which was a glass pane) may be used (Chhaba and Kooner, 1980a). The lines may be divided into 5 different categories depending upon the population levels of 'the insects. The selected entries may be grown in replicated trial and the plots may be surrounded with infector rows. When the pest population reaches to its peak the infector rows are cut down and after 10 to 15 days the pest population may be recorded with split cages randomly from 3 to 5 spots per plot. The average~ population of insect-pest(s) for each entry is calculated. The entries may be tested at "hot spot". The resistance of the best line(s) may be tested in the screen-house under artificial conditions and may be used in on-going breeding programme. This technique can be used for screening resistance against whitefly and jassids in the grain legumes like black gram and cowpea.

A pure culture of bruchids (*Callosobruchus* sp.) is maintained in the laboratory. The counted seed samples in pot/container are kept and 10 pairs of insect-pest(s) are released in each container covered with muslin cloth. The insects are removed after 24 hrs. Only 200 eggs per entry may be left and surplus eggs may be removed. The infected seeds arc divided into four replications in different containers and covered with muslin clolh. The numbers of nymphs that emerge from the eggs per plot are counted. The number of damaged seeds is also counted. The percentage of infesled grains and also damage in tenns of weight ean be assessed. The resislant lines can be selected and may be used in breeding programme. This technique ean effectively be used against bruchids in other grain legumes.

Sources or resistance to insect-pests in mungbean : The resistant lines LM 115 is resistant beanfly and aphids. The whitefly resistant (ML I, ML 3, ML 5, LM 170, ML 26) have also been reported to be resistant against MYMV. The whitefly is the vector of MYMV. The use of such lines in breeding programme will facilitate the incorporation of resistance against the virus and the vector simultaneously. ML 8 and ML 4 are not only resistant beanfly but also for bacterial blight and *cercospora* leaf spot and root, stem rots, respectively. A total of 3713 mungbean accessions were screened (Chiang and Talekar, 1980) for resistant to the beanfly, *Ophiomyia phaseo/i* Tryon) and two other agromyzid, *O. centrosemates* (de Meijere) and *Melanogromyza sojae* (Zehntner) in mass screening trials. No immune but three mungbean accessions were least affected and were highly resistant to all the three species.

'Breeding for Resistance to Insect-Pests in Black Gram

Whitefly and aphids cause considerable losses directly to blackgram and are also vector of MYMV and leaf crinkle virus, respectively and thereby cause considerable reduction in yield indirectly. The leaf eating larvae (*Amsacta moori* Butler and *A. olbistriga* walker) also cause considerable losses to blackgram. Pod borers, *Apion ampulum* and *Maruca testulis* feed on developing seeds and empty pods are Bruches species though attack in field and are carried to store where they continue to damage the grains. Root knot nematodes (meloidogyne incognita) also cause reduction in yield.

Efforts to screen germplasm in blackgram aginst insect-pests have been negligible and perhaps no attempt has been made to incorporate the genes for resistance. Chhabra *et al.* (1984) have reported LU 178, LU 190 as resistant to whitefly (*Bemisia tabaci*) and LU 178, LU 196 and LU 126 as resistant to jassids (*Emposca moti* and *E.terminails*). These lines are also resistant to MYMV, suggesting the possibility on incorporating the genes against these insect-pests and MYMV.

References

Ahmad, T. 1938. The tur podfly, *Agromyza obtusa* Mall., a pest of *Cajanus cajan. Indian J. Agric. Sci.*, 8: 63-76.

Amosu, J.O. 1974. The reaction of cowpea (*Vigna unguiculata*(L.)Walp.) to the root knot nematode (*Meloidogyne incognita)* in Western Nigeria. *Niger.Agric. J.*, 11:165-169.

Auclair, J.L., Maltals, J.B. and Cartier, J.J. 1957. Factors in resistance of peas to the pea aphid, *Acyrthosiphon pisum* (Harr.) Homoptera Aphididae. II. Amino acids. *Canadian Ent.*, 89: 457-464.

Bhosale, D.J. and Nawale, R.N. 1983. Relative susceptibility of pigeonpea germplasm to gram pod borer. *J. Maharashtra Agric. Univ.*, 8: 30-31.

Bindra, O.S. and Singh, H. 1972. *Melanogromyza obtusa* Malloch. *Pesticides*, 6(7): 11-14.

Chhabra, K.S. 1980. Mechanism of insect-pest resistance in pulse crop. *In*: Breeding Methods for the Improvement of Pulse Crops (ed., K. S. Gill), Ludhiana, India, pp. 142-160.

Chhabra, K.S. and Kooner, B.S. 1980. Screening of breeding material for insect-pest resistance in pulse crops. In: Breeding methods for the Improvement of Pulse Crops (ed. K. S. Gill). Ludhiana, India, pp.161-170.

Chhabra, K.S. Kooner, B.S. Saxena, A.K. and Sharma, A.K. 1984. Influence of biochemical components on the incidence of insect pests and yellow mosaic virus in blackgram. *Indian J.Ent.* 46: 148-156.

Chiang, H.S. and Talekar, N.S. 1980. Identification of sources of resistance to the beanfly and two other agromyzid files in soybean and Mungbean. *J.Econ. Ent.*, 73: 197-199.

Costa, A.S. 1969. White-flies as virus vector. *In* : Viruses, Veetors and Vegetation (ed., K. Maramorosch). Interscience, New York. pp. 95-119.

Costa, A.S. 1976, White-fly transmitted Plant diseases. *Ann. Rev. Phytopathol.*, 14: 429-449.

Dass, S. B. and Odak, S. C. 1987. Biochemical basis of resistance in pigeon pea pod walls to pod fly, Melanagromyza obtuse malloch (Diptera: Agromyzidae). *Crop Improv.* 14: 64-68.

Gangrade, G.A. 1963. Assessment of damage to tur (*Cajanus cajan*) in Madhya Pradesh by Tur podfly, *Agromyza obtusa* Malloch. *Indian J. Agric. Sci.*,33: 17-20.

Gangrade, G.A. 1964. Losses to tur (*Cajanus cajan*) by *Melanagromyza obtusa* Malloch. *Indian J. Ent.*, 26: 364-365.

Hare, W.W. 1959. Resistance to root-knot nematode in cowpea. *Phytopathology*, 49:318 (Abstr.).

Harrington, C.D. 1945. Biological races of the pea aphid. J. Econ. Entomol., 38: 12-22.

Hawtin, G.C., and Chancellor, G.J. 1979. Food Legume Improvement and Development. IDRC, Ottawa, Canada. Pub. 126e.

IITA, 1973. International Institute of Tropical Agriculture. Grain legume Improvement Program. Report for 1971 and 1972, Ibadan, Nigeria.

Jackal, L.E. 1982. A field screening technique for resistance of cowpea (*Vigna unguiculata*) to the pod-borer,maruca testulalis (Geyer) (Lepidoptera:Pyralidae). *Bull.Ent. Res.*, 72:145-156.

Karel, A.K. and Malinga,Y. 1980. Leafhopper and aphid resistance in cowpea varieties. *Trop. Grain Leg. Bull.*, 20; 10-11.

Kooner, B.S. and Chhabra, K.S. 1980. Pests of Pulse Crops and their control. *In*: Breeding Methods for the Improvement of Pulse Crops (eds., K. S. Gill), Ludhiana, India, pp-132-141.

Lateef, S.S., Bhagwat, V.R. and Reed, W. 1981. Screening of chickpea cultivars for borers (*Heliothis armigera*) susceptibility in pesticide free conditions at ICRISAT center. *Intl. Chickpea Newsl.* 5:13-14.

Lateef, S.S. 1985. Gram pod borer (*Heliothis armigera* Hub.) resistance in chickpeas. *Agril. Eco. and Environ.* 14:95-102.

Lateef, S.S. and Reed, W. 1981. Survey of insect pest damage in farmers fields in India. *Intl. Pigeonpea Newsl.* 1: 29-30.

Lateef, S.S., Reddy, L.J., Reed, W. and Faris, D.G. 1981. *Atylosia scarabaeoides*: a source of resistance to *Heliathis armigera. Intl. Pigeonpea News,* 1: 32-34.

Lemos, J.W.V. and da Ponte, J.J. 1978. Cultivars de feijaadecorela, *Vigna sinensis* (L.) Savi, résistentes a meloidoginose. *Bol.Soc.Coarense Agron.*, 19 :1-19.

Mani, A. and Sethi, C.L. 1984. Reaction of certain chickpea to root-knot nematode, *Meloidogyne incognita. Intl. Chickpea Newsl.* 10:15.

Nolte, H.W. 1959. Investigations on control of the pea moth (*L. nigricana* Steph.). *Albrecht-Thaer-Arch.*, 3: 146-157.

Ogbuji, R.O. 1978. Reactions of cowpea (*Vigna unguiculata)* to *Meloidogyne incognita* in Eastern Nigeria. *Pans*, 24; 32-34.

Orton, W.A. 1902. The wilt disease of cowpea and its control U.S.Dep. *Agric. Bur. Pl. Ind. Bull.*, 17:9-20.

Patel, G.J., Shah, H.M. and Patel, D.J. 1977. Screening of cowpea cultivars against root-knot nematodes. *Indian J.Nemat.*, 7:169-170.

Raman, K.V., Singh, S.R.and van Emden, hHF. 1978.Yield losses in cowpea following leafhopper damage. *J.Econ.Entomol.*, 71;936-938.

Rembold, H. 1981. Malic acid in chickpea exudate-a marker for *Heliothis* resistance. *Intl. Chickpea Newsl.* 4:18-19.

Rizvi, S.M.A. and Singh, H. M. 1980. *Autographa nigrisigna* a pest of chickpea. *Intl.Chickpea Newsl.* 3:15.

Sandhu, T.S. and Malhotra, R.S. 1980. Objectives, breeding approaches and achievements in lentil (*lens culinaris* Medik). In: Breeding Methods for the Improvement of Pulse Crops (ed., K.S.Gill), Ludhiana, India. pp. 260-271.

Saxena, H.P. and Raina, A.K. 1970. A bruchid resistant strain of bengal gram. *Curr. Sci.* 39:189-190.

Sharma, G.L. and Mathur, B.N. 1985. Reaction of chickpea cultivars to the root-knot nematode, *Meloidogyne incognita. Intl.Chickpea Newsl.* 12:31-32.

Sharma.N.K. and Sethi, C.L. 1976. Reaction of certain cowpea varieties to *Meloidogyne incognita* and Heterodera cajani. *Indian J.Nemat.,* 6 :99-102.

Singh, R.H. and Singh, K.M. 1978. Succession of insect-pests in early varieties of redgram, Cajanus cajan (L.) Millsp. *Indian J.Ent.,* 40:106.

Singh, S.R. 1977. Cowpea cultivars resistant to insect-pests in world germpasm collections, Trop. *Grain Leg. Bull.,* 9:3-7.

Singh, S.R. and Allen, D.J. 1980. Pests, diseases, resistance and protection in cowpeas. In: Advances in Legume Science (eds., R/J. Summerfield and A.H. Bunting). Royal Bot. Gardens, Kew. London. Pp 419-443.

Southgate, B.J. 1978. Bruchidae. In: Pests of Grain Legumes: Ecology and Control (eds., S.R.Singh), H.F. van Emden and T.A. Taylor). Academic Press, London. Pp. 219-229.

Sulyndin, A.F. 1947. The selective ability of *Bruchus* among pea varieties. *Agrobiology,* 2: 105-109.

Summerfield, R.J.Muehibaur, F.J. and Short, R.W. 1982. Description and culture of lentils. U.S. Dep. *Agric. Production Res. Report* No.181, pp. 22.

Thomason. I.J. and Mckinney, H.E. 1960. Reaction of cowpea, Vigna sinesis, to root knot nematode, Meloidogyne spp. *Pl. Dis, Reptr.,* 44: 51-53.

Ufer, M. 1948. The possibility of breeding *Bruchus* resistant peas. *Bodenkultur,* 2:28-35.

Varma, P.M. 1963. Transmission of plant viruses by whiteflies. *Natl. Instt. Sci. India, Bull.,* 24: 11-33.

Webber, H.J. and Orton, W.A. 1992. A cowpea resistant to root-knot *(Heterodera)*

Yadav, L.S. and Yadav, P.R. 1983. Pest complex of cowpea (*Vigena sinensis savi*) in Haryana. *Bull. Ent.,* 24: 57-68.

□□□

Ecologically Based Integrated Pest Management, 2012
© *D.P. Abrol and Uma Shankar (eds.), pp. 553-567*
New India Publishing Agency, New Delhi (India)
e-mail : info@nipabooks.com; website : www.nipabooks.com

29

Breeding for Insect Pests Resistance in Oilseed Crops

S.K. GUPTA AND D.R.C. BAKHETIA

In practical agriculture, resistance represents the ability of a certain variety to produce larger crop of good quality than an ordinary variety at the same level of insect population. During the 19th Century, the French wine industry survived a major set-back largely due to the use of resistant root stocks imported from America and which still maintain their resistance levels against the beetle pest (Painter, 1951). Another important example of resistant variety is the U-4 variety of cotton possessing high level of resistance to the jassid, *Emposasca* spp. (Parnell, 1925). The excessive use of pesticide on oilseed crops needs to be minimized owing to several reasons: (i) most of the pesticides are lipophilic and are likely to be carried in the oil (ii) in a few cases like *Brassica,* leaves and inflorescence are consumed as saag (pot herb), (iii) the oilseed crops provide the raw material for the vanaspati industry, cosmetics, etc., (iv) many of the oilseeds and oils are directly consumed without further processing, and (v) the oil cakes are used as cattle feed. Thus there is an urgent need to exploit the non-pesticidal methods of pest control, among which the use of resistant varieties assumes a great significance. The details of each oilseed crops are given below:

Rapeseed-Mustard

Among the oleiferous crops, rapeseed and mustard occupy an important place for their value as a source of oil and cake. The yield of these oilseeds is being limited by several factors but ravages caused by the insect-pests are quite considerable. The ill effects of chemical control of insect-pests have necessitated the search for suitable alternatives. Among these, role of resistant varieties in integrated pest management (IPM) of insect pests has been recognized as very useful and practical preposition. Under the prevailing situations, the resistant varieties comprise the foremost strategies of pest management because of their well documented advantages. In India, in the recent past, extensive work has been done on various aspects of HPR to the key pest (*Lipaphis erysimi*) but the work on the other major insect-pests of rapeseed and mustard has been negligible.

Insect Pest Complex

The number of insect pests associated with the rapeseed and mustard crops has increased from 24 (Rai, 1976) to about 40 (Bakhetia and Sekhon, 1989) owing to changed cropping patterns and introduction of new crops like *Brassica napus* and B. *carinata*. The mustard aphid, *Lipaphis erysimi* (Kaltenbach) is the key pest responsible for bulk of yield losses. The other major and minor insect pests in order of their severity are listed in Table 1.

Mustard Aphid

The mustard aphid, being a cosmopolitan, attains more importance on *Brassica* crops in which it can cause yield losses upto 96 per cent under varied agro-ecological situations (Baketetia, 1989; Landin, 1982). The continuous feeding of the mother aphids and their colonies drains out cell sap resulting plants with yellow, curled, crinkled and finally failed to grow.

Screening Techniques

Various screening criteria have been employed by different authors for evaluation of cruciferous genotypes in order to get the résistance sources. This aspect has been extensively reviewed recently by Bakhetia and Singh, 1992 (Table 2).

Table 1 : List of important insects pests of rapeseed-mustard

S.No	Name of the pest	Nature of damage
Key Pest		
1	Mustard aphid *Lipaphis erysimi* (Kalt)	Sucks sap from leaf, inflorescence ad pods
Major pests		
2	The saw-fly *Athalia Lugens proxima* (Klug)	Feeds on leaves of young seedlings
3	Painted bug *Bagradaa hilaris* (Burm.)	Sucks sap from leaves of seedlings, and also from pods and seeds
4	Pea leaf miner *Chromatomyia horticola* (Goureau)	Mines the leaves
5	Bihar hairy caterpillar *Spilosoma oblique* walker	Defoliates the vegetative parts
Minor Pests		
6	Cabbage butterfly *Pieris brassicae* (Linn.)	Defoliates the foliage
7	Flea beetle *Phyllotreta cruciferae* (Goeze)	Makes holes in young leaves
8	Green peach aphid *Myzus persicae* (Sulzer)	Sucks sap from the leaves flowers and pods
9	Diamond back-moth *Plutella xylostella* (Linn.)	Feeds on leaves and cuts holes
New Pests		
10	The larger moth *Crocidolomia binotallus* (Zeller)	Damages the inflorescence
11	Cabbage top borer *Hellula undalis* Fab.	Damages the inflorescence
12	Cabbages aphid *Brevicoryne brassicae* (Linn.)	Sucks sap from the foliage and inflorescence
13	The white fly *Bemisia tabaci* Linn.	Sucks sap from the leaves
14	Cabbages semi-looper *Plusia orichalcea* (Fab)	Defoliates the leaves etc.
15	Turnip moth *Agrotis segetum Dennis & Schiff*	Cuts the germinating Seedlings
16	Ants *Monomorium* sp	Takes away the seeds before germination
17	Pod fly *Dasineura brassicae* (Winnertz)	Damages the flowers and pods

Source: Kalia and Gupta (1997): Recent advances in oilseed brassicas.

Table 2 : Resistance source identification criteria used in *Brassica* species against the mustard aphid.

Technique	Resistance modality	Remarks
Seedling Survival	Non-preference, antibiosis, tolerance	useful in preliminary screening in screen house
Aphid	Injury	Resistance, tolerance suitable for specific screening on a smaller scale in screen-house and field too
Aphid Population	Non-preference, antibiosis, tolerance	A far more laborious method suited best for screening on small scale in the field or screen house
Fecundity and development	Antibiosis	Screening of select germplasm preferably in laboratory or under caged conditions
Grain yield	Tolerance	Best suited for large scale screening in the field or under specially designed cages
Honey-dew Secretion	Host-food	Most laborious and time consuming best suited for laboratory suitability, resistance testing of a few selected ones only

Source : Kalia and Gupta (1997): Recent advances in oilseed brassicas.

Seedling Survival

This technique was developed by Bakhetia and Bindra (1977) which advocates using the optimum level of aphid population per seedling or plant under the screen house conditions. This technique of resistance screening against the mustard aphid is widely accepted because of its higher efficiency and ease in application at seedling stage, thus has become an integral part of mass screening programme.

Aphid Injury

Aphid damage symptoms expressed as injury grades (0-4) were proposed by Pathak (1961) and Teotia and Lal (1970) while Rajan (1961) and Bakhetia and Sandhu (1973) used a different set of injury grades formed on the basis of injury symptom (Table 3).

Table 3 : Scoring methods used by different workers for aphid resistance screening in rapeseed-mustard

Injury symptoms as explained in different reports			
Injury Grade	**Pathak (1961), Teotia and Lal (1970)**	**Rajan (1961)**	**Bakhetia and Sandhu (1973)**
0	Normal	Plants unaffected by aphids	Free from aphid infestation
1	Yellowing in a few leaves	Apparently healthy plants with a few young ones in the flower	Normal growth, no yellowing or curling of leaves, a few aphids without injury symptoms, flowering and pod setting almost normal
2	Yellowing and curling of some leaves, medium growth no	Plant severely infested, inflorescene and pods full of	Average growth, flowering and pod setting; curling and yellowing of few leaves aphids, no visible symptoms of damage
3	Poor (stunted) Growth and flowers	Plants completely damaged and in different stages of decay	Curling and yellowing of some branches, below average growth, poor flowering with very little pod setting
4	Stunting, plants defoliated and	--	Very poor growth, heavy curling of leaves, stunting of plants, a little or no flowering and pods formation. Heavy aphid colonies
5	--	--	Severe stunting of plants, curling and yellowing of almost all leaves. No flowering and pod formation. Plants full of aphids.

Source : Kalia and Gupta (1997): Recent advances in oilseed brassicas.

Aphid Population

Pathak (1961) screened the Brassica germplasm at the flowering stage on the basis of the number of aphids/plant part under laboratory conditions which are given in the following grades:

Identified Aphid Resistant Strains/Varieties

There are numerous reports on testing of cruciferous germplasm for resistance to aphid and further resulted in identification of good number of lines showing promise against the mustard aphid under varied climatic conditions (Table 4).

Table 4 : List of lines/strains of cruciferous oilseeds reported as possessing resistance/ tolerance against the mustard aphid

Species	Strains/lines	Modality of Resistance	Reported by
B.alba	Lethbridge 22 A	R	Sachan and Singh (1983)
	Metapollaska		
B. campestris	IB 787	R	Prasad (1983)
	Pusa Kalyani	T	Singh *et al.* (1982)
B. hirta	ACC 751516	R	Bakhetia (1985)
B. juncea	Laha 101	HR	Lal (1969), Teotia and Lal (1970)
	T 6342	R	Bakhetia & Sandhu (1973)
			Asthana and Das (1978)
			Brar and Sandhu (1978)
			Chander (1993)
	RLM 198	R	Bakhetia (1975)
			Bakhetia & Bindra (1977)
			Brar and Sandhu (1978)
	RK 2, RK 9, RH 785	R/T	Bakhetia *et al.* (1982)
Rai-75	-1	R	Singh *et al.* (1982)
	T 16	R	Chand and Singh (1977)
	IB 680	R	Prasad (1983)
	RC 1425	T	Yadava *et al.* (1985)

Contd...

	(non-waxy)		Bakhetia (1985)
	B 85 (Glossymutant White flower)		Chatterjee and Sen-Gupta (1987)
	IM 76-103	HR	Ran et. al. (1989)
	RLC 1036, RLC 1037 RJ LM 198, PR 43, RW-white glossy, B 85, RLM 198,	R	Rohilla (1990)
	T 6342, RH 7361, RH 7847	R	Kumar et. al. (1990)
	T 6342, RE 5, RW-2-2, RW-32-2, B 85	R	Chander (1993)
B. napus	PI 171538	R	Jarvis (1970)
	GSL 1501, GS 47, GS 140, GS 825	R	Rohilla (1990)
	GSL 8851 & 8965	R	Chander (1993)
C. abyssinica	PI 247310	R	Jarvis (1969)
Eruca. sativa	ITSA	HR/T	Bakhetia & Bindra (1977) Narang (1982), Dilawari & Dhaliwal (1988)
	T 27	HR	Chander (1993)

HR=Highly resistant; T=Tolerant; R=Resistant; LS=Low Susceptibility; S=Susceptible; HS=highly susceptible
Modified after Bakhetia and Singh (1993)
Source : Kalia and Gupta (1997): Recent advances in oilseed brassicas.

Morphological Traits Versus Resistance

The morphological traits of plants imparting relative resistance to aphid injuries are not clearly understood (Table 5). More hairy plants of E. *sativa* were not preferred by the aphid for settlement, movement and oviposition and thus acted as a barrier (Rai and Sehgal, 1975; Narang, 1982). Chander (1993) recorded that T27 showed minimum aphid population in the field because of its antibiosis and not due to the presence of hairy structures around the inflorescence. It is concluded that possibly the role of biochemical constituents is of greater significance in the relative resistance of rapeseed-mustard varieties.

Table 5 : Association of morphological characters of plants with the aphid resistance

Characters	Species	Reaction	Author (s)
Smaller and hardy inflorescence with loose packing of buds	*B. juncea*	Resistance	Rai and Sehgal (1975)
Darkers leaves	*B. juncea* B. *campestris* *(Brown sarson)*	Resistance	Pathak (1961)
Non-waxiness (Glossy)	*B. rapa,* *B. napus*	Resistance	Srinivasachar and Malik (1972) Yadava *et al.* (1985) Angadi *et al.* (1987) Bakhetia *et al.* (1992) Narang (1982)
Profuse branching and wider angle of orientation	*B. juncea* *B. campestris*	Negative correlation with aphid population	
Hairiness Yellow or greenish-yellow leaves	*E.sativa* *Brassica* spp	Anti-xenosis Attraction	Narang (1982) Dilawari (1985)
White petal mutant	*B. juncea*	Anti-xenosis and negative correlation with population	Bakhetia *et al.* (1992)
Leaf colour, leaf thickness, plant height, inflorescence (type and colour), leaf size, leaf hairness	*Brassica* spp *E. sativa*	No. correlation with aphid population in the field	Chander (1993)

Source : Kalia and Gupta (1997): Recent advances in oilseed brassicas.

Biochemical Constituents Versus Resistance

Many workers have tried to attribute the aphid resistance phenomenon to the biochemical composition of the plants (Table 6 & 7). Total sugars and sulphur contents in Brassica species have been found to be co-related negatively with aphid reproduction and population build-up (Bakhetia *et al.,* 1982). Higher concentrations of total sugars and iron are known to impart resistance while nitrogen, zinc, copper and manganese seem unrelated (Narang, 1982). It has also been found that total protein, free amino acids, reducing sugars and total sugars were higher in susceptible strains while higher flavanoid contents in B. napus imparted resistance to the aphids (Malik, 1981).

Table 6 : Biochemical constituents versus aphid resistance in cruciferous species

S. No.	Biochemical constituents(s)	Biological effect	Plant species	Author(s)
1	Glucosinolates and sinigrin	Toxicant/growth inhibition/deterrent	*Brassica* spp.	Anand(1976) Malik (1981) Rani(1981) Kumar et. al (1990) Chander (1993)
2	Total sugars, flavanoids, phenols, ascorbic acid K,S, Fe	Growth inhibition and toxicant	*Brassica* spp	Rani (1981) Narang (1982) Gill and Bakhetia (1985) Chander (1993)
3	Proteins, free amino acids, ash N,P	Stimulants	*Brassica* spp	Malik (1981, 1988) Gill and Bakhetia (1985)
4	Free amino acids, N,Cu, Zn, Mn	No relation	*E.sativa* *Brassica* spp	Kundu and Pant (1967), Narang (1982) Weibull and Melin (1990) Chander (1993)
5	Gluconapin, Allyl isothiocyanate, volatile isothiocyanate	Stimulant, repellent, attractant	*Brassica* spp *E. sativa*	Malik (1981) Dilawari (1985) Chander (1993)

Source : Kalia and Gupta (1997): Recent advances in oilseed brassicas.

In the olfactometer studies, higher and lower release rates of allylisothiocyanate proved as repellent and stimulant, respectively (Dilawari 1985; Dilawari and Atwal 1989b). Similarly, Dilawari and Atwal (1987) had found that the number of probes increased and feed intake drastically reduced on the artificial media containing higher levels of glucosinolates.

Table 7 : Morphological, anatomical and biochemical factors of resistance in rapeseed-mustard against the leaf miner.

S.No.	Character/chemical	Species	Response/effect	References
		Morphological		
1	Plant apparency wider angles	Brassica spp.	Easy host finding	Singh (1985)
2	Thinner and non-hairy leaves	Brassica spp.	Host selection	Singh *et al.* (1986)
3	Yellow colour	Brassica spp.	Attraction	-do-
		Anatomical		
1	Thick cuticle, thick epidermis, closely packed nesophyll tissue	Brassica spp.	Reduced larval feeding and adult oviposition	-do-
2	Tissue proliferation	*B. juncea*	Destruction of eggs and larvae	-do-
		Biochemical		
1	Carotenoid pigments	*B. juncea*	Visual cues attraction	
2	Glucosinolates, volatile ITC	Brassica spp.	Oviposition stimulant/ deterrent larval feeding and growth inhibitor, olfactory attractant etc.	Singh and Bakhetia (1983)
3	Phenolics and sinaptic acids	Brassica spp.	Feeding inhibition oviposition deterrent toxicant	Singh (1985) Singh *et al.* (1986)
4	Absence of histidine and arginine	Brassica spp.	Oviposition inhibition	
5	Progoitrin	Brassica spp.	Growth and feeding inhibition	

Source : Kalia and Gupta (1997): Recent advances in oilseed brassicas.

Resistance to Groundnut Pests

It has been reported that the bunch varieties had higher population of, and showed more injury owing to, leaf-webber, aphid and grey weevil, whereas the spreading and semi-spreading types harboured lower populations (Bakhetia, 1969-1971, 1977, 1978; Brar and Sandhu, 1975). This difference may be due to more of foliage and more of succulence of the bunch types which provide conditions favourable for providing with better shelter and preferred food.

Resistance to Sesame Leaf-Roller

It has been reported that the branching pattern and maturity of the plants are positively correlated with their resistance to leaf-roller. Profuse branching and early maturing varieties provide conditions which are more conducive to pest development. The late maturing varieties NP 29, C 1036 and Chanda 3 have been reported to the less susceptible to the pest.

Breeding for Resistance to Insect-Pests

A breeding programme for resistance to insects depends on their behavior and the nature of the crop involved as well as on the information available at the beginning. Painter (1951) suggested the following steps for such a programme:

i) Discover possible sources of resistance to insects among varieties and strains locally available.

ii) Search for new germplasm that may carry resistance.
   a) New selections out of older varieties.
   b) Plant introductions out of older varieties.
   c) Varieties of related species of plants, if necessary.

iii) Determination of some of the basic properties of the plants responsible for resistance; at least a separation of the three classes: preference, antibiosis and tolerance.

iv) Hybridization to combine genes for resistance with desirable agronomic characters.

v) Study of genetics of resistance to insects where possible.

vi) The testing of resistance in advanced-generation hybrids.

vii) The study of resistance of released varieties in plots and on farms to evaluate as an insect control method.

In case where selection for resistance to a particular pest is difficult, perhaps because assessment is complex or environmental effects mask the genetic expression of resistance. In such cases, it may be possible to use indirect selection to achieve the objectives by selecting for early recognizable characters which is highly correlated with resistance or pleiotropic effects. Either way they can provide convenient method for the selecting insect resistance strains.

Summary and Conclusion

The more important oilseed crops comprise groundnut, rapeseed-mustard and sunflower which occupy a pride place in the present-day agriculture.

These are valued for their edible oils for human consumption and the cakes for animals. In India, the former two crops contribute largely to the oilseed production while sunflower is gaining importance. The important cruciferous oilseeds grown on commercial scale world over include *Brassica campestris*, B. *napus*, B. *juncea* and B. *carinata*.

Resistant varieties are very useful in pest management. The varieties reported to be resistant to a particular pest (or pests) should be introduced directly for cultivation in certain areas. Such resistant varieties should be used extensively in a hybridization programme aimed at incorporating pest resistance into the well adapted varieties of the region. Similarly, the species known to carry pest resistance may be used in inter-specific or intergeneric crosses through the genetic and cytogenetic techniques known for chromosomal manipulations. Mutation breeding is also an important tool for developing new varieties with desired characters of pest and disease resistance.

Among the alternative approaches, use of host plant resistance (HPR) in IPM programme has been given considerable importance in the past two decades, which has resulted in the development and standardization of suitable techniques for germplasm screening in the screen house and fields as well.

Despite these endless scientific endeavours, it is unfortunate that resistant/tolerant parents have not been evolved which could be possibly used in conventional breeding programmes to develop new resistant varieties against the mustard aphid. However, the identification of resistance imparting gene would solve the problem through more inter-generic transfer of genes from Brassica species. The more recent advancements in biotechnology have explored the possibility of transferring the foreign genes like NPT II, DHFR (mouse), napin and NOS into B. *napus* and B. *oleracea* plants directly or through *Agrobacterium tumefaciens* mediated genes transfers.

References

Anand, I.J. 1976. Oil quality vis-à-vis aphid resistance in Brassicae. Paper presented in the seminar-cum-workshop on oilseed research in India. PKV Nagpur, April, 5-9, 1976.

Asthana, K.S. and Dass, S.R. 1978. Varietal evaluation of rapeseed-mustard to aphids (L. erysimi). *Proc. Bihar Academy Agric. Sci.* 24: 184-186.

Bakhetia, D.R.C. 1989. Current status of breeding for aphid resistance in rapeseed-mustard. In Bakhetia D.R.C, Kurl, S.P. and Rajendran, T.P.(Eds). The mustard aphid. Aphidological Society, India. MMPG College, Modinagar.

Bakhetia, D.R.C. and Bindra, O.S. 1977. Screening technique for aphid resistance in Brassica crops. *SABRAO* J. 9: 91-107.

Bakhetia, D.R.C. and Sandhu, R.S. 1971. Preliminary studies on the resistance of some Brassica species to the leaf miner, *Phytomyza horticola* (Meigen) infestation. Proc. 58th Indian Sci. Congress Part II, p. 753 (Abstract).

Bakhetia, D.R.C. and Sandhu, R.S. 1977. Susceptibility of some *Brassica* species and *Crambe abyssinica* to the leaf miner, *Phytomyza horticola* (Meigen). *Crop Improv.* 4: 221-223.

Bakhetia, D.R.C. and Sekhon, B.S. 1989. Insect pests and their management in rapeseed-mustard *J. Oilseeds Res.* 6: 269-299.

Bakhetia, D.R.C.,Sekhon, B.S. and Chatterjee, S.D. 1992. Management of mustard aphid: Review of the work from 1979-80 to 1989-90. Summing up concluding seminar on Indo-Swedish Collaborative Research Programme on Rapeseed-mustard Improvement. Organized by ICAR, Swedish Embassy and Swedish University of Agricultural Sciences, Uppasala, Sweden.

Bakhetia, D.R.C. and Singh P. 1982. Host plant resistance in oilseeds, pp: 208-248. In: Dhaliwal, G.S. and Dilawari, V.K. (Eds) Adv. Host. Plant Resistance to Insects. Kalyani Publishers, New Delhi.

Bakhetia, D.R.C., Sushma Rani and Ahuja, K.L. 1982. Effect of suphur nutrition of some Brassica plants on their resistance response to the mustard aphid, *L. erysimi* (Kalt). Calicut Univ. Res. J. (Special Conference Number), May 3-5, 1982, p. 36.

Brar, K.S., Rataul, H.S. and Labana, K.S. 1975. Differential reaction of mustard aphid *Lipaphis erysimi* (Kalt) to different rapeseed and mustard varieties under natural and artificial infestation. *J. Res. Punjab Agric. Univ.* 13: 14-18.

Chander, H., 1993. Identification and characterization of sources of resistance in some crucifers against the mustard aphid, *Lipaphis erysimi* (Kaltenbach), Ph.D. Dissertation, Punjab Agricultural University, Ludhiana 140 pp.

Dilawari, V.K. 1985. Role of glucosinolates in host selection, development and multiplication of mustard aphid, *L. erysimi* (Kalt). Ph.D. Dissertation, Punjab Agric. Univ. Ludhiana, 111 p.

Dilawari, V.K. and Atwal, A.S. 1987. Effects of cruciferous glucosinolates on probing pattern and feed uptake by mustard aphid, *L. erysimi* (Kalt). *Proc. Indian Acad. Sci.* (Animal Sciences) 96: 695-703.

Dilawari, V.K. and Atwal, A.S. 1989a. Response of mustard aphid, *L. erysimi* (Kalt) to visual stimulus -colour. *J. Res. Punjab Agric. Univ.* 26: 231-236.

Dilawari, V.K. and Atwal, A.S. 1989b. Response of mustard aphid, *L. erysimi* (Kalt) to visual stimulus -colour. *J. Res. Punjab Agric. Univ.* 2: 103-108.

Dilawari, V.K. and Dhaliwal, G.S. 1988. Population build-up of mustard aphid, *Lipaphis erysimi* (Kaltenbach) on cruciferous cultivars. *J. Insect. Sci.* 1: 149-153.

Downey, R.K. 1987. Breeding Brassica oilseeds crops. *IRDC: MR 153e*: 321-325.

Gill, R.S. and Bakhetia, D.R.C. 1985. Resistance of some Brassica *napus* and *B.campestris* strains to *Lipaphis erysimi* (Kalt.). *J. Oilseeds Res.* 2: 227-239.

Jarvis, J.L. 1969. Differential response of introductions of *Crambe* to the turnip aphid. *J.Econ. Ent.* 63: 1498-1502.

Kumar, V., Singh, D., Singh, H. and Singh, H. 1990. Combining ability analysis for resistance parameters to mustard aphid (*Lipaphis erysimi* (Kalt.). *Indian J. Agric. Sci.* 57: 227-279.

Kundu, G.G. and Pant, N.C. 1967. Studies on the *Lipaphis erysimi* (Kaltenbach) with special reference to insect plant relationship. I. Susceptibility of different varieties of Brassica and Eruca sativa to the mustard aphid infestation. *Indian. J. Ent.* 29: 241-251.

Landin, J. 1982. Life history, population dynamics and dispersal in the mustard aphid *Lipaphis erysimi*: A literature study, 18pp, Swedish University of Agricultural Sciences, Research Information Centre, Uppasala, Sweden.

Lindley, G. 1831. A guide to the orchard and kitchen garden. Longmans Press, London, 601 pp.

Malik, R.S. 1981. Morphological, anatomical and biochemical bases of mustard aphid, *Lipaphis erysimi* (Kalt.) resistance in cruciferous species. Sver. Utsad. *Tidsk.* 91: 25-35.

Narang, D.D. 1982. Studies on the basis of resistance in *Brassica campestris* var. yellow sarson, *B. juncea* and *Eruca sativa* to mustard aphid, *Lipaphis erysimi* (Kaltenbach). Ph. D. Dissertation, Punjab Agricultural University, Ludhiana 103 p.

Painter, R.H. 1951. Insect Resistance in Crop Plants. The University Press of Kansas, Lawrence, pp. 520.

Parnell, F.R. 1925. Breeding jassid resistant cottons. Emp. *Cotton Growers Rev.* 2: 330-336.

Pathak, M.D. 1961. Preliminary notes on the differential response of yellow and brown sarson and rai to mustard aphid (*Lipaphis erysimi* Kaltnbach). *Indian Oilseeds J.* 5:39-43.

Prasad, S.K. 1983. Varietal susceptibility of rapeseed and mustard cultivars to the aphid (L. erysimi Kalt.). *Indian J. Ent.* 45; 501-503.

Rai, B.K. 1976. Pests of Oilseed Crops in India and their Control, ICAR, New Delhi, pp. 121.

Rai, B. and Sehgal, V.K. 1975. Field resistance of Brassica germplasm to mustard aphid, *Lipaphis erysimi* (Kalt.). *Sci. & Cult.* 41: 444-445.

Rajan, S.S. 1961. Aphid resistance of tetraploid toria. *Indian Oilseeds J.* 5: 251-255.

Ram, S., Gupta M.P. and Maurya, R.P. 1989. Mustard varieties resistant to aphid (Lipaphis erysimi Kalt.) in India. *Trop. Pest. Managm.* 35: 150-153.

Rani, S. 1981. Role of sulphur nutrition in *Brassica species* for aphid resistance, M.Sc. Thesis, Punjab Agricultural University, Ludhiana, 74 pp.

Sachan, G.C. and Singh C.P. 1983. Relative susceptibility of some exotic *Brassica* and *Sinapsis* species to mustard sawfly, aphid and flea bettle. Annual Report Res. Dir. Ext. Stn. G.B. Pant Univ. Agric. Tech. Pantnagar.

Singh, P. and Bakhetia, D.R.C. 1983. Comparative development and host selection mechanism of the pea leaf miner, Phytomyza horticola Goureau on rapeseed and mustard. *Proc. 6th Int. Rapeseed Conf.* 17-19 May 1983, Paris, pp. 1135-1141.

Singh P., Bakhetia D.R.C., Sidhu H.S. and Setia R.C. 1986. Plant apparency and physical defences in Brassica species as index for resistance against the leaf miner, Chromatomyia horticola Goureau. In: Namopeth B. and Subhadvabandhu S. (Eds). New Frontiers in Breeding Researches; Kasetsart University, Bangkok, Thailand, pp. 755-772.

Singh, P., Mavi, G.S., Bakhetia, D.R.C. and Sidhu, H.S. 1985. Identification of leaf miner, Chromatomyia horticola Goureau punctures in *Brassica* strains. *Annals Biol.* 1: 226-228.

Singh, R.N., Dass, R., Saran, G. and Singh, R.K. 1982. Differential response of mustard varieties to *Lipaphis erysimi (Kalt.) Indian J. Ent.* 44: 408.

Teotia, T.P.S. and Lal., O.P. 1970. Differential response of different varieties and strains of Brassicae to aphid *Lipaphis erysimi* (Kalt.). Labdev *J. Sci. Technol.* 8B: 219-226.

Weibull, J. and Melin, G. 1990. Free amino acid content of phloem sap from *Brassica* plants in relation to performance of Lipaphis erysimi. *Ann. Biol.* 116: 417-423.

Yadava, A.K., Singh, H. and Yadava, T.P. 1985. Inheritance of non-waxy trait in Indian mustard and its reaction to aphids. *J. Oilseeds Res.* 2: 339-342.

❑❑❑

Section

V

IPM IN FIELD CROPS VEGETABLES AND FRUIT

Ecologically Based Integrated Pest Management, 2012
© D.P. Abrol and Uma Shankar (eds.), pp. 569-595
New India Publishing Agency, New Delhi (India)
e-mail : info@nipabooks.com; website : www.nipabooks.com

30

Integrated Pest Management in Cereal Crops

UMA SHANKAR AND D.P. ABROL

Introduction

Cereal crops are of global importance and the lifeline of the humane civilization. Nutritionally, they are important sources of dietary protein, carbohydrates, Vitamin B complex, Vitamin E, iron, trace minerals and fiber. Major cereal crops produced worldwide include wheat, rice, maize and millets. Among them, wheat, maize and rice together comprise at least 75% of the world's grain production. More than 70 per cent of the cereal crop produced in developed countries is fed to livestock whereas, in developing countries, 68-98 per cent of the cereal crop is used for human consumption (Chaven and Kadam, 1989). Rice is the most important and widely cultivated crop in the world and is the staple food for about 60% of the world's human population. It is basically a crop of humid tropics, but it varies widely in physiological adaptability, hence grown successfully both in tropical and temperate conditions. Wheat (*Triticum aestivum* L.) is the another important and widely cultivated food crop in the world and important rabi crop of our country. In India, wheat is the second stable food crop after rice which contains more protein than any other cereals. Maize is an important *kharif* cereal in hilly, sub-mountained and even in plain regions of our country where it forms staple diet of the people.

Agricultural production in India is limited by many factors like soil, water, genetic potential of the crop and the organisms that feed on or compete with food plants. The global losses due to various categories of pests vary with the crop, the geographical location and the weather. Total yield losses from different pests of all crops have been estimated to be US$ 500 billion worldwide. Despite the plant protection measures adopted to protect the principal crops, 42.1% of attainable production is lost due to attack by pests. However, if no control measures were used to protect the crops, the figure would be 69.8%. Animal pests account for 15.6% loss of production, pathogens 13.3% and weeds 13.2 % (Oerke *et al.*, 1994, Liebhold and Tobin, 2008).

Insects not only cause direct loss to the agricultural produce, but also indirectly due to their role as vectors of various plant pathogens. In addition to direct losses caused by insects, there are additional costs in the form of pesticides applied for pest control, valued at US $10 billion annually (Sharma *et al.*, 2000). Extensive and indiscriminate usage of chemical pesticides has resulted in environmental degradation, adverse effects on human health and other organisms, eradication of beneficial insects and development of pest-resistant insects.

The increase of *Bt* crops has decreased the use of pesticides. The efficacy of *Bt* crops against major pest species has been associated with an estimated 136.6 million kg global reduction of insecticides used during 1996-2006 (Naranjo, 2009). Even so, although non-target organisms are more abundant in *Bt* crops than in crops managed with insecticides, certain not-target organisms may be even more abundant in insecticide-free crops than in *Bt* crops (Marvier *et al.*, 2007). Therefore, attention must be drawn to the fact that *Bt* is not infallible, as it has been recently demonstrated by the studies of Tabashnik *et al.* (2008).

To realize the impact of chemical pesticides, sustainable pest management strategies have become vital to systematically review the recent advancements in insect control and to develop better crop protection programmes. Introduction of high yielding varieties and agro inputs during the green revolution period increased the productivity of land with a concomitant increase in the proportion lost to insect pests in India and other developing countries in Asia (Dhaliwal and Arora, 1994). Trend of crop losses between the 1950s and 1994 showed that the crop loss in paddy increased from 10% to 25% (accounting for 150% increase). Widespread outbreaks of the rice brown planthopper, *Nilaparvata lugens* during 1970s caused by the excessive use of insecticides has triggered the development of IPM strategies for pest management. The losses caused by insect pests like *Spodoptera, Helicoverpa,* whitefly and aphids are so enormous that farmers use insecticides excessively and as a result disturb the ecological balance. The current annual loss due to insect pests and diseases in agricultural sector in India is around Rs. 150 billion (US $ 3.3 billion, 1US $ = Rs.45) and in addition over 20 million man days are lost due to the vector borne diseases (TIFAC, 2004).

In order to cope with the burgeoning pest problems in the 1970s, the concept of integrated pest management (IPM) based on ecological principles was introduced into Asia from industrialized countries and over time the new knowledge gained has resulted in management systems that have mostly pacified the once turbulent pest outbreak situation (Matteson *et al.*, 1984; Gallagher *et al.*, 1994). In India, research on integrated pest management was started in 1974-75 on two crops, rice and cotton, under Operational Research Projects (ORP) (Swaminathan, 1975). But it was only in the mid 1980s that the Government of India re-oriented its plant protection strategy. India became a member country of the FAO initiated Inter - Country Program in 1980, but IPM activities have been intensified only since 1993.

Integrated Pest Management

IPM is a dynamic and constantly evolving approach to minimize crop losses in which all the suitable management tactics and available surveillance and forecasting information are utilized to develop a holistic management programme as part of a sustainable crop production technology (Dhaliwal and Arora 1996).

Integrated Pest Management (IPM) is a system designed to provide long term management of pests, instead of temporarily eradicating them. It is the coordinated use of pest and environmental information with available pest control methods to prevent unacceptable levels of pest damage by the most economical means with the least possible hazard to people, property, and the environment. The main goal of IPM is to manage pests and the environment so as to balance costs, benefits, public health, and environmental quality. IPM systems use all available technical information on the pest and its interaction with the environment.

IPM, is an ecological approach to suppressing pest populations in which all available necessary techniques are utilized in a unified program, so that pests are kept at acceptable levels in effective, economical and environmentally friendly ways. Because pest problems are often symptomatic of ecological imbalances, the goal is to attempt to plan and manage ecosystems to prevent organisms from becoming pests. Crop losses from pests vary on an average between 30 and 45%, but can reach 100%. The value of pesticides in protection of crops can be questioned when it is noted that estimated percentage losses in food production attributed to pests are as great today as they were 50 years ago when organic pesticides began to be widely used. Further, it has been estimated that pesticide use could be reduced by 35-50% without crop yields or causing an appreciable increase in the price of food (Pimentel 1991). With increasing population pressure and need for more food from same area, pesticide will be important component of IPM. However, the safe and judicious use of pesticide will be the priority. Environmental issues are debated for use of pesticide and transgenic crops as dominant component of pest management. The scientist extension workers and policy planners must learn to work with farmers who have better understanding of their land and what is needed to improve. The future prospect and challenges of IPM are (Dhawan and Peshin, 2009):

1. Emphasis should be placed on the development of pesticides that are active at lower doses, more specific for the target organisms, less toxic to the user, consumer, wildlife, biocontrol agents, and less persistent in the environment. Compatibility of pesticides with other pest management options should be improved. Pesticides must be integrated with other pest control technologies including genetic resistance in plants, cultural practices, biological control, and biotechnology.

2. Transgenic cops are one of the alternative to pesticide dominated IPM and therefore it is imperative to use on basis of IPM principles.

3. Biological control is an emerging technology to control pests, diseases and weeds. Genetic engineering will play a vital role in production of transgenic biocontrol agents having biocontrol potential and ecological acceptability.

4. Integration of IPM must be done in the context of agro-ecosystem management.

5. Knowledge based system on pest management should designed to improve IPM communication.

It is therefore important for farmers to be able to identify beneficial fauna, to know how they function in the crop ecosystem, and to reduce pesticides which destroy them. The Farmer Field School (FFS) model of training was designed to train farmers how to apply this Ecology-Based model of IPM (Kenmore *et al.*, 1995). In the FFS, a method called 'agroecosystem analysis' is used to assess all beneficials, pests, neutral insects and disease, and then determine if any intervention like a pesticide spray is needed. Economic Threshold Levels are discussed in the FFS, but crop protection decisions are based on conserving beneficial insects/spiders.

Tools of IPM

Monitoring

Proper scouting or monitoring to determine the pest status population or damage is prerequisite for need based application of insecticides. Counting the number of dead hearts/white earheads by stem borer, damaged leaves by leaf folder, hispa or rice skipper and working out their percentage have been directly used. Keep tracks of the pests and their potential damage. This provides knowledge about the current pests and crop situation and is helpful in selecting the best possible combinations of the pest management methods.

Host plant resistance

Host plant resistance to insect pests is the key component in the pest management system. It is the most economical method and compatible with other methods of pest control. Changed field conditions, including less insecticide use by farmers in some regions (Heong *et al.*,1998; Pincus, 1994), and new rice cultivars (Heinrichs, 1994), may have reduced yield losses to pests. The use of host-plant resistance in an IPM program can be a fairly straightforward matter of avoiding the use of varieties with high levels of susceptibility to an arthropod pest (Eigenbrode and Trumble, 1994). Rice has a relatively large and diverse germplasm collection (Heinrichs and Quisenberry, 1999) which has been used to good effect in breeding programs for arthropod resistance, notably for Asian rice gall midge (*Orseolia oryzae*), brown planthopper (*Nilaparvata lugens*), and green leafhopper (*Nephotettix virescens*). For these pests, high levels of resistance, conditioned by one or a few genes, are present in the germplasm collections at relatively high frequencies (Heinrichs, 1986; Alam and Cohen, 1998; Heinrichs and Quisenberry, 1999).

With the development of genetic transformation techniques, it has become possible to bring about quick and dramatic improvements in the tolerance to many Lepidopteran and other insect pests. As a result, much emphasis is now placed on the transgenic resistance to pests, particularly in crops like cotton, corn and rice where the Bt genes represent a major step in the direction of HPR. The first Bt toxin gene was cloned in 1981 and since then concerted efforts have been undertaken in several crops and consequently, genes conferring resistance to different insects have been inserted into potato, cotton, maize, rice, tobacco, alfalfa, lettuce, soybean, tomato and walnuts etc. (Benet, 1994; Fedrici, 1998; Griffiths, 1998). Bt rice has a potential to eliminate yield losses caused by Lepidopteran pests of up to 2-20 per cent of Asia's annual rice yield of 523 million tonnes (High *et al.*, 2004). Gene pyramiding, which comprises stacking of multiple genes and leads to simultaneous expression of more than one toxin in the transgenic variety, has been advocated as one of the solutions though it is also associated with problems such as development of cross- and multiple resistances (Pratap and Gupta, 2009).

Cultural pest control

Cultural methods to control insects involve crop husbandry practices that have a dual purpose of crop production and insect pest suppression. Farmers have developed these practices through observations which include crop production practices that make crop environment less susceptible to pests. Deep summer ploughing, flooding, synchronous transplanting, trap crop, crop rotation, cover crop, row and plant spacing, water management, fertilizer application, planting and harvesting dates, destruction of old crop debris are a few examples. Moreover, this is based on pest biology and development.

Mechanical control

Mechanical methods include collection and destruction of eggs, egg masses, larvae and pupae of stem borer, leaf folder, rice hispa and rice skipper, etc. and is a cheaper method for managing the leaf eating insects of rice.

a) *Rouging:* Rouging is removal of pest infested plants or plant parts by hand. Stem borer, leaf folder and skipper's larvae can be controlled by rouging.

b) *Bund shaving:* Mowing and shaving of bunds is a common practice that directly affect the weed and rodent population. A number of insects pass a part of life in the soil e.g. army worms pupate, grasshopper oviposits and mole crickets tunnel in rice bunds.

c) *Oil film on paddy water:* A traditional method to control many insect pests is to pour kerosene on to flooded field. Kerosene works best against more mobile insect pests like leaf and plant hoppers, root weevils, armyworm, leaf folder and caseworm larvae and grass hoppers. Fields are drained after 6 hours to prevent phytotoxicity.

d) *Light traps/ pheromone traps:* All the major pests of rice can be trapped through light. Although mercury bulbs are more advantageous in trapping the insects but single electric bulb of 200 watts can also be used (Fig. 1). If trap is not available then use a tub with kerosinized water below the source of light.

Biological control

Biological control is an emerging technology to control pests, diseases and weeds. These include augmentation and conservation of natural enemies of pests such as insect predators, parasitoids, pathogen and weed feeders. In IPM programes, native natural enemy population are conserved and non-native agents are released with utmost caution. Genetic engineering will play a vital role in production of transgenic biocontrol agents having biocontrol potential and ecological acceptability.

Large number of beneficial insects particularly the natural enemies have either reduced their activity or become extinct with the advent of modern agriculture and climate change. Most rice pests are controlled by a complex and rich web of generalist and specialist predators and parasites that live in or on the rice plant, paddy water, or soil. Beneficial insects/spiders comprise the majority of species in healthy ecosystems. For example, in Indonesian tropical wet rice ecosystem, 64% of all species identified were predators (306 species) and parasitoids (187 species); neutrals (insect detritivores, plankton feeders) comprise 19% (Settle *et al.*, 1996). Rice pests constitute only 17% of species. Beneficials are extremely effective in controlling major rice pests; very substantial reduction of pesticide applications does not threaten rice yield. Early season insecticide applications destroy that ecological balance. Insecticide wipes out predators along with their food supply, leaving the field open for pest buildup (Barrion *et al.*, 1994; Cohen *et al.*, 1994; Heong and Schoenly, 1998; Schoenly *et al.*, 1996; Settle *et al.*, 1996). Insecticide suppression of natural enemies, particularly spiders, predacious water striders in the genera *Microvelia* and *Mesovelia*, and the mirid bug *Cyrtorhinus lividipennis* Reuter, has been confirmed as the key factor in the emergence of BPH as a secondary pest (Heinrichs and Mochida,1984; Kenmore *et al.*, 1984; Ooi, 1986).

Chemical control

The present strategy for management of rice pests' emphasis the need based application of insecticides instead of prophylactic or fixed time bound schedules advocated earlier. Pesticides are used to keep the pest population below economically damaging levels when the pests cannot be controlled by other means. Effective and economic suppression of insect pests in rice ecosystem by the judicial use of pesticides on the basis of economics threshold levels (ETL) is very essential. ETLs for the chemical control of rice stem borers, and rice leaffolder in the traditional Basmati rice growing area of the Punjab, Pakistan. On the basis of cost benefit ratio an action threshold of 5% dead-hearts (DH) and 3% folded leaves for stem borers and rice leaffolder, respectively were developed and use of insecticides to control these insects on need basis despite calendar based insecticides application was stressed (Asghar *et al.*, 2009). The main tools for insecticide decisions are economic thresholds which are pest densities that trigger a corrective action before the damage reaches the critical economic injury level (Norton and Mumford 1993, Morse and Buhler, 1997). Action thresholds are surrogates for an economic threshold that are determined empirically if the damage function is not known (Bandong and Litsinger 1988, Smith *et al.*, 1988). ET for rice pests have been worked out, based on insect population as well as damage.

Name of insect pests	Economic threshold level (ETL)
Stem borer	5-10% dead hearts 1 egg mass/m^2 at vegetative stage 1 egg mass or 1 adult/m^2 at flowering
BPH/WBPH	5-10 insects/hill, 20 insects/hill
Green leaf hopper	20 Insects/hill
Leaf folder	1-2 larvae/hill
Root weevil	2 Grubs/plant at initial stage
Cut worm	1 larva/hill
Gundhi bug	1-2 bugs/m^2
Gall midge	5% silver shoot
Other leaf eating insects	1 larva/hill

Application of cartap hydrochloride granules in the nursery seed beds at 15 kg/ha 5 days before uprooting the seedlings proved better and provided adequate control of *Scirpophaga incertulas* and *Hydrellia philippina* upto 40 DAT (Dash and Mukherjee, 2003), of stem borers @ 1kg a.i./ha (Khan and Khaliq, 1989), of stem borer and leaf folder (Krishnaiah *et al.*, 1996) and kept population of *Sogatella furcifera* low (Lal, 2000).

IPM in Rice

Rice (*Oryza sativa* L.) is an important staple food crop for more than two third of the population of India and more than 65 per cent of the world population (Mathur *et al.*, 1999). India is the largest rice growing country and accounts for more than 40% food grain production. Rice ecosystem represents one of the highly diversified agro-ecosystem in the world. There are some 800 insect pest species that have been recorded to feed on rice (Grist and Lever, 1969). In tropical Asia there are some 18-20 species that are considered to be pests of major importance and regular occurrence (Pathak 1968).

About 300 species of insects have been reported to attack rice crop in India out of which 20 have been found to be the major pest (Pathak, 1975; Arora and Dhaliwal, 1996). Rice crop is attacked and damages by large number of pest insects from nursery to harvest but only a few of them are considered key pest, which cause severe yield losses. Among these major pest insects rice stem borers, (*Scirpophaga incertulas* and *S. innotata*) and rice leaf folder (*Cnaphalocrosis medinalis*) considered as dominant pest insects of rice, throughout the rice growing areas all over India. They induce highly discernible damage symptoms (Chaudhary *et al.*, 1984; Heong *et al.*, 1994; Hafeez *et al.*, 2006). Among the various methods used by the farmers to protect the rice crop from these economically important pests the chemical control remains the only mean of economical and rapid method for suppression of insect pests infestation. This is because other methods of pest control are laborious, time consuming and less effective (EJF, 2002). However the indiscriminate use of insecticides has caused many side effects, including, loss of biodiversity, biological imbalance, resulting in changes in community structure the problem of secondary pests, insecticide resistance, residual toxicity, proliferation and resurgence of selected insect pests and environmental pollution (Roger and Kurihara, 1988; Wilson and Tisdell, 2001; Tahir and Butt, 2009). Rice stem borers are probably the most serious group and usually 1-4 species are important in any given area. The dynamics of harmful pests may lead to plant injury on the standing crop which may or may not be visible to the observer. Injury leads to damage which may or may not lead to yield loss or reduction of crop value in economic terms (Litsinger, 1991). Pest assessment studies frequently show that crops vary greatly between sites and between years in their response to attacks even by similar densities of insects (Litsinger *et al.*, 2005). Chemical control is widely adopted for the management of these pests (Matteson, 2000). However, in the last decade, the role of entomopathogenic nematodes (EPNs) as a safe alternative to the use of insecticides in IPM of different crops, including rice, has gained worldwide attention (Georgis, 1987; Kaya, 1990; Klein, 1990; Mason and Wright, 1997; Rahman *et al.*, 2000).

Kinds of Rice Insect Pest: A Primer for Identification

A. Defoliators

Insects that feed on plant leaves. A portion or the whole leaf blade may be consumed early, late, or at all stages of the crop growth.

I. Seedling to Vegetative

1. Rice whorl maggot

Whorl maggot is a small grey fly with strong preference to young transplanted rice plants. It causes damage during the tillering stage of rice, for about 40 days after transplanting, but thereafter little damage occurs. Damage symptoms are visible when the leaf grows out of the whorl. The edge of the leaf is discolored, whitish or yellowish. The extent of damage varies from slight feeding causing a "pinched" or narrow area or pinholes to a severely damaged leaf which is completely broken due to extreme feeding across the entire width of the leaf.

2. **Rice caseworm** (*Nymphula depunctalis)*

   Moth is immaculate white with light brown and black spots on the wings. Eggs are pale, yellowish green, laid in one or two long rows on under surfaces of leaves drooping into the water. Larvae are yellowish green, translucent with threadlike gills on body and freshly hatched larvae feed on the surface of the tender leaves but later instar larvae, with their heads and thoraces protruding from within the case, feed on the surface of the older leaves. The larva consumes the green leaf tissue, leaving only the papery upper epidermis. Frequently several larva attack a plant and cut off most of the leaf tips for making cases and may consume the remaining leaves. Uneaten leaves can be seen floating on the paddy water surface.

3. **Asian rice gall midge** (*Orseolia oryzae)*

   The orange red fly is mosquito-like with long, slender legs and bead-like antennae. This insect is nocturnal and is attracted to light. Eggs are shiny white with pinkish, red or yellow shading and are laid openly on leaf blades. Soon after hatching, the larva moves down between the leaf sheath and the stem until it reaches the growing point of the apical or side buds and feeds at the base of the growing plant. A gall begins to form and eventually results in the development of a tubular gall or "onion leaf". Plant damage is caused by the transformation of tillers into tubular galls which do not bear panicles. Early infestation also causes excessive tillering but these new tillers often become infested and few bear panicles.

B. General defoliators

1. **Rice hispa** (*Dicladispa armigera)*

   Adult beetle is small, metallic blue-black and covered with short and long spines in the thorax and forewings. Both adults and larvae feed on the leaf tissue between the epidermal membranes. The adults scrape the upper surface, leaving only the lower epidermis. Adult feeding damage shows white streaks along the long axis of the leaf. Larvae produce irregular longitudinal white mines as they tunnel between the epidermal layers. The affected leaves may wither and die, giving the field a burnt appearance.

2. **Rice leaf folders** (*Cnaphalocrocis medinalis)*

   The moths of the 3 species: *C. medinalis, M. patnalis* and *M. exigua* are yellow brown which can be differentiated with dark line markings on their wings. Damage to the rice plant is caused by the caterpillars which fold the leaf blades into tubular structures and feed on the green leaf tissues (Fig. 2). The affected leaves dry up giving seriously infected fields a scorched appearance. The discoloration and folding of the leaves affects the general vigour of the rice plants and reduces their photosynthetic ability. The damaged leaves also serve as an avenue for fungal and bacterial infection.

3. **Rice thrips**

Rice thrips are minute, soft-bodied, elongated insects having narrow wings with long marginal hair and asymmetrical rasping and piercing mouthparts. Adults and nymphs prefer to feed on young plants and seedlings. Damaged leaves show fine white to silvery scraped areas. Due to severe rasping, the leaves rolland may even wither. Leaf damage increases in dry weather, particularly when there is drought stress. Upland rice being prone to drought commonly suffer from thrips infestation.

4. **Rice skippers** (*Pelopidas mathias)*

Moths of rice skippers, namely, *P. mathias* Fabricius (Fig. 3). and *P. guttata* Bremer and Grey are brown in color with transparent white marks on the forewings and their antennae have hook-like tips. The hind wings of *P. mathias* have no transparent spots but *P. guttata* bear four white transparent spots. The head of *P. mathias* larva has red oblique bands on each side while *P. guttata* larva has a brownish frontal W-band on its head. The larvae of two species of rice skippers have similar feeding habits. They skeletonize the leaf by eating the entire leaf margins and leave the midrib uneaten.

5. **Rice short-horned grasshoppers** (*Oxya sp.* and *Acrida sp)*

Several genera and species of short horned grasshoppers (Fig. 4) of varying shapes are attacking the rice plants. The common genera are easily identified. *Oxya* and *Acrida* species have no deep furrow along midline of head but present in *Atractomorpha. Oxya* have large rounded eyes with bold heads while *Acrida* and *Atractomorpha* have small elongate eyes and strongly slanted faces. Both nymphs and adults feed on leaf tissue consuming large sections from the edges of the leaf blades.

II. Sap-feeders

A. Plant sap-feeders

1. **Green leafhoppers** (*Nephotettix* spp.)

Green leafhoppers are the most common leafhoppers (Fig. 5) in the rice plant. Four species transmit virus and mycoplasma like diseases of rice such as tungro, yellow dwarf, waika, transitory yellowing, gall dwarf, and dwarf. Leafhoppers can be distinguished by the shape of the head and the presence or absence of black lines or bands between the eyes. For instance, the temperate species, *N. cincticeps* (Uhler) has bluntly rounded head and without central black spots on the forewings. The tropical Asian species *N. nigropictus* (Stal) has a transverse black band on the head and central black spots on the forewings On the other hand, N. virescens (Distant) has a pointed head without black bands . Both nymphs and adults feed by extracting plant sap with their needle-shaped mouthparts.

2. **Zigzag leafhopper** (*Recilia dorsalis)*

   The zigzag leafhopper can be readily recognized by the brownish zigzag pattern on its wings. It feeds well on rice and some grassy weeds, and is an important vector of viral diseases.

3. **Brown planthopper** (*Nilaparvata lugens)*

   The adult brown planthopper is light to dark brown (Fig. 6). The male is distinctly smaller than the female. The adult planthoppers occur in both the short- and long- winged forms. The eggs, characterized by broad flat egg caps, are laid in batches in the leaf sheath or midrib of the leaves. There are five nymphal instars. Newly emerged nymphs are white, but later turn brown. Removal of plant sap and blockage of vascular bundles by feeding sheaths cause the plant to wilt and die. This condition is referred to as hopperburn. Brown planthopper transmits grassy stunt and ragged stunt.

4. **White-backed planthopper** (*Sogatella furcifera)*

   The adult white-backed planthopper has a prominent white band on its thorax (Fig. 7). Male occurs only in long-winged form while female can either be short or long-winged. Eggs are laid in rice stem or leaf sheaths. Newly emerged nymphs are white but show coloration as they mature. Feeds equally on some grasses as well as on rice.

B. Grain sap-feeders

Rice bugs (*Leptocorisa acuta)*

Adults are brown and slender with long legs and antennae (Fig. 8). The dark reddish brown, disc-shaped eggs are laid in rows on rice foliage. Nymphs are green with white bands on its reddish antennae and their bodies elongate as they mature. Mature nymphs feed at higher rate than the adults. Rice bug feeding at milking stage results to empty grain while feeding at dough stage gives rise to broken/chaffy grains.

Stem borers

Yellow stemborer (*Scirpophaga incertulas* Walker), Pink stemborer (*Sesamia inferens*), Dark-headed stemborer (*Chilo polychrysus*) and Striped stemborer (*Chilo suppressalis*)

The rice stem borer, which infest the rice from seedling to maturity act as a major constraints for rice production (Pathak, 1975 and Dale, 1994). Adults are harmless but the larvae (Fig. 9) bore into the plant stem and do the damage while feeding inside the stem. Moths are nocturnal, attracted to light and are strong fliers. Eggs of stem borers are laid mostly on the leaf blades except for pink which is hidden on the inner side of the leaf sheath. The larvae of *S. incertulas* cause dead hearts during vegetative stage and white ear heads during reproductive stage. Even though rice plant can compensate if dead heart infestation does not exceed 10 per cent, it cannot compensate for white ear loss.

III. Root feeders

Root feeders : Soil-inhabiting insects (partly or totally) that feed on underground parts. Root pests are favored by well-drained and unflooded rice soils.

Oriental mole cricket (*Gryllotalpa orientalis)*

Adults and nymphs have enlarged and spade-like front legs which enable them to tunnel in the soil. Adults and nymphs feed on sown seeds and roots. Young plants can die from root loss producing bare spots and missing plant stand in the field

White grubs (*Leucopholis irrorata)*

Adults are gray brown with stout and smooth bodies. Larvae are white with large yellow brown heads. They feed on the root mass trimming the roots. Young larvae roll into C shape when disturbed while older larvae have an enlarged area at the tip of the abdomen. Leaves of damage plants turn orange-yellow, wither and die due to root loss.

Integrated Pest Management in Basmati Rice (Crop and Area specific)

A. Cultural method

- Remove/destroy stubbles after harvest and keep the field free from weeds.
- Trim and plaster the bunds of rice field to expose the eggs of grasshoppers and to eliminate the bugs breeding in grasses.
- Form the buds narrow and short to reduce the damage by rodents.

- Use resistant varieties wherever available.
- Provide effective drainage wherever there is problem of BPH.
- Clip the tip of seedling before transplanting to prevent the carryover of egg masses of rice yellow stem borer from nursery to main field.
- Organise synchronized planting wherever possible.
- Leave 30 cm rogue space at every 2.5 m to reduce damage by BPH and rodents.
- Avoid use of excessive nitrogenous fertilizers.
- Use irrigation water judiciously (Alternative webbing x drying reduce BPH and case worm).
- Remove the egg masses of stem borer in the main field.

B. Mechanical methods

- Dig out the rat burrows and destroy the rats and young ones at the beginning of the season.
- Set up-bow traps to kill rodents.
- Pheromone traps have been found very useful in the management of Yellow Stem Borer, *Scirpophaga incertulas* for which they are used for monitoring of the species or management through mating disruption or mass trapping of male (male annihilation technique). For the monitoring install the pheromone traps in the field at the rate of 3 trap per acre at a distance of 60 m in a triangular pattern and record the males trapped daily to access the peak population.

C. Biological methods

Most of the paddy pests are attacked by an array of natural enemies. In the world, there are 220 cases of successful biological control in 60 different countries and involves 110 species of the pests. Subba Rao *et al.* (1983) reported 77% egg parasitism by *Tetrastichus* spp, *Telenomos spp* (Fig. 10) and *Trichogramma spp* in India. Bharati and Kushwaha (1988) recorded the pupal parasitoids of rice leaf folder and reported *Xanthopimpla flavolineata* as a major parasitoid. More than 100 species of parasitoids and 72 species of predators have been reported from various hoppers (Manjunath *et al.*, 1989). Spiders are the dominant pradators found in the rice ecosystem. Wolf spider, *Lycosa pseudoannulata and Oxyopes sp.* (Fig. 11) killed up to 90% of 130 borer larvae in one day. This spider is also capable of attacking stem borer moths.

- Release *Trichogramma japonicum* on 30 and 37 DAT twice 5 ml/ha/ release against stem borer.

- Release *Trichogramma chilonis* on 37, 44 and 51 DAT (Thrice) @ 5 ml/ha/release against leaf folder.
- Release of *Platygaster oryzae* parasitized galls @ 1 per 10 m2 in the mainfield on 10 DAT against gall midge.
- Set up owl perches to reduce rat damage.

D. Plant products

- Spray neem seed kernel extract 5% (25 kg/ha), neem oil 3% (15 lit/ha) to control brown planthopper.
- Spray botanical powder formulation *viz.*, NSKE, *Vitex negundo* (Notchi), *Prosopis juliflora* and *Ipomoea carnea* leaf extract 5% to control earhead bug and black bug.

E. Chemical methods

- In BPH prone areas/season avoid use of synthetic pyrethroids, methyl parathion and quinolphos and use recommended chemical at recommended doses.
- Use insecticides based on FTLs.

Integrated Pest Management in Maize and Wheat

Maize (*Zea mays* L.) is an important staple food in many tropical countries in Africa, Asia and Latin America (Paroda and Kumar, 2000) and occupies third place in the world after wheat and rice. The important reason for Maize low productivity is improper management of insect pests' infestation. More than 130 insect pests have been recorded causing damage to maize in India. However, only about a dozen of these insect-pests are among the key factors contributing to low yield of maize. Globally losses due to insect pests of maize are reported to be approximately 9%, which is equivalent to 52 million MT and worth approximately $5.7 billion annually. A variety of insect pests attack maize crop right from seedling to maturity. These insect pests includes lepidopterous pests (borers, cutworms, armyworms, earworms, and grain moths) are the most damaging world over, followed by the coleoptera (root worms, wireworms, grubs, grain borers, and weevils). Next in importance is the group of insects that serve as carriers (vectors) for disease agents or pathogens, among which leafhoppers and aphids are of prime importance. Maize cultivation and yield realization are highly affected by the degree of insect-pest management.

In India wheat is a principal *rabi* cereal crop grown all over the country under a wide range of agroclimatic conditions in an area of about 25.12 m ha with an annula production of 75.80 mt. Over hundrered different insects pests have been reported to be associated with wheat crop, of these 85 are insects, 2 mites, 12 nematodes, 7 birds and 3 mammals. However a few of them are regular and serious pests. Three species of aphids namely *Sitobion aveneae, S. miscanthii* and *Rhopalosiphum maidis* are the most predominat and causing a sizeable reduction n the yield.

Insect pests of maize and their management

Maize Cutworm : *Agrotis ipsilon* Rott.

Cutworm belongs to notucid moths which have a common habit of biting through the stems of seedlings at ground level and eating the leaves or the entire seedlings. *Agrotis ipsilon* is a polyphagus pest and has been reported feeding of many field crops viz. cereals, millets, pulses, oilseeds and vegetables etc. The pest is active from October-April in plains and probably migrates to the mid hills for further breeding in summer. The young larvae feed on the epidermis of the leaves. As they grow, their habit changes. During day time they live in cracks and holes in the ground and come out at night and fell the plants (Fig. 12) by cutting their stems, either below the surface or above the ground. The cut branches/leaves/stems are sometimes seen to have been dragged into the holes where the leaves are eaten at leisure. They seem to be very voracious eaters and they fell more plants than they can consume.

Management

- Flooding (heavy irrigation) the field may force larvae up to the surface and destruction by the predators.
- Heaps of green-grasses may be kept at suitable places in the field in the evening, so that the caterpillars seek shelter below them. The heaps along with the caterpillars may than be destroyed in early morning.
- The moths are strongly attracted to artificial light. Therefore, light traps of electric or petromax lamps placed just above a broad flat basin full of kerosin water, should be put on the night following the first shower of the monsoon and continued throughout the period of emergence for about one month.
- Soil treatment by mixing carbofuran 3G @ 20 kg/ha in the soil at time of last ploughing.
- In case soil application could not be done than spray endosulfan @ 0.07% or chlorpyriphos @ 1.5 ml/litre of water.
- Use of biopesticides viz., *Heterorhabditis indica, metarhizium anisopliae, Beauveria bassiana, Steinernema carpocapsae, S. carpocapsae + B. bassiana,*

S. carpocapsae+ *M. anisopliae* and *H. indiaca*+ *M. anisopliae* are found effective in providing protection at later stage of seedling growth and also registered higher yield compared to control (Bhagat *et al.*, 2008).

Stem borers: *Chilo partellus* Swinhoe, *Sesamia inferens* Walk. (Lepidoptera: Pyralidae)

It is a most destructive pest of maize and sorghum. The adult moth is medium sized, straw coloured insect with numerous shining brown spots. They are active in night, when they mate and lay eggs on the underside of the leaves of plants particularly early crops of maize and sorghum grown for fodder. As it opens more short holes visible, indicating an earlier attack and the plants also show dead hearts. The full grown caterpillars of the last generation hibernation in stubble, stalks etc. and remain there till the next spring. Short holes in the central whorl leaves later on and production of dead hearts, frass and weakening of plants are the typical symptoms of the stem borer. The extent of loss caused by the pest varies from 20-25%.

Management

- The potential for carry over may be reduced by destroying the stubble, stalks, weeds and other alternate hosts of the stem borer by ploughing the field after harvest.
- Removal and destruction of dead hearts and destruction of infested plants showing early pin holes damage has been found to be successful practice in reducing the pest incidence.
- Destruction of crop residues and chopping of stems harboring diapausing larvae could be very effective in reducing borer population.
- Hand picking or light trapping of adult moths and collection of their eggs for destruction.
- *Trichogramma cllilonis* has been found parasitising egg masses of *C. partelllus* upto 70%. *Apanteles flavipes* is the most dominant and most widely distributed larval parasitoid of *C. partellus* in India with a parasitisation ranging from 5.3 to 42.8%.
- Caterpillars of this insect feed on plants particularly tender parts during the night and hide during the day time. They damage leaves and ear head also. The pest can be managed through Bt and NPV. Take up two applications of NPV at 10 days interval at 250 LE/ha along with crude sugar 2.5 kg + cotton seed kernel powder 250 g on the ear heads to reduce the larval population of *Helicoverpa* sp.

- Spray the crop with 2-3 weeks after sowing or as soon as the borer injury to the leaves is noticed with any of the synthetic pyrethroids like Fenvalerate @ 100ml/ha or Cypermthrin @ 100ml/ha or Deltamethrin @ 200ml/ha.

Blister beetle: *Mylabrus pustulata* Thunberg.

Blister beetle, a polyphagous pest attacking maize, millets, bhindi, arhar, aak, etc., and is a serious pest in foot hills of Poonch, Rajouri and Udhampur and even in Jammu plains (Fig. 13). The beetle is pollen feeding and feed on the silk of the maize cobs and flowers of other crops and thus preventing pollination and setting of grains. The adult beetle lays eggs in soil in masses which hatch in about 2 weeks. The young ones are reported to feed on eggs of grasshoppers and are thus beneficial. The beetle emerges from pupae and remains active from July to October. The adult beetles of the pest are only damaging stage.

Management

- Collection by means of hand nets and destruction.
- Raising 1 or 2 rows around the fields of trap crops viz., bhindi or arhar.
- Spraying the crop with carbaryl 50 WP @ 1.2 kg/ha.

Insect Pests of Wheat and their Management

Wheat aphid, Sitobion avenae *(F.) (Aphididae: Homoptera)*

Wheat aphids are worldwide in distribution and reported in India from Delhi, Haryana, Himachal Pradesh, Jammu and Kashmir (Gupta, 2001), Madhya Pradesh, Punjab, Rajasthan, Uttar Pradesh and West Bengal. Aphids are soft bodied insects and major pest of cereal crops in the spring. Heavy infestations can cause a reduction of the number of grains per ear and thus a noticeable reduction of the yield. Honeydew secretion by the aphids further aggravates the sooty mould development which hampers the photosysthetic activities of plants. Yield losses of 5-15% are common but losses can exceed 30% during epidemics. They also act like a vector of the barley yellow dwarf virus (BYDV).

Management

- For the management of aphids, foliar spray of imidacloprid 200SL @ 20g a.i./ha on border rows at the start of the aphid colonization be given.
- A well timed application of a persistent pyrethroid insecticide such as Lamda cyhalothrin can reduce the incidence of BYD and increase yields.
- Spraying formothion or oxydemeton methyl @ 250 ml a.i./ha.

Shootfly, Atherigona naqvii (Muscidae: Diptera)

The maggots bore into the shoot of young plants, a week after germination to about one month and as a result the central shoot dries up resulting in 'dead hearts'. If it is a little later the mother plant may produce side tillers. But the tillers also may be attacked. The infestation often goes as high as 60%.

Management

- Early sowing of wheat will reduce the attack.
- In late sown crops (end of December to 1st week of January) spraying of cypermethrin 0.002% twice at fortnightly intervals in seedling stage was found effective. The most vulnerable period of crop is being up to 60 days of germination.

Termites: Microtermes obesi *and* Odontotermes Obesus *Ramb. (Isoptera: Termitidae)*

These are social insects like honey bees and ants live in colonies. They build up nests consisting of small chambers connected with narrow passages make superficial tunnels on the ground to attack growing plants, timber wood, dry leaves and grasses. In the plants, generally the roots and stems are damaged with the result that they don't grow properly, wilt soon and dry up. An attacked plants when cut or dug up shows irregular galleries on the bark which are made by termites eaten up the woody tissues. A new termite colony is usually started at the beginning of rains when sexually mature winged males and females which emerge in vast numbers come out of their colonies. The male and female termites shed their wings and copulate; the female burrows into the ground to build a new nest. The queen may lay 30,000 to 80,000 eggs/day and 15 million eggs throughout her life span of five years.

Management

- Donot apply raw or undecomposed farmyard manure; if possible irrigate the field when infestation is serious.
- Treat the seed with chlorpyriphos @ 5 ml/kg of seed before sowing.
- Mounds of termites if any in the area should be treated with 0.5% chlorpyriphos after breaking open the earthen structure and making holes with an iron bar. The insecticidal emulsion should be used @4 litres / m^3 of the mound.
- In the termite prone areas, (especially the northern states) seed treatment with chlorpyriphos @ 0.9g a.i or endosulfan @ 2.4g a.i/kg seed, be taken up for their management.

- Fipronil (Regent 5FS @ 0.3 g a.i./kg seed) is also very effective.
- In the standing crop, the broadcasting of the insecticide treated soil 15 DAS be practiced. For this, endosulfan 35EC @ 2.3 l or chloropyriphos @ 3 l mixed in 50 Kg soil be used for one hectare field.
- Entomopathogenic formulation of *Steinernema carpocapsae* and *H. Indica* are promising in controlling the termite population in laboratory (Shankar and Kaul, 2010).

References

Ahmad, H., Shankar, U., Monobrullah, Md. and Khan, R.B. 2006. Screening of rice cultivars against rice leaf folder, *Cnaphalocrocis medinalis* (Guenee). *Environment and Ecology,* 24 S (3): 673-675.

Alam, S.N. and Cohen, M.B. 1998. Durability of brown planthopper, *Nilaparvata lugens,* resistance in rice variety IR64 in greenhouse selection studies. *Entomologia Experimentalis et Applicata,* 89: 71-78.

Arora, R. and Dhaliwal, G.S. 1996. Agroecological changes and insect pest problems in Indian agriculture. *Indian J. Ecol.* 21 (2): 109-122.

Asghar, M., Suhail, A., Afzal, M. and Khan, M.A. 2009. Determination of Economics Threshold Levels for the Stem Borers (*Scirpophaga* sp.) and Leaffolder (*Cnaphalocrosis medinalis*) of Rice (*Oryza sativa*) in the Kallar tract of Punjab, Pakistan. *Int. J. Agric. Biol.,* 11(6): 717-720.

Bandong, J.P. and Litsinger, J.A. 1988. Development of action control thresholds formajor rice pests. In: Teng P.S. and Heong K.L. (eds.), *PesticideManagement and Integrated Pest Management in Southeast Asia,* CICP/USAID, College Park, MD, pp. 95-102.

Barrion, A.T. and Litsinger, J.A. 1994. Taxonomy of rice insect pests and their arthropod parasites and predators. In: Heinrichs E.A. (ed.). *Biology and Management of Rice Insects.*Wiley Eastern Ltd., New Delhi, 779 p., pp. 13-359.

Benet, J. 1994. DNA-based techniques for the control of rice insects and diseases: Transformation, gene tagging and DNA fingerprinting. In: Teng, P.S., Heong, K.L. and Moody, K. (eds), *Rice Pest Science and Management.* International Rice Research Institute, Los Banos, Phillipines, pp. 147-172.

Bhagat, R.M., Khajuria, M.K., Shankar, U., Monobrullah, Md. and Kaul, V. 2008. Efficacy of bio-pesticides and insecticides in controlling maize cutworm in Jammu. *Journal of Biological Control,* 22(1): 99-106.

Bharati, L.R. and Kushwaha, K.S. 1988. Parasitoids of leaffolder (LF) pupae from Haryana, India, *International Rice Research Newsletter,* 13 (5):31.

Chaudhary, R.C., Khush, G.S. and Heinrichs, E.A. 1984. Varietal resistance to rice stem borers in Asia. *Insect Sci. Applic.,* 5: 447-463.

Chaven, J.K. and Kadam, S.S. 1989. Nutritional improvement of cereals by fermentation. *CRC Critical Reviews in Food Science and Technology,* 28(5): 349.

Cohen, J.E., Schoenly, K.G., Heong, K.L., Justo, Jr., H.D., Arida, G.S., Barrion, A.T. and Litsinger, J.A. 1994. A food web approach to evaluating the effect of insecticide spraying on insect pest population dynamics in a Philippine irrigated rice ecosystem. *Journal of Applied Ecology,* 31: 747-763.

Cohen, M.B., Alam, S.N., Medina, E.B. and Bernal, C.C. 1997. Brown planthopper, *Nilaparvata lugens,* resistance in rice cultivar IR64: Mechanism and role in successful *N. lugens* management in Central Luzon, Philippines. *Entomologia Experimentalis et Applicata,* 85: 221-229.

Dale, D. 1994, Insect pests of rice plants-their biology and ecology. In : *Biology and Management of Rice Insects* Eds.(Heinrichs, E.A., Indian. Rice Res. Inst., Philippines. Wiley Eastern Ltd. pp. 363-485.

Dash, A.N. and Mukerjee, S.K. 2003, Insecticidal control of major insect pests of rice. Pest Mgmt. Econ. Zool., 11: 147-151.

Dhaliwal, G.S. and Arora, R. 1994. *Trends in Agricultural Insect Pest Management.* Commonwealth Publishers, New Delhi.

Dhaliwal, G.S. and Arora, R. 1996. *Principles of Insect Pest Management.* National Agricultural Technology Information Centre, Ludhiana.

Dhawan, A.K. and Peshin, R. 2009. Integrated Pest Management: Concept, Opportunities and Challenges. In: *Integrated Pest Management: Innovation Development Process Vol-1,* 2009. Peshin, R. and Dhawan, A. K. (eds.), Springer Verlag. pp. 51-82.

Eigenbrode, S.D. and Trumble, J.T. 1994. Host plant resistance to insects in integrated pest management in vegetable crops. *Journal of Agricultural Entomology,* 11:201-225.

EJF, 2002. *Death in Small Doses: Cambodia's Pesticides Problems and Solutions,* pp: 1–37. Environmental justice Foundation, London, UK.

Fedrici, B.A. 1998. Broad-scale leaf pest-killing plants to be true test. *California Agriculture* 52: 14-20.

Gallagher, K.D., Kenmore, P.E. and Sogawa, K. 1994. Judicial use of insecticides deter planthopper outbreaks and extend the role of resistant varieties in Southeast Asian rice. In: Denno, R.F. and Perfect, T.J. (eds), *Planthoppers Their Ecology and Management,* Chapman and Hall, London, 772 p., pp. 599-614.

Georgis, R. 1987. Nematodes for biological control of urban insects. Pp. 816-821. *In*: 194th Meeting, American Chemical Society, New Orleans, Vol. 27.

Griffiths, W. 1998. Will genetically modified crops replace agrochemicals in modern agriculture? *Pesticide Outlook,* 9: 6-8.

Grist, D.H. and Lever, R.J.A.W. 1969. *Pests of Rice.* Longmans, Green and Co. Ltd, London and Harlow, 520 p.

Gupta, A. 2001. Population dynamics of wheat aphid (*Sitobion avenae* Fab.) in Jammu sub tropical area. *MSc. Thesis.* SKUAST-Jammu. pp95.

Heinrichs, E.A. 1986. Perspectives and directions for the continued development of insect-resistant rice varieties. *Agriculture, Ecosystems and Environment,* 18: 9-36.

Heinrichs, E.A. 1994. Impact of insecticides on the resistance and resurgence of rice planthoppers. In: Denno, R. F. and Perfect, T.J. (eds.) *Planthoppers: Their Ecology and Management*. New York: Chapman and Hall., pp. 571- 598.

Heinrichs, E.A. and Mochida, O. 1984. From secondary to major pest status: the case of insecticide-induced rice brown planthopper, *Nilaparvata lugens*, resurgence. *Prot. Ecol.*, 7:201-18.

Heinrichs, E.A. and Quisenberry, S.S. 1999. Germplasm evaluation and utilization for insect resistance in rice. In: Clement, S.L. and Quisenberry, S.S. (eds), *Global Genetic Resources for Insect-Resistant Crops*. CRC Press, Boca Raton, Florida, pp. 3-24.

Heong, K.L. and Schoenly, K.G. 1998. Impact of insecticides on herbivore-natural enemy communities in tropical rice ecosystems. In: Haskell P.T. and McEwen P. (eds), *Ecotoxicology: Pesticides and Beneficial Organisms*, Kluwer Academic Publishers, Boston MA, pp. 381-403.

Heong, K.L., Escalada, M.M. and Mai, V. 1994. An analysis of insecticide use in rice: case studies in the Philippines and Vietnam. *Int. J. Pest Manag.*, 40: 173-178.

Heong, K.L., Escalada, M.M., Huan, N.H. and Mai, V. 1998. Use of communication media in changing rice farmers' pest management in the Mekong Delta, Vietnam. *Crop Prot.*, 17(5):413-25.

High, S.M., Cohen, M.B., Shu, Q.Y. and Altosaar, I. 2004. Achieving successful deployment of Bt rice. *Trends in Plant science*, 9: 286-292.

Kaya, H.K. 1990. Entomopathogenic nematodes in biological control. Pp. 93-116. *In*: Soil Ecology (Gaugler R. and Kaya H.K., eds). CRC Press, Boca Raton, Florida, USA.

Kenmore, P.E., Carino, F.O., Perez, C.A., Dyck, V.A. and Gutierrez, A.P. 1984. Population regulation of the rice brown planthopper (*Nilaparvata lugens* Stal) within rice fields in the Philippines. *Journal of Plant Protection in the Tropics*, 1: 19-37.

Kenmore. P.E., Gallangher, K.D. and Ooi, P.A.C. 1995. Empowering farmers: Experiences with integrated pest management. *Entwicklung and Landlicher Raum*, 1/95: 27-28.

Khan, L. and A. Khaliq, 1989. Field evaluation of some granular insecticides for the control of rice stem borers. *Pak. J. Sci. Indust. Res.*, 32: 824- 826.

Klein, M.G., 1990. Efficacy against soil-inhabiting insect pests. Pp. 195-214. *In*: Entomopathogenic nematodes in Biological control (Gaugler R. and Kaya H.K., eds). CRC Press, Boca Raton, Florida, USA.

Krishnaiah, N.V., Reddy, A.A. and Ramaprasad, A.S. 1996, Studies on buprofezin and synthetic pyrethroids against hopper's in rice. *Indian J. Pl. Protec.*, 24: 50-60.

Lal, R. 2000. Management of rice leaf folder, *Cnaphalocrocis medinalis* Guenee by using certain neem based formulations. *Pesticide Res. J.*, 12(1): 141-144.

Liebhold, A.M. and Tobin, P.C. 2008. Population ecology of insect invasions and their management. *Annu. Rev. Entomol.*, 53:387-40.

Litsinger, J.A. 1991. Crop loss assessment in rice. In: Heinrichs E.A. and Miller T.A. (eds), *Rice Insects: Management Strategies.*, Springer-Verlag, NY, pp. 1-65.

Litsinger, J.A., Bandong, J.P., Canapi, B.L., dela Cruz, C.G., Pantua, P.C., Alviola III A.L. and Batayan, E. 2005. Evaluation of action thresholds against chronic insect pests of rice in the Philippines: I. Less frequently occurring pests and overall assessment. *International Journal of Pest Management,* 51:45-61.

Manjunath, T. M., P. S. Rai and G. Gowda. 1978. Parasites and predators of *Nilaparvata lugens* in India. *PANS* 24(3): 265-269.

Marvier, M., McCreedy, C., Regetz, J. and Kareiva, P. 2007. A meta-analysis of effects of Bt cotton and maize on nontarget invertebrates. *Science,* 316: 1475-1477.

Mason, J.M. and Wright, D.J. 1997. Potential for the control of *Plutella xylostella* larvae with entomopathogenic nematodes. *Journal of Invertebrate Pathology, 70*: 234-242.

Mathur, K.C., Reddy, P.R., Rajamali, S. and Moorthy, B.T.S., 1999, Integrated pest management of rice to improve productivity and sustainability. *Oryza,* 36(3): 195-207.

Matteson, P.C., 2000. Insect pest management in tropical Asian irrigated rice. *Annual Review of Entomology, 45*: 549-574.

Matteson, P.C., Altieri, M.A. and Gagne, W.C. 1984. Modification of small farmer practices for better pest management. *Annual Review of Entomology,* 29:383-402.

Morse, S. and Buhler, W. 1997. IPM in developing countries: The danger of an ideal. *Integrated Pest Management Reviews,* 2: 175-185.

Naranjo, S.E. 2009. Impacts of Bt crops on non-target organisms and insecticide use patterns. *CAB Reviews: Perspectives in Agriculture, Veterinary Science. Nutrition and Natural Resources,* 4: 1-23.

Norton, G.A. and Mumford, J.D. 1993. *Decision Tools for Pest Management.* CAB International: Wallingford UK.

Oerke, E.C., Dehne, H.W., Schönbeck, F. and Weber, A. 1994. Crop Production and Crop Protection-Estimated Crop Losses in Major Food and Cash Crops. Elsevier, Netherlands.

Ooi, P.A.C. 1986. Insecticides disrupt natural control of *Nilaparvata lugens* in Sekinchan, Malaysia. In *Biological Control in the Tropics. Proc. Reg. Symp. Biol. Control, 1st, 1985, Serdang, Malaysia,* ed. MY Hussein, AG Ibrahim, pp. 109-20. Serdang: Univ. Pertanian Malaysia.

Paroda, R.S. and Kumar, P. 2000. Food production and demand situations in South Asia. *Agricultural Economics Research Review,* 13(4): 1-25.

Pathak, M.D. 1968. Ecology of common insect pests of rice. *Annual Review of Entomology,* 13:257-294.

Pathak, M.D. 1975, Insect Pests of Rice. Indian Rice Research Institue, Philippines. pp. 68.

Pimentel, D. 1991. CRC *Handbook of Pest Management in Agriculture,* 2nd ed., Vol. IIIrd, CRC Press, Boca Raton. FL.

Pincus, J. 1994. Discussion points on pesticide policy and rice production in Indonesia: combining reform from above and action from below. *Proc. Göttingen Workshop*

Pestic. Policies, 28 Feb.-4 March, Göttingen, Ger., ed. S Agne, G Fleischer, H Waibel, pp. 61-73. Göttingen: Inst. Agrarökon. Univ. Göttingen.

Pratap, A. and Gupta, S.K. 2009. Biotechnological Interventions in Host Plant Resistance. In R. Peshin, A.K. Dhawan (eds.), *Integrated Pest Management: Innovation-Development Process*, Vol-I, Springer Verlag.pp.

Rahman, P.F., Sharma, S.B. and Wightman, J.A. 2000. A review of insect parasitic nematodes research in India 1927- 1997. *International Journal of Pest Management, 46:* 19-28.

Roger, P.A. and Kurihara, Y. 1988. Floodwater biology of tropical wetland ricefields. *In*: '*Proc. First Int. Symp. Paddy Soil Fert.*, pp: 275-300. The Hague, Netherlands: International Soil Science Society, The Netherlands.

Schoenly, K.G., Cohen, J.E., Heong, K.L., Arida, G.S., Barrion, A.T. and Litsinger, J.A. 1996. Quantifying the impact of insecticides on food web structure of rice-arthropod populations in a Philippine farmer's irrigated field: a case study. *In*: Polis G. and Winemiller K. (eds), *Food Webs: Integration of Patterns and Dynamics*, Chapman and Hall, New York NY, pp. 343-351.

Settle, W.H., Ariawan, H., Astuti, E.T., Cahyana, W., Hakim, A.L., Hindayana, D., Lestari, A. S., Sartanto, P. 1996. Managing tropical rice pests through conservation of generalist natural enemies and alternative prey. *Ecology*, 77: 1975-1988.

Shankar, U. and Kaul, V. 2010. Entomopathogenic Nematodes: A Tool in IPM for Vegetable Crops. *In:Quality Seed Production of Vegetable Crops* , 2010, Sharma, J. P. (ed.), Kalyani Publishers, Ludhiana, India.pp. 207-225.

Sharma, H.C., Sharma, K.K., Seetharama, N. and Ortiz, R. 2000. Prospects for using transgenic resistance to insects in crop improvement. *Electronic Journal of Biotechnology*, available on line at *http://www.ejb. org/content/vol3/issue2/full/3.*

Smith, J., Litsinger, J.A., Bandong, J.P., Lumaban, M.D. and dela Cruz, C.G. 1988. Economic thresholds for insecticide application to rice: profitability and risk analysis to Filipino farmers. *Journal of Plant Protection in the Tropics*, 6:67-87.

Subba Rao, C., N. Venugopal and S. A. Razvi. 1983. Parasitism, a key factor in checking rice pest population. *Entomon*, 8: 97-100.

Swaminathan, M.S. 1975. ICAR, Operational Research Projects, purpose and approach. *Indian Farming*, August.

Tabashnik, B., Gassmann, A.J., Crowder, D.W. and Carriere, Y. 2008. Insect resistance to Bt crops: evidence versus theory. *Nature Biotechnology*, 26: 199-202.

Tahir, H.M. and Butt, A. 2009. Predatory potential of three hunting spiders inhabiting the rice ecosystems. *J. Pest Sci.*, 82: 217-225.

Technology Information, Forecasting & Assessment Council (TIFAC). 2004. *Technology for Futuristic Pesticides*. Published by Technology Information, New Delhi, Retrieved on 30 May 2007 from *www.tifac.org.in*

Wilson, C. and Tisdell, C. 2001. Analysis, why farmers continue to use pesticides despite environmental, health and sustainability costs. *Ecol. Econ.*, 39: 449-462.

Fig. 1 : Light trap installed near rice field

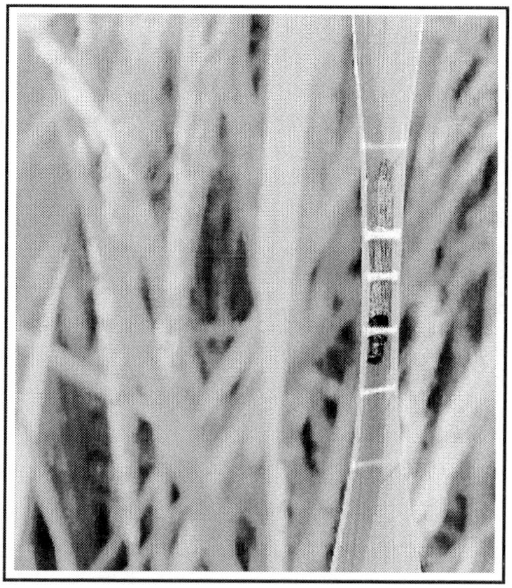

Fig. 2 : Rice leaf fold larva inside the leaf blade

Fig. 3 : Rice skipper moth

Fig. 4 : Shoot horned grasshopper

Fig. 5 : Green leafhopper

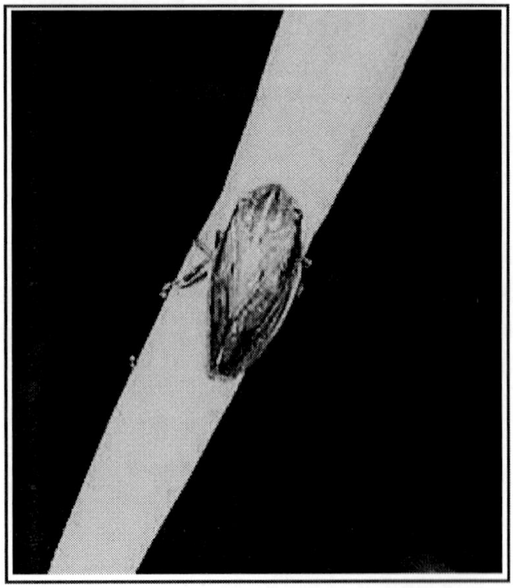

Fig. 6 : Brown planthopper

Fig. 7 : White-backed planthopper

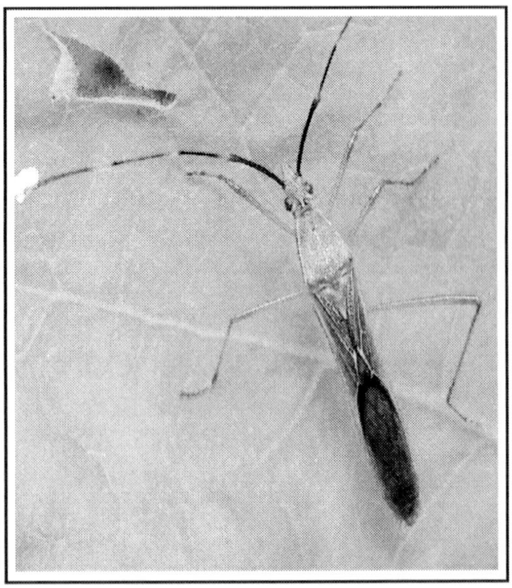

Fig. 8 : Rice gundhi bug

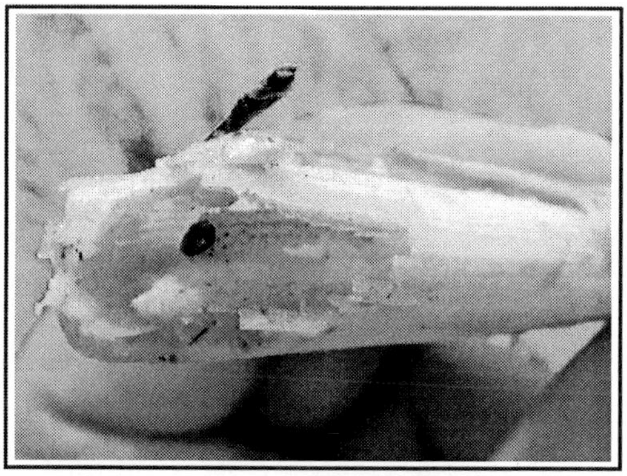

Fig. 9 : Yellow stemborer larva inside plant stem

Fig. 10 : Egg. parasitism by *Telenomus* spp.

Fig. 11 : Spider feeding on rice leaf folder larva

Fig. 12 : Felled maize plant by cut worm

Fig. 13 : Blister beetles feeding on maize silk

Photographs courtesy by Dr. Uma Shankar

Ecologically Based Integrated Pest Management, 2012
© D.P. Abrol and Uma Shankar (eds.), pp. 597-617
New India Publishing Agency, New Delhi (India)
e-mail : info@nipabooks.com; website : www.nipabooks.com

31

Integrated Disease Management of Field Crops under Hill Agro-Ecosystem

S.K. SINGH, DEEPAK KUMAR AND P.K. SINGH

The global importance of cereal crops to the human diet and moreover to the written history of man and agriculture cannot be overstated. Cereal grains are the fruit of plants belonging to the grass family (*Gramineae*) most important group of food crops produced in the world. Nutritionally, they are important sources of dietary protein, carbohydrates, complex vitamin B, vitamin E, iron, trace of minerals, and fiber. It has been estimated that global cereal consumption directly provides about 50 per cent of protein and energy necessary for the human diet. Some cereals, notably wheat, contain proteins that form gluten, which is essential for making bread. Major cereal crops produced worldwide include wheat, rice and maize. Wheat, rice, maize, sorghum, and millet are produced in large quantities in India.

Integrated Disease Management

Disease control is, managing the plant disease severity below the economic threshold following economically viable, eco-friendly and easily operational procedures. The position of IDM in the overall agricultural system indicates clearly that it is an important sub-system and interacts with other such systems that fall under the cropping system.

Surveillance of Disease

Surveillance of crop health is an important component of integrated disease management system. Disease surveillance includes the monitoring of disease prevalence and severity. Surveillance helps in the identification of potentially important diseases in early stage of development. Surveillance focuses primarily on decision making for the selection of proper timing of pesticide application and prevents from unnecessary uses of chemical in the agriculture system. Thus surveillance, along with other logistic support, may be an effective and important tool of disease forecasting.

Resistant Cultivars

The use of resistant varieties may be the main component in disease control in several crops. Varietal replacement provides a spatial and temporal discontinuity of susceptible host against the pathogen. This approach reduces the disease severity and the extent of the losses. The use of resistant varieties is the cheapest, safest and the most effective approach of plant disease management. The important resistant varieties developed to counter the diseases epidemics of various crop shown in Table 1.

Table 1 : Important resistant varieties of crops at present under cultivation.

Crop	Varieties	Resistant to
Rice	Vikramarya	Rice tungro virus
	Ajaya	Bacterial leaf blight
	Tulsi, Rasi, IR 64	Rice blast
	Suraksha	Multiple disease resistance
Wheat	PBW 550	Rust, Blight and Karnal bunt

Resistance may be defined as a inherent capability of host plant to reduce the chance and degree of infection and disease development. Any variety that expresses a high degree of resistance to certain specific race(s) is considered to be vertical resistance (Vander plank, 1963). Vertical resistance to specific races is generally governed either by a single (monogenic) dominant gene or by a few (oligogenic) dominant genes. Horizontal resistance reduced the rate of disease spread and is evenly spread against all races of the pathogen.

Hot Spot Testing

In India, various types of climate and weather provide an opportunity to evaluate the host resistance against the various pathogens at hot spots. The pathologists have confirmed several resistance stocks by repeated testing them under these hot spot condition.

Gene Deployment

Gene deployment strategies are applicable for long distance spreading of diseases like the black, brown and yellow rusts of wheat, rice tungro disease and several other pathogens. The spatial and temporal deployment of resistant genes has immensely benefited India in many crops diseases including the wheat diseases.

Cultural Control

The proper and judicious use of fertilizers or nutrients, sowing time adjustment, proper soil and water management, crop rotation, sanitation and proper weed management help in reducing the losses due to different diseases and resulting better yield.

Avoidance of Pathogen

Smut and ergot of bajra are more severe in wet areas where rain occurs for longer period during the flowering stage of the crop. A number of soil borne diseases can be prevented if the same crop is not grown in the same field year after year.

Crop Rotation

Crop rotation helps to control many soil borne diseases, such as wilts of arhar and linseed, smut of bajra, bunt of wheat and root rot of vegetables crops (Table 2).

Table 2 : Important diseases of crop plant controlled by crop rotation.

Disease	Pathogen	Crop interval (years)	Rotation crop
White rot of pea	*Sclerotium cepivorum*	8-10	Non host
Anthracnose of pea	*Colletotrichum pici*	2+	Non host
Seedling blight of pea	*Rhizoctonia solani*	1+	Cereals and corms
Potato scab	*Streptomyces scabies*	1+	Soyabean
Flag smut of wheat	*Urocystis tritici*	1	Except wheat
Soil rot of tomato	*R. solani*	1	Pangola grass
Sorghum stalk rot	*Fusarium moniliforme*	2	Winter wheat

(*Source :* Paroda, 1997).

Sanitation

Field sanitation is another effective tool for controlling diseases. Burning of disease stubbles or previous crop refuse is common practice in this method. Removal of crop debris and the irrigation water regulation from fields with disease crops also help in reducing the number of primary inoculum and further spread of the disease. This method is found very effective in case of soil borne diseases, caused by species of *Fusarium* sp., *Verticillium* sp., *Rhizoctonia* sp., *Sclerotium* sp. and *Sclerotinia* sp.

Rogueing

Roguing of diseased is used for management of simple interest type of disease as loose smut of wheat, yellow vein mosaic of okra, panama disease of banana etc. This is the practice of careful removal and destruction of infected plants or plant parts from the field at an early stage which minimize the foci of infection and preventing wide spread of pathogen.

Heat Treatment and Solarization

Hot water treatment of seeds is most successful and useful to eliminate *Ustilago segetum tritici* from wheat seeds, *Alternaria brassicae* from cabbage seeds. Hot water treatment is effective for inactivating the virus inoculum present in planting materials like bulbs and suckers etc. Heat treatment of planting sets in sugarcane eliminates the red rot fungus effectively. Soil solarization, covering the soil with polythene sheets, exposing the soil by deep ploughing in summer and solar radiation eliminates several soil borne disease and nematodes in the soil.

Trap Crops

Decoy or trap crop are the non host crops which are sown with the main crop with the purpose of reduction of inoculum potential of soil borne pathogens. This is achieved by stimulating the dormant propagules of the pathogens in the absence of the host.

Quarantine and Exclusion of Pathogen Inoculum

The aim of quarantine is preventing new disease and pathogen to a newer area and avoiding contact between the pathogen and the host. Seed certification and growing crops in the region unfavourable for a particular pathogen are the means for preventing the further spread of the pathogen. Quarantine is the legal restriction on movement of agricultural commodities from one area to another area for the purpose of exclusion, prevention or delay in the spread of the plant pests and diseases in an unaffected area. In 1914, Govt. of India passed the destructive insect pest act (DIP act) prohibiting or restricting the import of plant and plant materials, insects, fungi etc. to India from foreign countries.

Biological Control

The natural mortality of the pathogen by parasites, predators or competitor fungus is the most important way to minimize the pest outbreaks. The use of natural parasites, predators or competitor fungus is another important component of IDM, which can be useful in managing the pathogens and crop diseases. In this method of disease control care has to be taken so that the organism used for control of the pathogen should not be pathogenic to crop concerned or to the other crops that are likely to be grown in the cropping sequence. The example of antagonistic micro-organism enlisted in Table 3.

Table 3 : Management of plant pathogens through biocontrol agents.

Diseases	Pathogens	Antagonist (s)
Chickpea wilt	*Fusarium oxysporum* f. sp.*ciceri*	*Trichoderma harzianum*
Cotton root rot	*Rhizoctonia solani*	*T. harzianum*
Spot blotch of wheat	*Bipolaris sorokiniana*	*Chaetomium globosum*
Cucumber fruit rot	*Rhizoctonia solani*	*T. harzianum*
Damping off of vegetables (Bean, Cotton, Solanaceous Vegetables, Tobacco)	*Pythium* spp. *Rhizoctonia solani* *S. rolfsii*	*T. harzianum* *T. harzianum* *T. harzianum*

(*Source :* Paroda, 1997).

Transgenic Plants

Biotechnological approach provides a new tool in plant disease management through the resistant genes transfer into agronomically superior genotypes for the management of diseases. For the management of viral diseases coat protein mediated resistance appears to be very promising. Chitinase gene of plant and bacterial origin has also been introduced into *Rhizobium meliloti*, *Pseudomonas flourescencs.* These bacteria, colonizing around the roots carrying the chitinase gene and suppresses wilt disease caused by *Fusarium oxysporum*. Cloning of plant defense genes conferring resistance to varieties of bacteria, fungal and virus diseases has need urgent attention for management of crop diseases. Transfer of Xa-24 gene of *Oryza longistaminata* to rice cultivars which confer resistance against different races of bacterial blight.

Major Field Crops

The major field crops grown in the tropical and sub-tropical areas are wheat, rice and maize. All the cereals commonly grown in this area are prone to a number of fungal, bacterial and viral diseases. These diseases are the main constraints in the cultivation of these high value cereal crops. Few diseases of major importance are being discussed below.

Important Wheat Diseases and their Management

Wheat is most extensively grown cereal crops. Wheat is the second major staple Food Crop, In India, it is the second important crop, next to rice, with a total production of 64.5 million tones grown over 24 million hectare area.

One of the major constraints in increasing production of wheat are diseases that effect crop yield as high as 10 per cent. Wheat is attacked by a number of fungi, bacteria, viruses etc. But the fungal diseases are the most predominant biotic stress as compared to other microorganisms. The most important fungal diseases that affect wheat crop are stem, leaf and stripe rusts, loose smut, karnal bunts, foliar Blight, powdery mildew, flag smut and hill bunt. In wheat, brown rust, yellow rust, black rust, loose smut, karnal bunt and powdery mildew were observed to be important diseases in different states. Karnal bunt and foliar blights were more damaging whereas stem rust became relatively less important after the green revolution (Chander *et al., 2003)*.

Wheat Rusts

The three important rust diseases occurs on wheat viz., stem or black rust (*Puccinia graminis* Pers. f. Sp. *tritici* Schof.), leaf or brown rust (*Puccinia recondita* Rob. ex. Desm. F. sp. *tritici*) and stripe or yellow rust (*Puccinia striiformis* West). Rust diseases are responsible for causing huge economics losses. In India leaf rust are most important diseases and prevalent all over the country. The disease in north-western region appears in late January and assumes epidemic growth rate in the month of February- march when the crop is in anthesis or grain filling stage but in case of stem rust it is mainly prevalent in central and peninsular region of country but it appear very late with low intensity in the northern belt. Stripe rust is restricted only in northern parts of the country particularly in foot hills of Himalayas and adjoining plains of north-western region. Spores of both *Puccinia recondita* f. sp. *Brown* and *P. striiformis* (yellow rust of wheat) are thought to be primarily dispersed by wind and rainfall has the potential to spread both brown rust and yellow rust of wheat (Geagea *et al.,* 2000).

Rust diseases are responsible for huge economics losses. In 1972-73, stem and stripe rust which appears in epidemic form in Punjab, Haryana and western Uttar Pradesh were responsible for a loss of nearly 0.8 to 1.5 million tones of wheat. In 1980, leaf rust epidemic in Uttar Pradesh and parts of Bihar caused a loss of one million tones of wheat.

Symptoms

In case of stem rust the pustules appear a narrow, dark brown elliptical, parallel with the axis of the stem, on the leaf, leaf sheath and on glumes of the spike. The epidermis covering the pustules rupture irregularly, exposing masses of reddish brown uredospores. The pustules of leaf rust are small, oval shaped, orange to orange brown in colour, scattered on leaf and leaf sheath. Uredinaks pustules are may also developed on awns, glumes, peduncles, and internodes. Stem rust has been a serious disease of wheat, barley, oat and rye, as well as various important grasses. Infections in cereals occur mainly on stems and leaf sheaths, but occasionally they may be found on leaf blades and glumes as well. The first macroscopic symptom usually appears as a small chlorotic fleck, after a few days of infection. About 8-10 days after infection, a long pustule is formed by rupturing the host epidermis from pressure of a mass urediospores of brick-red colour produced in the infection. The powdery masses of urediospores appear similar to rust spots on a weathered iron surface. With age, the infection ceases production of brick-red urediospores and produces a layer of black teliospores in their place, causing the stems of heavily infected plants to appear black in the late season (Leona and Szabo, 2005).

Integrated Management

Grow resistant varieties of wheat recommended for particular area and grow different varieties at the farm level. Apply balanced dose of nitrogen and potash. Spray Dithane M 45 @ 0.2% or propiconazole or tebuconazole @ 0.1% along with 0.1% sandovit. Kalappanavar (2008) reported that the fungicide propiconazole performed best followed by triadimefon and hexaconazole and in the categories other than fungicides, neem leaf extract, *Trichoderma harzianum* and Panchgavya were the succeeding treatments effective against leaf rust of wheat. The yield of the plots sprayed with propiconazole, triadimefon and hexaconazole were significantly superior with low intensity of the leaf rust on yield and higher 1000-grain weight. Parsons (2000) found that the phosphorus (P) fertilization had a moderately suppressive effect on leaf rust and the increased yield due to production of about 50% more ear heads per unit area apparently from more prolific tillering and potassium (K) fertilization reduced leaf rust severity and improved yield by increasing kernel weight. P and K fertilization along with fungicide application is an important management tool for reducing disease and increasing winter wheat yield.

Smuts and Bunts

Five bunt and smut diseases namely Loose smut (*Ustilago segetum* (Pers.) Roussel *tritici*, Flag smut (*Urocystis agropyri* (preuss) schroet), Common bunt (*Tilletia caries* (DC) Tul.) and *Tilletia floetida* (Wallr. Liro.), Dwarf bunt (*T. controversa* Kuhn) and Karnal bunt (*Neovossia indica* (Mitra) Mundkur) attack wheat crops.

Loose smut is more severe in north than south India and can causes yield losses up to 2-4 percent in north western India. Karnal bunt has assumed importance in recent years due to its quarantine importance. Reduction in yield due to karnal bunt in India is not beyond 0.2 to 0.5 percent but 3-5 percent infected grains can cause blackening of whole meal. Hill bunt and flag smut are of minor importance because of their restricted distribution in isolated pockets of Rajasthan, Haryana and Punjab.

Symptoms

In case of loose smut entire inflorescence, except the rachis, is replaced by masses of smut spores. These black teliospores often are blown away by the wind, leaving only the bare rachis and remnants of other floral structure.

Karnal bunt is not easily detected prior to harvest, since it is usual for only a few kernels per spike to be affected by the disease. Following harvest, disease kernels can be easily detected by visual infections. A mass of black teliospores replaces a portion of endosperm, and the pericarp may be intact or ruptured. Disease kernels give off a fetid or fishy smell when crushed. Karnal bunt of **wheat** is caused by the fungus *Tilletia indica*, which partially converts kernels into sori filled with teliospores. Despite minor overall yield and quality losses, the disease is of considerable international quarantine concern (Goates and Jackson, 2006).

Integrated Management

Flag Smut

Flag smut of wheat incited by *Urocystis agropyri* could be managed by growing resistant cultivars like WH 293 or HD 2329. Seed treatment with tubeconazole @ 2.0 g/kg, or carboxin and carbendazim @ 2.5 g/kg seed reduced cent per cent disease incidence. Application of poultry manure @ 4–6 t per ha managed flag smut. $N_{160}P_{50}$ kg per ha and early or late sowing were found effective in management of the disease (Kumar and Singh, 2004).

Loose Smut

Sowing of resistant varieties like Kalyan Sona, Sonalika and PU 18 is mainly adopted for the management of the disease. Seed treatment with

Vitavax @ 2-2.5 g/kg seed before sowing can manage the disease by reducing the seed borne primary inoculum. Roguing out suspected disease ears at the time of ear emergence and destroyed them properly by burning or burring under the ground. Soaking of seeds and solar heat treatment of seeds is practicable in hot summer areas of plains. During bright sunny days in the last weeks of May or first week of June soak the seed in water for about four hours in the morning and dry it under hot sun in the afternoon. Storage of properly dried seed for use in the next season is also reduce the disease incidence.

Karnal Bunt

Sowing of resistant varieties and use of disease free seeds are the important control measure for the management of the disease. Treatment of the seeds before sowing with Vitavax @ 2-2.5 g/kg seeds minimize the disease intensity through elimination of seed borne infection. Avoidance of excessive irrigation at the time of flowering also reduce the disease intensity.

Common Bunt

Treatment of the seeds before sowing with Vitavax @ 2-2.5 g/kg seeds reduce the disease intensity by elimination of seed borne primary inoculums. Cultivation of common bunt resistant varieties like Kalyan Sona, Panjamo-62 and PU-18.

Foliar Blight Complex

A number of pathogens causing blight, blotch and spots are associated with wheat in India. *Bipolaris sorokiniana* Shoem. is the major pathogen followed by *Alternaria triticina* prasada and prabhu. *Drechslera tritici repentis* and *Aternaria alternata* are also associated with wheat but are not very prominent. Both the pathogen are widely prevelant in Bihar, West Bengal, Uttar Pradesh, Rajasthan and Orissa. Of late *B. sorokiniana* has made appearance in North western region of the Indo-Gangatic plains also.

The importance of foliar blights must be expressed in terms of yield losses. Recent study on yield losses due to *B. sorokiniana* revealed that yield of wheat cultivars UP 2338, UP 262, HP 1633 may be reduced upto 36.24, 20.93 and 22.26 per cent respectively due to spot blotch disease.

Symptoms

Alternaria blight symptoms appear as small, oval, discoloured lesion, irregularly scattered on leaves. As lesion enlarged, they become irregular and dark brown to grey surrounded by yellow margin. Later several lesions coalesce and cover large areas exhibiting bunt appearance. The lesion on leaves caused by *B. sorokiniana* are light brown to almost black, which are circular linear or elliptical. A well developed lesion is typically elliptical with abundant sporulation.

Integrated Management

Cultivation of resistant varieties is an important component of disease management. Hot water treatment of seeds at 52°C for 10 minutes before sowing provides good control of the disease by reducing primary inoculum. Seed treatment with fungicide vitavax @ 2.5 g/kg of seeds also effective for the management of the disease in the field. Use of balanced NPK fertilizers and irrigation reduce the intensity. Spraying of plants with Dithane M 45 @ 0.2% or propiconazole or tebuconazole @ 0.1% at 25-35 days interval starting from the the first symptoms appearance in the field.

Powdery Mildew

Powdery mildew of wheat is caused by *Erysiphe graminis tritici*. It is widely distributed in the humid and sub humid wheat growing regions of the world. It is basically important disease of northern hilly region of India. This disease is also occurs in the north-western region of the Indo Gangatic Plains where stripe and brown rust are the potential threat to wheat production. It is believed that heavy attack of powdery mildew in hilly and north-west plain regions result in yield losses through reduced grain size.

Symptoms

All the green part of the plant, the upper surface or undersize of the leaves, sometimes both surface together, the stem or even the ears are affected by the pathogens, exhibiting white to grey coloured powdery masses on the plant surface. The cottony powdery mass contains the conidia and conidiophores of the fungus. As the season progresses, small black cleistothecia developed on the powdery mass. The infected leaves are twisted, crinkled and deformed, infected plant become yellow and photosynthetic rates are severely affected.

Integrated Management

Sowing of resistant varieties is the major control measure. Burning of the previous crop refuses in the field after harvesting the crop could reduce the disease intensity. Spraying of the fungicide mixture of 16 parts Dithane M-45 and 4 parts Karathane 25 WP @ 2 kg/ha in 800 litres water. Two to three sprays of the mixture should be given at an interval of 10-15 days if losses are expected to very high.

Rice Diseases and their Integrated Management

In India rice is grown in various agro-climatic regions ranging from Kashmir to Kanyakumari as upland, medium and low land rice. Total area coverage under rice in India is 42.4 million hectares. A brief account of the important rice diseases and their management strategies are given below.

Rice blast : (*Pyricularia grisea*)

The disease is considered to be most important one among all the rice diseases not only India but all the rice growing area of the world. The loss due to this disease may reach up to 50 per cent and sometimes the complete loss of the crop has been recorded in different areas under epiphytotic condition.

Symptoms

Spot appear as distinct large, indefinite, spindle shaped, grey centred and lavender blue, purple grey horizon with yellow margin on leaf blade (Fig. 7). On the stem, one or two nods just below the panicles, grayish black or dark lesions appear. The grains are mostly half filled or unfilled and generally grains are partially chaffy, brittle or unfilled; panicle show white appearance.

Integrated Management

Cultivation of resistant varieties like Vandana, Pathara, Hira, Savitri, Gayatri, Ratana, Annada, Udaya, Sarasa etc. help to get effective control of rice blast diseases. Use of healthy seeds for nursery raising and seed treatment with thiram @ 2.5 g or bavistin 2 g/kg of seed gives good control by avoiding the primary inoculum. Burning of affected straw and stubbles after harvest is also a good practice of disease management. High dose of nitrogenous fertilizer (> 60 kg/ha) increases the disease intensity so, it should be avoided and phosphorus and potash fertilizer (60 kg/ha) should be use to reduce the epidemics of the disease. To avoid vulnerable phase for blast development, planting of crop adjusted in such a way that active tillering stage should not fall during 3rd week of September to 2nd week of November. In endemic area the disease appears every year spray the crop with 0.1 % Hinosan 50 EC (1 ml/ litre) or Edifenphos (0.06%) or copper oxychloride (0.3%). Singh *et al.* (1975) found hinosan @ 0.1 per cent superior to other treatments. Two treatments of azoxystrobin and trifloxystrobin at 250 and 125 g/ha, respectively, were more effective for management of blast and neck blast and reduced incidence and severity by 90-100% and by 75-90%, respectively. Antibiotic Kasugamycin @ 20-30 mg/l to be found very effective in preventing secondary spread.

Brown spot : (*Bipolaris oyzae*)

Symptoms

The spot are small, definite, some spot may be oval to oblong, light brown or straw centre and dark reddish brown margin. Dark brown to black, oval or oblong spots on the glumes, whole surface of grain black and velvety.

Integrated Management

Treatment of the seeds with thiram or bavistin 2 g/kg of seeds. Add murate of potash to correct the potash deficiency in the soil and apply balanced dose of fertilizer to minimize the disease in the field. Three to four spray of Dithane M-45 @ 2.5 g/l of water or bavistin or propiconazole 0.1 per cent at 15 days interval just before the appearance of first symptoms of the disease. Cultivation of moderately resistant varieties like Bala, Krishna, Sabarmati and IR 24. Seeds should not be collected from the diseased fields and plants.

Sheath Blight

Paracer and Chahal (1963) first described the disease in detail from Gurdaspur, Punjab. Later, this disease was reported from Uttar Pradesh (Kohli, 1966), Orissa (Das, 1975) and Tamil Nadu (Kanniyan and Prasad,1976) as well. Sheath blight a fungal disease of rice caused by *Rhizoctonia solani* Kuhn., is prevalent in upland and low land rice. It is gaining importance in Andhra Pradesh, Assam, Kerala, Punjab and Tamil Nadu and its incidence has also been recorded in Jammu and Kashmir, Orrisa, Uttar Pradesh and West Bengal. The extent of damage due to sheath blight is more than 25 per cent. Roy (1979) reported that yield losses vary from 10 to 36 per cent depending upon the crop growth stage at which the disease first appears.

Symptoms

On leaf blade greenish grey irregular lesions with dark line on margins, lesion large and develop irregularly on older plants. The symptoms appear on stem and sheath with large greenish grey or dark edged whitish lesions which are oval, oblong, cobra patches, irregular and necrotic. Small, white or brown, globular sclerotial bodies are loosely attached to the surface.

Integrated Management

The disease control strategy should be rather preventive to realize maximum yields. Using seeds from healthy crop. Deep ploughing and use of limited but adequate nitrogenous fertilizers. Avoiding close planting and follow field sanitation i.e. burning of disease straw, stubbles and chaff and eradication of graminaceous weeds and water hyacinth in and around the field. Soil amendment with green manure and rotation of rice with dicot crops. Avoiding field to field irrigation. Some of promising ones with moderate resistant to sheath blight are T141, CR 1014, Ptb 21, Ptb 33, Swarndhan, IET 6056 IET 6105, CR-44-11, Krishna and Saket-I. The disease can be controlled by need based application of fungicides such as Bavistin or MBC (0.1%) and delayed top dressing of nitrogenous fertilizer when 10 per cent or more tillers are affected during early planting of mid tillering. Second spraying of Bavistin or MBC (0.1%) is done when 10 per cent or more tillers are affected during panicle initiation to booting stage. Bhuyan *et al.* (1994) reported the antagonist activity of *Trichoderma viride, Trichoderma harzianum and Aspergillus terreus* against *R. solani.* Baby and Manibhushanrao (1996) evaluated *Trichoderma spp.* and *Gliocladium virens* as well as some vesicular arbuscular fungi against sheath blight *(R.solani*) of rice in field and reported that *Tirchoderma spp.* (*T. harzianum and T. longibrachiotum*) and *Gliocladium virens* were equally effective in controlling the disease.

Bacterial blight of rice : (*Xanthomonas oryzae pv. Oryzae*)

This is also most important disease of rice and appears in rice field cultivated under low land areas. The losses due to bacterial blight may reach 30 to 50 per cent. In certain years rice fields infected with blight in different states are reported to suffer yield loss up to 60 per cent.

Symptoms

The disease appears in two phases viz., wilt or kresek phase and leaf blight phase.

Kresek Phase

This is most destructive phase of the disease which result from the early systemic infection. The leaves roll completely, droop and turn yellow or grey ultimately the tillers wither away. Affected shoot may be completely killed in severe attack.

Leaf Blight Phase

The symptom is characterized by the appearance of straw coloured stripe with wavy margins generally are both side of the leaves. This stripe usually start from the tip and extend down wards the leaves turn straw yellow. Yellowish bacterial ooze appears on the surface which dry into bead like encrustations if there is no rain for a few days. This blight phase usually appears 4-6 week after transplanting. The grain are partially filled or become chaffy in the disease plants.

Integrated Management

Use disease free seeds from the reliable sources. Drain the standing water in the field from time to time. Incorporate balanced dose of NPK fertilizers after soil testing and avoid use of heavy dose of nitrogen fertilizer aggravates the disease. Spray the crop with a mixture of copper oxychloride and streptomycin. Apply 7.5 g streptomycin or Agrimycin and 500 g Copper oxychloride (fytolan or blitox-50) in 500 litre water. Seed soaking for 12 hours and treating in hot water at 53^{0}C for 30 minutes will make the seeds free from bacterium. Cultivation of resistant varieties like Ratna, Pusa 2-21, IR 20, Prasad, Govind, IR 24, Jaya, Vijaya, Shakti, Rudra, Shankar, Parijat, Keshari, Annapurna, IR 36, Mahalaxmi.

Maize Diseases and their Integrated Management

Maize is not only utilized as a staple food by the large population in rural areas, but is also a crop per excellence for industrial use. During the last few years, there has been progressive escalation in demand for the maize grains for the value added products. Among the factors adversely affecting productivity, incidence of diseases in the preharvest stage is prominent.

Foliar Diseases of Maize

In India about twenty foliar diseases are known to occurs on maize. However, Turcicum leaf blight, Maydis leaf blight and common rust are considered to be important in terms of geographical distribution and potential to cause significant yield reductions.

Turcicum leaf blight (*Exserohilum turcicum* (Pass) Leon. and Suggs.)

Turcicum or northern corn leaf blight (NCLB) incited by the ascomycete *Setosphaeria turcica*, anamorph *Exserohilum turcicum*, is a ubiquitous foliar disease of maize distributed throughout the world of economic importance (Dorothea *et al.*, 1998, Welz and Geiger, 2000 and Khampila *et al.*, 2008). Turcicum leaf blight (TLB) of maize caused by *Exserohilum turcicum* (Pass) Leonard. Suggs. is a major production constraint of maize crop in Karnataka (Harlapur *et al.*, 2008) . This disease is prevalent in the areas where maize is planted in the highlands (more than 1000 m altitude), Where temperate conditions prevail and in winter plantings in the plains as the cool/moderate conditions favour disease development. Shivankar and Shivankar (2000) reported that inoculated plots showed high incidence of turcicum leaf blight in AMC1 (39.23%), AMC2 (35.46%), Manjari composite (33.61%) and Kargil 633 (31.42%). The inoculated plots recorded lowest grain yield with significant losses in the grain yield of from 10.50 to 18.77 per cent.

Symptoms

An early symptom is the easily recognizable; the symptoms produced as slightly oval, water-soaked, small spots produced on the leaves. These lesions grow into elongated, spindle-shaped necrotic lesions. First symptoms appear on lower leaves and gradually progress upwards and can lead to complete burning of the foliage. *Turcicum* leaf blight (or northern leaf blight) occurs worldwide and particularly in areas where high humidity and moderate temperatures prevail during the growing season. Under high humidity and moderate temperature the whole area becomes necrotic and plants appear dead. Lesions may also appear on leaf sheaths and husk. The highest disease severity (55.89%) was observed in Belgaum district and the lowest disease severity (27.64%) in Koppal district (Harlapu1r *et al.*, 2008).

Management

Harlapur *et al.* (2007) reported that Neem seed kernel extract @ five per cent concentration showed maximum inhibition of growth (56.64%) followed by Aloe vera @ 10 per cent (53.50%). Among the bio-agents, *Trichoderma harzianum* was effective in inhibition of growth (65.17%) and in the fungicides tested, mancozeb @ 0.25 per cent completely inhibited the growth. Carboxin power @ 0.1 per cent and propiconazole @ 0.1 per cent were the other effective fungicides. The development of cultivars resistant to *E. turcicum* seems to be the most efficient method to control NCLB damage (Khampila *et al.*,2008). Vieira *et al.* (2009) reported that the hybrids of popcorn UEM129, UEM145, UEM140, UEM116 and UEM101 presented the lowest levels of disease and were considered the most resistance genotypes. Okori *et al.* (1999) reported that the infection by maize streak virus (MSV) masked the symptoms expression of turcicum leaf blight (TLB).

Maydis Leaf Blight: (*Drechslera maydis* Nishikado)

The disease is distributed all over the world in warm temperate to tropical (20-30^0 C) maize producing areas. The disease symptoms appear as small and diamond shaped lesions limited by adjacent veins of the leaf. The disease is favoured by high temperature. The fungus mainly over winters in the plant debris, although it seed borne nature is established.

Symptom

The young lesions are observed as small diamond shaped lesion. As they mature the elongation of lesion takes place. Growth is limited to adjacent veins, so finally the shape of lesion is rectangular of 2 to 3 cm long. Lesions may coalesce, producing a complete burning of large areas of the leaves. The symptoms described above correspond to the "O" strain of the fungus. In the early 1970s the "T" strain caused severe damage to maize cultivars. Lesions produced by the T strain are oval and larger than those produced by the O strain. A major difference is that the T strain affects husks and leaf sheaths, while the O strain normally does not affects husks and leaf sheaths. Maydis leaf blight (or southern maize leaf blight) is prevalent in hot and humid maize-growing areas. The fungus requires slightly higher temperatures than *E. turcicum* for infection. Both species are often found on the same plant.

Management

Sprying of any copper fungicides@ 0.3 per cent or Dithane M 25 @ 0.25 per cent after the appearance of the disease gives good result. Kumar *et al.* (2009) found that the garlic, clove, neem leaf and tulsi leaf and bael leaf extract was highly effective in inhibiting the growth of *Helminthosporium maydis* from 37 to 65 per cent.

Common rust: (*Puccinia sorghi* Schw.)

This disease is prevalent in all over the world from moderate to severe form in temperate to semi temperate environments and in the winter season crop in tropics and high land tropics. Symptoms appear as circular to elongated brown powdery erumpent pustules on both the leaf surfaces which turn into black coloured teleutosori on maturity under favourable conditions for disease development these pustules extent t leaf sheath and husk. This disease is favoured by moderate temperature (17-25 ^{0}C). Shah and Dillard (2006) reported that for processing sweet corn, every 10% increase in rust severity reduced yield by 2.4 to 7.0%; the corresponding reduction for fresh market sweet corn was between 3.0 and 6.2%.

Management

The major approach to management of these diseases has been the utilization of host resistance. Resistance varieties viz., Ganga 11, Deccan 105 and inbred lines CM 104, Cm 105 have been reported to durable resistance to this disease. Complete control was obtained by a schedule of four spray applications of carbamate group of fungicide (Dithane M-45) @ 0.2 % a.i., on a susceptible cultivar Basi in case of common rust. Reuveni *et al.* (1994) reported thata single foliar spray of phosphate salts solution induced systemic resistance against common rust as expressed by 98% reduction in the number of pustules of *P. sorghi*.

Stalk Rot

This group of disease is broadly dividisible into two categories i.e., pre flowering and post flowering types.

Pythium stalk rot: *Pythium aphanidermatum* (Eds) Fizt.

This stalk rot is prevalent in the maize growing area of the countries with sub tropical climate. In India it occurs in considerable proportion in the north India.

In this case the rot is confined to a single basal internodes leading to destruction of pith parenchyma and subsequent weakening of stalk the infected plant topple and remain green and do not die upto two weak after attack. White fluffy growth of the fungus on the breaking point is very common under high humidity conditions.

Management

Planting should be done timely between 10 and 20th of July, which is prevalent in North India. Plant population should not exceed 50,000/ha. Maintenance of good field drainage, and Removal of previous crop debris to minimize the primary inoculum. Growing of resistant varieties viz., hybrids Ganga Safed 2, high starch and composite Suwan 1 is recommended for the disease prone area. Application of Captan @ 120g a.i.,/100 litre of water as soil drench at the base of the plants when crop is 5 to 7 week old can effectively check this disease.

Bacterial stalk rot: (*Erwinia chrysathnemi* pv. *zae*)

Prasad (1930) firstly reported the bacterial stalk and sheath rot of maize from Pusa Bihar. The disease is severe in areas of high rain fall like UP, Bihar, Andhra Pradesh, Rajasthan and Jammu and Kashmir.

The disease affects younger plants as well as older plants. The rotting starts from the lower nodes and moves up and down causing dark soft decay of the stalk. Stalk rotting causes yellowing and drying of leaves. The symptom developed as soft rot of stalk tissues followed by rapid wilting and plant death. The typical characteristics of disease are the omission of fermenting odour from the site of infection. The vascular bundle of the plant usually remains intact.

The resistance to disease has been identified as hybrids Ganga safed-2 and inbred lines CM 104 and CM 600. The effective control of the disease by using bleaching powder ($CaOCl_2.H_2O$ containing 33% of chlorine) @ 3.3 g/ 10 litre of water could be obtained by drenching of the basal stalk region. Avoidance of water logging and proper drainage also help in reducing disease incidence. Sharma *et al.* (1982) and Thind and Soni (1983) reported that two application of calcium hypochlorite (Klorocin) as drenching near the root region @ 25 kg/ha at flowering stage after 10 days interval gives good control.

Fig. 1 : Brown rust of wheat

Fig. 2 : Yellow or stripe rust

Fig. 3 : Loose smut of wheat

Fig. 4 : Flag smut of wheat

Fig. 5 : Spot blotch of wheat

Fig. 6 : Powdery mildew of wheat

Fig. 7 : Rice blast

Fig. 8 : Leaf spot of rice

Fig. 9 : Sheath blight of rice

Fig. 10 : Banded leaf and sheath blight Maize
Photographs courtesy by Dr. S.K. Singh SKUAST-J

Referances

Agrios, G.N. 2005. Plant Pathology. Academic Press, California.

Baby, U.I. and Manibhusharrao, K. 1996. Fungal antagonists and VA Mycorrhizal fungi for bio- control of *Rhizoctonia solani*, the rice sheath blight pathogen, pp. 1-9. In : *Recent Development in Biocontrol of Plant Pathogens* (Eds. K. Manibhushan Rao and A. Mahadevan). Today and Tomorrow Printers and Publishers, Karol Bagh, New Delhi.

Bhuyan, S.A., Das, B.C. and Bore, L.C. 1994. Antagonistic effect of *Trichoderma viride, Trichoderma harzianum* and *Aspergillus terrus* on *R. solani. Journal of Agri. Science Society of North East India.* 7: 125-127.

Chander, S., Aggarwal, P.K., Kalra, Naveen and Swaruparani, D.N. 2003. Changes in Pest Profiles in Rice-Wheat Cropping System in Indo-Gangetic Plains. *Annals of Plant Protection Sciences.* 11 (2).

Chatrath, R., Mishra, B., Ortiz Ferrara, G., Singh, S.K. and Joshi, A.K. 2007. Challenges to Wheat Production in South Asia. *Euphytica.* 157 (3) : 447-456.

Das, S.R. 1975. Occurrence of sheath blight disease of rice in Western Orrisa. *Indian Journal of Mycology and Plant Pathology.* 5: 54.

Dorothea S. B.; H. G. Welz and H.H. Geiger, 1998. Molecular marker analysis of European Setosphaeria turcica populations. *European Journal of Plant Pathology.* 104(6): 611-617.

Geagea, L., Huber, L., Sache, I., Flura, D., McCartney, H.A. and Fitt, B.D.L. 2000. Influence of Simulated Rain on Dispersal of Rust Spores from Infected Wheat Seedlings. *Agricultural and Forest Meteorology.* 101 (1) : 53-66.

Goates, B.J. and Jackson, E.W. 2006. Susceptibility of Wheat to Tilletia indica During Stages of Spike Development. *Phytopathology.* 96 (9) : 962-966.

Harlapur, S.I., Kulkarni, M.S., Wali, M.C. and Kulkarni, S. 2007. Evaluation of Plant Extracts, Bio-agents and Fungicides Against Exserohilum turcicum (Pass.) Leonard and Suggs. Causing Turcicum Leaf Blight of Maize. *Karnataka Journal of Agricultural Sciences.* 20 (3).

Harlapur, S.I., Kulkarni, M.S., Wali, M.C., Kulkarni, S., Hegde, Y. and Patil, B.C. 2008. Status of Turcicum Leaf Blight of Maize in Karnataka. *Karnataka Journal of Agricultural Sciences.* 21 (1).

Joshi, L.M., Singh, D.V. and Srivastava, K.D. 1986. Problems and progress of wheat pathology in South Asia. Malhotra Publishing House, New Delhi.

Kalappanavar, I.K., Patidar, RK., Kulkarni, S. 2008. Management Strategies of Leaf Rust of Wheat Caused by Puccinia recondita f. sp. Tritici Rob. ex. Desm *Karnataka Journal of Agricultural Sciences.* 21 (1).

Kannaiyan, S. and Prasad, N.N. 1976. Efficacy of certain fungicides in the control of sheath blight disease of rice. *Madras Agriculture Journal.* 63: 407-408.

Khampila, J., Lertrat, K., Saksirirat, W., Sanitchon, J., Muangsan, N. and Theerakulpisut, P. 2008. Identification of RAPD and SCAR Markers Linked to Northern Leaf Blight Resistance in Waxy Corn (Zea mays var. ceratina). *Euphytica.* 164 (3) : 615-625.

Kohli, C.G. 1966. Pathogencity and host range studies of paddy sheath blight pathogen. *Journal of Research*, Punjab Agriculture University. 3:37-40.

Kumar, S., Rani, A. and Jha, M.M. 2009. Evaluation of Plant Extracts for Management of Maydis Leaf Blight of Maize. *Annals of Plant Protection Sciences*. 17 (1).

Kumar, V.R. and Singh, B. 2004. Management of Flag Smut in Wheat. *Annals of Plant Protection Sciences*. 12 (1).

Leona, K.J. and Szabo, L.J. 2005. Stem rust of small grains and grasses caused by *Puccinia graminis*. *Molecular Plant Pathology*. 6 (2) : 99-111.

Okori, P., Adipala, E. and Kyetere, D.T. 1999. Induced Resistance to Turcicum Leaf Blight in Maize when Doubly Infected with Maize Streak Virus Disease. *International Journal of Pest Management*. 45 (2) : 111-116.

Paracer, C.S. and Chahal, D.S. 1963. Sheath blight of rice caused by *Rhizoctonia solani* Kuhn, a new record in India. *Current Science*. 32: 320-329.

Reuveni, R., Agapov, V. and Reuveni, M. 1994. Foliar Spray of Phosphates Induces Growth Increase and Systemic Resistance to *Puccinia sorghi* in Maize. *Plant Pathology*. 43 (2): 245-250.

Roy, A.K. 1979. Identification of strains of *Corticium sasakii* in India. *International Rice Research Newsletter*. 4: 17.

Sweeney, D.W., Granade, G.V., Eversmeyer, M.G. and Whitney, D.A. 2000. Phosphorus, Potassium, Chloride, and Fungicide Effects on Wheat Yield and Leaf Rust Severity *Journal of Plant Nutrition*. 23 (9) : 1267-1281.

Sharma, G., Saxena, S.C. and Sharma, G. 2001. Evaluation of biocontrol agents against *Rhizoctonia solani*-the causal agent of banded leaf and sheath of maize. *Annals of Plant Protection Sciences*. 9: 144-145.

Shah, D.A. and Dillard, H.R. 2006. Yield Loss in Sweet Corn Caused by *puccinia Sorghi*: A Meta-analysis. *Plant Disease*. 90(11): 1413-1418.

Shivankar, S.K. and Shivankar, R.S. 2000. Losses in Grain Yield Due to Turcicum Leaf Blight Disease in Maize. *Agricultural Science Digest*. 20 (3).

Singh, G., Chohan, J.S. and Sandhu, K.S. 1975. Fungicidal control of rice blast. *Indian Phytopathology* 28 : 422-423.

Thind, T.S. 2005. Diseases of field crops and their management. Daya Publishing House, New Delhi.

Vieira, R.A., Tessmann, D.J., Hata, F.T., Souto, E.R. and Mesquini, R.M. 2009. Resistance of Popcorn Hybrids to Exserohilum Turcicum, Causal Agent of Northern Leaf Blight of Maize. *Scientia Agraria*. 10 (5) : 391-395.

Welz, H.G. and Geiger, H.H. 2000. Genes for Resistance to Northern Corn Leaf Blight in Diverse Maize Populations. *Plant Breeding*. 119 (1) : 1-14.

❑❑❑

Ecologically Based Integrated Pest Management, 2012
© D.P. Abrol and Uma Shankar (eds.), pp. 619-650
New India Publishing Agency, New Delhi (India)
e-mail : info@nipabooks.com; website : www.nipabooks.com

32

Integrated Pest Management in Vegetable Eco-system

UMA SHANKAR AND SVS RAJU

Introduction

Vegetables contain large quantities of minerals, vitamins and essentials amino acids required for normal functioning of human metabolic processes and plays an important role in food and nutritional security. India is the second largest vegetables producing countries next to China and occupies 7.981 million ha area with an annual production of 129.077 million tons (Anonymous, 2009). Due to expanding urbanization and deforestation, most of the natural resources are depleting and shrinking with a rapid pace. With the limited land and depleting water resources, India has to feed the burgeoning population without destroying the ecological balances. In this context, we must have to look after the vertical expansion through increase in the intensity, diversification, mixed and intercropping of vegetable crops. Besides this, vegetables are also more prone to insect pests and diseases mainly due to their tenderness and softness as compared to other crops.

The extent of crop losses in vegetables varies with the crop type, crop location, damage potential of the insect pests involved and cropping season. Insect pest and disease infestation is one of the major limiting factors (Dhandapani *et al.*, 2003) in increasing the yield potential in brinjal, tomato, okra, cucurbitaceous and cole vegetable crops. The crops are susceptible to the attack of various insect pests and diseases in different stages with varying degree of damage which causes considerable losses by reducing potential yield and quality of the produce. In view of the lucrative returns from vegetables in short span, farmers with limited land holdings and resources are forced to follow monoculture and intensive cultivation, which also exacerbates the survival of various insect pests and pathogens from one season to another.

Changing Pest Scenario and Yield Losses in Vegetable Ecosystem

With the introduction of high yielding varieties and hybrids vegetable seeds resulted in dramatic changes in insect pest scenario leading to minor pests assuming the status of major pests. The serpentine leaf miner (*Liriomyza trifolii*) attacks tomato, brinjal, melons, leafy vegetables and cucurbits while the spiraling white fly damage okra. The mealy bug, *Coccidohystrix insolita,* and leafhopper, *Empoasca motti*, attack brinjal and bitter gourd, respectively (Balasubramanian, 2004) cabbage leaf webber, *Hellula undalis* and red spider mites on okra, brinjal and cucurbits. Overall, the insect pests inflict crop losses to the tune of 30-40 per cent in vegetable production (Srinivasan, 1993) and in many cases, there is 100 per cent yield loss due to viral diseases vectored by insects (Shivalingswami *et al.*, 2002).

Table 1 : Yield losses due to insect pests in vegetables

Major crops and their relative insect pests	Yield losses (%)
Crucifer crops	
Diamondback moth (*Plutella xylostella*)	16.87-98.83
Cabbage butterfly (*Pieris brassicae*)	68.5
Crucifer leaf webber (*Crocidolomia binotalis*)	28.09-50.88
Cabbage borer (*Hellula undalis*)	30-58
Crucifer aphid (*Lipaphis erysimi*)	36.5
Mustard saw fly (*Athalia lugens proxima*)	36.5
Chillies	
Chilli thrips (*Scirtothrips dorsalis*)	11.8-90
Chilli mites (*Polyphagotarsonemus latus*)	34.14
Brinjal (Egg plants)	
Fruit and shoot borer (*Leucinodes orbonalis*)	48-66
Tomato	
Fruit borer (*Helicoverpa armigera*)	22.39-37.79

Okra/ Lady's finger	
Shoot and fruit borer (*Earias vitella*)	22.79-54.04
Jassids/Leafhoppers (*Amrasca bigutulla bigutulla*)	54-66
White flies (*Bemesia tabaci*)	54.04
Fruit borer (*Helicoverpa armigera*)	22.1
Cucurbit crops/Summer vegetables	
Fruit flies (*Bactrocera cucurbitae*)	
Bitter gourd	60-80
Cucumber	20-39
Muskmelon	76-100
Snake gourd	63
Sponge gourd	50

(*Source :* Shivalingaswamy *et al.*, 2002)

Injudicious Use and misuse of Pesticides

Vegetable growers by and large depend on chemical pesticides to counter the problem of insect pests. It accounts for 13-14 per cent of total pesticides consumption, as against 2.6 per cent of cropped area (Sardana, 2001). The continuous, excessive and injudicious use of pesticides has triggered several ecological consequences like resistance (currently, more than 500 species of insects) and resurgence of insect pests, pesticide poisoning, environmental pollution, destruction and elimination of natural enemies, effect on non target organisms, disruption in food cycle and food web and residues in food commodities (Balasubramanian, 2004). Insects are not only responsible for massive direct losses of productivity but also cause massive indirect losses due to their role as vectors for various plant pathogens. These losses occur despite the extensive use of pesticides and fungicides. In the absence of such crop protection measures losses would be much more serious. This is of special importance in developing countries where, billions of people are living and working in farming areas are exposed to pesticides. Farmers spray methyl parathion over cauliflower to give extra white appearance. Okra is dipped in copper sulphate ($CuSO_4$) to give greener colour. To give a shiny colour of tomato and brinjal, the fruits are dipped in carbaryl WP (Balasubramanian, 2004).

Integrated Pest Management

The concept and impetus for IPM grew out of the discontent with using a purely insecticidal approach to insect control in many areas in the 1950's. Overuse of insecticides resulted in insects that were resistant to insecticides, resurgence of pests after treatment and occurrence of secondary pests that became a problem only after the natural enemies that had been keeping them in check were killed by insecticides. "Integrated control" was developed emphasizing use of selective insecticides so that natural enemies were conserved in the system. This "integration" of control techniques was expanded in later years to include other management options such as resistant crop varieties, crop rotation and other tactics and to include weed and diseases as pests in addition to insects. It was only after World War II that chemical pesticides and other agrochemicals were widely adopted as part of the "green revolution". Chemical pesticides had many advantages, and became an integral part (along with chemical fertilizers, mechanization, and high-yielding crop varieties) of modern agriculture. A large proportion of the productivity increases seen during the "green revolution" were due to the widespread use of new pesticides. Often these agrochemicals were applied on a routine or prophylactic basis, resulting in immediate yield increases but contributing to future environmental, agricultural, and socio-political problems. By 1962, when "Silent Spring" by Rachel Carson was published, serious concerns about the disadvantages of pesticide use were widely raised. Carson and others suggested that pest control methods other than chemical pesticides should be used in order to protect wildlife, human health, and the environment. Public pressure led to government legislation restricting pesticide use in many countries, causing agriculturists to reconsider the heavy use of persistent pesticides such as DDT. This new approach, termed Integrated Pest Management or IPM, applies a holistic solution to the problem of managing harmful insects, weeds and plant diseases.

Integrated pest management is a systems approach to pest control that combines biological, cultural, and other alternatives to chemical control with the judicious use of pesticides. The objective of IPM is to maintain pest levels below economically damaging levels while minimizing harmful effects of pest control on human health and environmental resources. IPM is a dynamic and constantly evolving system in which all the suitable control tactics and available surveillance and forecasting information are combined into a holistic management programme delivered to the farmer at requisite interval as part of the sustainable crop production technologies. In another words, IPM aims at combining all available methods or tools of insect pest control in a judicious manner that minimizes insecticide use and disturbance to the ecosystem but becomes a multidisciplinary one. A broader definition was adopted by the FAO Panel of Experts (Food and Agriculture Organization, 1975): "Integrated Pest Control is a pest management system that, in the context of the associated environment and the population dynamics of the pest species, utilizes all suitable techniques and methods in as compatible a manner as possible and maintains the pest population at levels below those causing economic injury." This definition has been cited frequently and has served as a template for others.

Prerequisites in Integrated Pest Management

Understanding of an Agro-ecosystem

IPM technology manages the pest population in such a manner that economic loss is avoided and adverse side effects of chemical pesticides are minimized. The IPM packages encompass various management strategies for containing the pest and disease problems. Pest monitoring is one of the important components of IPM to take proper decision to manage pest problem well in time. It can be done through agro-ecosystem analysis (AESA), field scouting, light, sticky and pheromone traps. The economic threshold level (ETL) of important pests and disease are determined to take appropriate control measures when pest population crosses ETL. Certain pests are recorded by (DPPQS, 2003) as the pests of national significance. AESA is an approach in which extension functionaries and farmers to analyze the field situations with regard to pests, defenders, soil conditions, plant health, the influence of climatic factors and their relationship for growing healthy crop. Such a critical analysis of the field situation will help in taking appropriate decision on management practices. An agro-ecosystem is a set of model having uniform pattern or cultivation of crops and exposed for the biotic and abiotic components to act upon (Figure 1). Many agroecosystems are unfavorable environments for natural enemies due to high levels of disturbance.

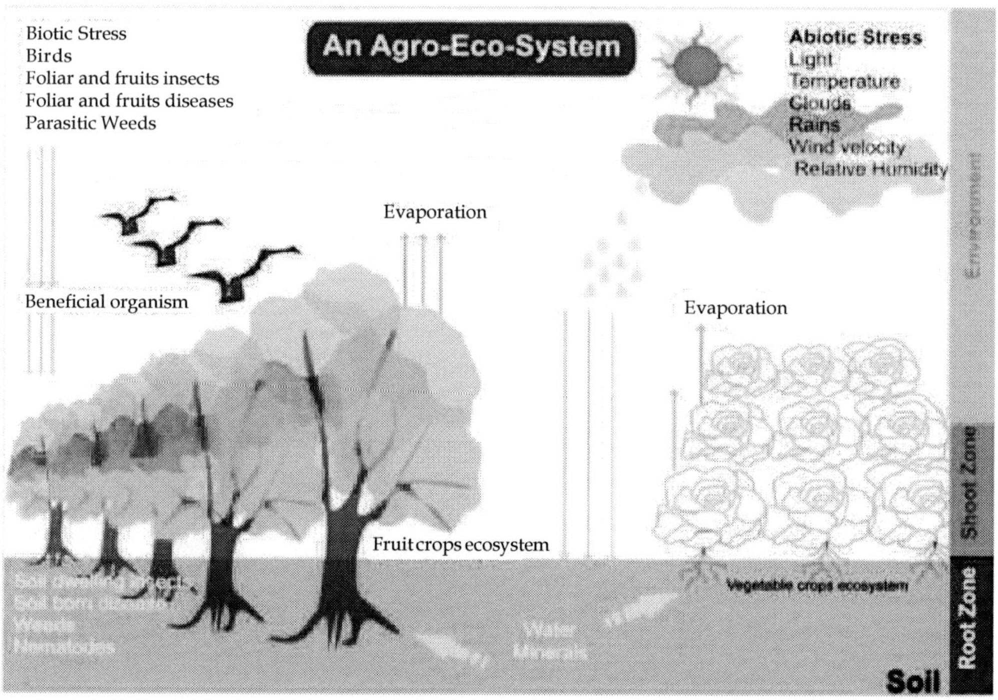

Fig. 1 : Model depicting an agro-ecosystem

Habitat management, a form of conservation biological control, is an ecologically based approach aimed at favouring natural enemies and enhancing biological control in agricultural systems (Landis *et al.*, 2000). Understanding the existing eco-system and their relationship in an environment is the fundamental aspects in IPM programme. Although continuous flowering not only encourages but also conserve the natural enemies fauna in short duration vegetable crop. Shennan (2008) discussed the biotic interactions in agro-ecosystems and how they may be manipulated to support crop productivity and environmental health by provision of ecosystem services such as weed, pest and disease management, nutrient cycling and biodiversity conservation. Important elements for understanding biotic interactions include consideration of the effects of diversity, species composition and food web structure on ecosystem processes; the impacts of timing, frequency and intensity of disturbance; and the importance of multitrophic interactions.

Understanding of the ecology of the cropping system

IPM requires an understanding of the ecology of the cropping system including that of the pests and their environment. For example, temperature is the primary factor determining the rate at which insects develop; higher temperatures increase the rate of development. Temperature can be important when determining the frequency of insecticide applications. In sweet corn for example, the interval between insecticide applications for the European corn borer can be increased when the weather is cool because eggs are not hatching as rapidly and the rate of larval development is slower; with warmer temperatures the situation is reversed. Degree-day models can aid in determining how fast insects are developing and the timing of applications. Weather conditions can also influence insect pest populations. For example, wind-driven rain is a major source of mortality to nymphs of the potato leafhopper on potatoes and thrips on onions and frame leaves of cabbage.

Sampling and Monitoring

To estimate the severity of pest infestations, the crop must be sampled at least a week. Sampling may involve examining plants and recording the number of insect pests or the amount of damage observed, or traps may be used to capture the adult pest species to estimate pest abundance. Sampling is conducted at regular intervals throughout the season or during critical stages of crop growth. Monitoring includes detecting, identifying, and determining the level of pest populations on a timely basis. Insect traps can often be used to detect pests and identify them when scouting should be intensified or control measures should be taken. Scientifically based, accurate, and efficient monitoring methods are available for many pests on vegetable crops.

Forecasting

Weather data and other information help predict when specific pests will most likely occur. Weather-based pest forecast models for diseases and insects of many crops have been developed.

Economic threshold level

Use thresholds to determine when pest populations have reached a level that could cause economic damage. The decision to use an insecticide, or take some other action, against an insect infestation requires an understanding of the level of damage or insect infestation a crop can tolerate without an unacceptable economic loss. The level of infestation or damage at which some action must be taken to prevent an economic loss is referred to as the 'action threshold'. Thresholds levels should be adjusted based on market, environmental conditions, variety, etc.

Table 2 : Economic Thresholds (ETL) for major vegetable pests

Insect pests	Economic Threshold Level (ETL)
Diamondback moth (*Plutella xylostella*)	2 larvae plant at 1-4 weeks after transplanting or 5 larvae plant at 5-10 weeks after transplanting
Cabbage Leaf webber (*Crocidolomia binotalis*)	0.3 egg mass per plant
Tomato fruit borer (*Helicoverpa armigera*)	8 eggs per 15 plants or one larva per plant
Pea aphids (*Acyrthosiphon pisum*)	3-4 aphids per stem tip
Onion thrips (*Thrips tabaci*)	13-15 thrips per plant 15 days after transplanting
Cut worms (*Agrotis ipsilon*) in potato	One larvae per ten plants
Chilli thrips (*Scirtothrips dorsalis*)	2 thrips per leaf
Chilli mites (*Polyphagotarsonemus latus*)	1 mite per leaf
Brinal Fruit and shoot borer (*Leucinodes orbonalis*)	0.91% on shoots and 0.67% on fruits
Okra fruit borer (*Earias vitella*)	5.3% infestation of fruits
Okra spider mite	2 mites per leaf
Okra Jassids (*Amrasca devastance*)	4.46 nymphs per plant 2.3 nymphs per leaf
Okra white flies (*Bemesia tabaci*)	4 adults per leaf
Cucrbits leaf hopper (*Amrasca bigutulla*)	2.5 nymphs per leaf
Cucrbits spotted leaf beetle (*Epilachna* sp.)	1 insects per sq m

(*Source :* Shivalingaswamy *et al.*, 2002; Singh and Singh, 2002; DPPQS, 2003)

Record keeping

Records kept from year to year on insect pest occurrence in fields can be valuable tools for avoiding insect pests in the future.

Strategies in IPM

Plant Resistance to insects (PRI)

The importance of host plant inherent resistance or partial resistance as a key component in IPM is increasingly recognized (van Emden, 1991). Host plant resistance to insects (PRI) is an underutilized pest management strategy in vegetable production. Increased pressures to reduce pesticides, and changes in technology now increase the economic viability and probable role of PRI in vegetable pest management. In the short term, existing varieties can be screened more extensively and quantitative information provided to producers. In the long term, variety specific recommendations for the use of chemical controls, and other management tactics in conjunction with PRI will be beneficial (Eigenbrode, 1994). In several cereal and forage crops, host plant resistance to insects (PRI) has been an extremely successful technique for suppressing pest populations or damage. In contrast, there has been much less use of this method for the management of insect pests in commercial vegetable production (Smith, 1989). A resistant variety may be less preferred by the insect pest, adversely affect its development and survival, or the plant may tolerate the damage without an economic loss in yield or quality.

Cultural Control

There are many agricultural practices that make the environment less favorable to insect pests. Crop rotation, location and alteration in sowing / planting date reduces the attack of red pumpkin beetle, fruit fly, shoot and fruit borer. Early planting of cucurbits in November escapes the attack of red pumpkin beetle whereas flowering beyond October in bitter gourd suffers less from fruitfly. Sowing of okra during second week of June retain less population of borers, thereby incurred maximum healthy yield whereas July planted brinjal face more attack of shoot and fruit borer. Thus, synchronization of most susceptible stage of the crop with the inactive period of insect reduces the infestation and chemical intervention. Summer ploughing is an effective practice to destroy the soil dwelling inactive stages of insect. Deep ploughing of the field after the harvest reduces the activity of fruitfly, red pumpkin beetle and cut worm as they are exposed for sun drying and for bird and other predator. Trap crops are planted to attract and hold insect pests where they can be managed more efficiently and reduce their movement onto crops. Trap crop must be distinctly attractive to the pest than the main crop. It provides protection either by preventing the pest from reaching the main crop or by concentrating them in certain part of the field where they can be economically destroyed. Mustard as a trap crop and coriander and tomato as deterrent crop along with cabbage has been successfully utilized for the control of diamondback moth, aphid and leaf webber. African marigold functions as good source to trap the adults of *Helicoverpa armigera* besides it also attracts the adults of leafminer for egg laying on the leaves. Maize plants applied with bait spray trap and kill fruit fly adults when sown in combination with bitter gourd.

Table 3 : Insect Pests resistant varieties of vegetables

Crops	Insect pests	Resistant varieties	Sources
Tomato	Fruit borer	Atkinson, Angur Lata, Punjab Chuhara Sabou Prabha, Hisar 1370, No. 128, No.133, No. 135	Gill *et al.*, 1980 Kashyap *et al.*, 1982
Eggplant or Brinjal	Shoot and fruit borer	Black Pendi, Thorn Pendi, H-165, H-407 H-4, Punjab Chamkila, Pusa Purple Cluster, Pusa Purple Long, S-4, S-6	Panda *et al.*, 1971 Gill and Chadha, 1979
		PPL, Pusa Kranti, SM- 41, H-4	Raut and Sonone, 1980
		Arka kusmakar, Kalaynpur long green, Muktakeshi, Nimbkar Green, Pusa Kranti, Purple Long, PPL	Mehto and Lall, 1981
		Barahmasi, Gulabi Doria, Pusa Purple Cluster, Pusa Purple Round, PPL, Nurki,Punjab bahar	Kale *et al.*, 1986
		SM 17-4, PBr 129-5, Punjab Barsati, ARV 2-C, Pusa purple round, Punjab Neelam.	Singh, 1991
	Aphid, jassid, whitefly	Kalyanpur, Punjab chamkila, GB-1, GB-6	Singh, 1991
Lady's Finger (okra)	Jassids	Climpson Spineless, Early long Green, Pusa Sawani, Crimpson Smooth Long	Teli and Dalaya, 1981; Sandhu *et al.*, 1974
	Shoot and fruit borer	Bhindi red-I, Bhindi red-II, Red wonder-I, Red wonder-II	Srinivasan and Naranaswami, 1961; Mote, 1978
		AE-3, AE-22, AE-52, AE- 57, AE-67 AE-79, Wonderful pink	Kashyap and Verma, 1983
		Parkins long green, Climpson spineless, White snow, Cell round, Climpson spineless, Rashmi	Patil *et al.*, 1986
Cucurbit vegetables			
(Louki/ bottle gourd)	Fruitfly	M B- 29	Nath, 1966
	Red pumpkin beetle	NB-19, NB-25, NB-29, NB.-3	Nath and Thakur, 1965
Melons	Mites	Durgapur cell-1, Early gold, Kocha-4, Hara madhu, MM-1, Punjab Sanchari, Pusa sarbati	Dhooria and Sukhija, 1986
Onion	Thrips	Bombay white	Saxena, 1970
		Bombay white, Pusa ratnar, White big round	Kaur *et al.*, 1978

Physical and Mechanical Control

The use of physical barriers such as row covers or trenches prevents insects from reaching the crop. Row covers can help prevent early-season damage to cucurbits by cucumber beetles, and plastic-lined trenches are effective in trapping dispersing Colorado potato beetles in the spring and fall. Other methods include handpicking of pests, sticky boards or tapes for control of flying insects in greenhouses, and various trapping techniques.

Biological Control

Biological control is defined as the reduction of pest populations by natural enemies (predators, parasitoids, and pathogens). Predators, such as lady beetles and lacewings, are free-living species that consume a large number of preys during their lifetime. Parasitoids (certain wasps and flies) are species whose immature stage develops on or within a single insect host, ultimately killing the host. Pathogens are disease-causing organisms including bacteria, fungi, and viruses. Pathogens kill or debilitate their host and are relatively specific to certain insect groups.

Natural enemies can be either generalists, who utilize many different species of insects as food, or specialists, which are fairly selective in their prey. Lady beetles are considered generalists because they feed on many types of prey; whereas, species of *Trichogramma* are microscopic wasps specialized to parasitize eggs of moths. There are three broad and somewhat overlapping types of biological control: classical, augmentative, and conservational. The conservation of natural enemies is probably the most important and readily available biological control practice available. Natural enemies occur in all vegetable production systems. They are adapted to the local environment and to the target pest, and their conservation is generally simple and cost-effective. Lacewings, lady beetles, syrphid fly larvae, and parasitized aphid mummies are almost always present in aphid colonies. Fungus infected adult flies are often common following periods of high humidity. These natural controls are important and need to be conserved when making pest management decisions. The following factors influence the impact of insecticide applications on natural enemies:

Spectrum of activity : Broad-spectrum insecticides will more adversely affect natural enemies than materials that are selective for specific insect species or life stages.

Residual (half life) activity of insecticide : Insecticides that remain toxic to pests for a long time and stay on the treated surface will have a similar effect on natural enemies.

Coverage and formulation of insecticide : Full coverage sprays will generally have a greater impact on natural enemies than directed sprays, systemics, or bait formulations. Spot or edge treatments directed at localized pest infestations will be less detrimental than those applied to the entire field.

Dosage and frequency of application : Higher rates and repeated applications will have a more detrimental impact on natural enemies.

Susceptibility of natural enemy to pesticides : Some natural enemies are inherently more resistant to insecticides than others and some populations of natural enemies have been selected to possess higher levels of resistance. Applications of some fungicides can also reduce the incidence of fungal diseases of insects. Fungicides and herbicides may also have lethal and sub-lethal effects on natural enemies.

Augmentation is the supplemental release of natural enemies. Relatively few natural enemies may be released at a critical time of the season or literally millions may be released. For example, the recommended release rates for *Trichogramma spp.* in sweet corn is 30,000 per acre. In contrast, entomopathogenic nematodes are released at rates of millions and even billions per acre for control of certain soil-dwelling insect pests. Augmentation may also involve modifying the cropping habitat to favor natural enemies. Many adult parasitoids and predators benefit from sources of nectar and the protection provided by refuges such as hedgerows, cover crops, and weedy borders. The modification of surrounding habitats must, however, be done carefully. Refuges may benefit some pests, but may also be hosts for certain plant diseases, especially plant viruses that could be vectored by insect pests.

Chemical Control

When other management tactics fail to keep pest infestations from causing economic damage, insecticides are needed. The decision to use an insecticide should be based on an action threshold. A field should be scouted regularly to determine if it is at the action threshold. When an insecticide is required, the choice of material, formulation, and rate and method of application will vary depending on several factors, including the pest(s) to be controlled and the stage of crop development.

Insecticides are formulated into emulsifiable liquid concentrates, flowable liquids, and dry concentrates such as wettable powders, soluble powders, dusts, and granules. A sticker-spreader should be added to spray mixtures for use on some cole crops, cucumbers, onions, and tomatoes, so that the mixture will spread more evenly over the plant surfaces. Insecticides will be effective only if the insect consumes or comes in contact with them; good coverage is thus essential for efficient pest control. Always use enough water and adequate pressure to cover plants completely, especially if insects are concealed (e.g., thrips on onions). Some insecticides are more effective than others for a particular pest, and certain insecticides must be applied at specific times. For example, to be most effective *Bacillus thuringiensis* (Bt) products must be applied early in the pest's development when it is more susceptible to the toxin. Coverage is also important. Aphids, for example, are frequently most abundant on the undersides of leaves. If the insecticide is applied only to the tops of leaves, aphid control will not be adequate. Thorough coverage is also important for stomach poisons, such as the Bt products, which must be ingested by the pest.

Managing Resistance

Insecticide resistance is a measurement of an insect's ability to tolerate the toxic effects of particular insecticide, resulting in repeated failure of the product to achieve an effective level of control when used according to the label recommendations (Luckmann and Metcalf, 1982). A major obstacle to the continued, effective use of insecticides is the development of resistance.

Insect pest populations develop resistance to insecticides through genetic selection. Some individuals in a population have genetic traits that allow them to survive an insecticide application. Some of the progeny of these surviving individuals will possess these traits. When additional applications of insecticides are made, more of the resistant individuals survive. Over time and in the absence of appropriate resistance management, the pest population may consist mostly of resistant individuals.The initial use of pesticides has been very effective in reducing pest infestations and increasing agricultural production and productivity. However, over time targeted pests have developed resistance to pesticides necessitating increasing applications or resulting in rising populations of pests or both. After a point, resistance of pests may grow to such an extent that application of pesticides is no longer economic. Increasing the frequency of the applications or amounts of insecticide will not improve control. Switching to different or new insecticides within the same class or even different classes may not help because pests resistant to one insecticide may also be resistant or rapidly develop resistance to others. This is referred to as cross-resistance. For example, some insect pests that are resistant to chlorinated hydrocarbons quickly develop resistance to pyrethroids. Repeated applications to control a target pest may lead other non-target species subjected to these same applications to develop resistance. The following steps are recommended to minimize the development of resistance to insecticides: apply insecticides only when needed; follow recommended action thresholds where available; use the minimum recommended rate that provides control; rotate insecticides that are in different chemical classes and have different modes of action during the season. Ideally, this rotation should be done across pest generations, that is, each generation should be treated with a different class of insecticide. Also, use insecticide synergists if available and effective.

Insect pests of Brinjal

Eggplant (*Solanum melongena*) is one of the most important vegetables in South Asia, which accounts for almost 50% of the world's area under eggplant cultivation (Alam *et al.*, 2003). Eggplant fruit and shoot borer (EFSB), *Leucinodes orbonalis* Guenee (Lepidoptera: Pyralidae) is one of the most serious insect pests of eggplant (Alam *et al.*, 2003). Larvae of this pest bore inside tender shoots. This can result in reduction of plant growth and subsequently the yield (Mall *et al.*, 1992). Farmers in South Asia rely primarily on the application of chemical insecticides to combat EFSB. This has resulted in tremendous misuse of pesticides. For example, surveys conducted in Bangladesh indicate that farmers spray insecticides daily to control EFSB (BARI, 1995). To address this serious issue, AVRDC-The World Vegetable Center developed, validated, and promoted an IPM strategy for the control of EFSB in South Asia from 2000-2005 (Alam *et al.*, 2006). Yield losses of up to 90% have been reported in India (Kalloo, 1988) and similar losses are also common in neighboring Bangladesh (Ali *et al.*, 1980).

1. Brinjal Shoot and Fruit Borers (BSFB), *Leucinodes orbonalis* (Lepidoptera: Pyraustidae)

Brinjal shoot and fruit borer inflicts a serious damage to the fruit causing drastic reduction in the marketable quality fruit yield. The damage starts soon after transplanting the crop and continues till harvest of the fruits. The adults are small moths having brown to black spots on the wings. The adult female lays eggs on the ventral surface of the leaves, flower buds and on young fruits. After hatching within the week, small pinkish larva initially bores into the terminal shoots resulting in wilting and drying of the shoot and at later stage, it damages the young fruits by making holes and feeds inside which makes the fruits unfit for consumption and marketing (Fig. 2).Wilting reduces plant growth, which causes a reduction in fruit number and size. Over 90% of fruits can be infested in high pest pressure years (Dhandapani *et al.*, 2003). Regular pruning and prompt disposal of pest damaged eggplant shoots, especially up to the first harvest, is an important component in the BSFB IPM strategy (Rahman *et al.*, 2002; Talekar, 2002; Miller *et al.*, 2003; Satpathy *et al.*, 2005). Often the entry hole encourages the secondary bacterial infection and rotting of fruits. As its larvae are concealed within the twigs as well as in fruits it is very difficult to control.

Management

- Monoculture or continuous cropping of brinjal on the same piece of land should be avoided.
- As soon as the insect is detected, the affected twigs should be clipped/ removed along with the insect and destroyed. Fruits showing any boring symptoms should also be picked and destroyed (Fig. 3).
- The moths can be mass trapped and killed by installation of pheromone traps @ 100 per ha at 10 m spacing (Cork *et al.*, 2001, 2003; Fig. 4)
- Apply ash dust and avoid the excessive dose of nitrogenous fertilizers in brinjal crops to reduce the infestation of BSFB.
- Long and oblong varieties with bunchy fruiting/flowering behavior are less susceptible to BSFB.
- Spraying the crops alternatively with profenophos 50 EC @ 2ml /litre of water at 15 days interval and cypermethrin (0.5 ml/litre of water) starting from 20 days after transplanting to control the pest.
- The safer insecticides like malathion, endosulfan, deltamethrin etc. may be applied in alternate or in association with *Bt* formulations as per need (Raju *et al.*, 2007).
- Before spraying, all the affected twigs and fruits should be clipped off or removed from the field (Shankar *et al.*, 2010; Alam *et al.*, 2003, 2006).

2. Jassids, *Amrasca biguttula biguttula* (Hemiptera:Cicadellidae)

Both nymphs and adults suck the sap from the lower surface of the leaves. The infested leaf curl upward along the margins, which may turn yellowish and show, burnt up patches (Fig. 5). They also transmit mycoplasma disease like little leaf (Fig. 6) and virus disease like mosaic or leathery leaves (Fig. 7). Fruit setting is adversely affected by the infestation.

Management

- Seed treatment with thiamethoxam 70 WS @ 3g/kg seed at the time of sowing.
- Foliar spray of imidacloprid (0.006 %) during the vegetative stage of the crop.

3. Red Spidermite, *Tetranychus cinnabarinus* and *Paratetranychus indicus* (Acari: Tetranychidae)

Nymphs and adults of phytophagous mite suck cell sap and leaves exhibiting yellowish white symptoms leading to their wilting, burning and shedding (Fig. 8). It also infests floral buds (Fig. 9) and tender fruits. The loss in yield is related to the population level and stage of infestation. Castanera and Arroyo (1979) found that initial infestation of 20 or 60 mites/plant significantly reduces the yield. The red spider mite infests brinjal particularly in low humid dry condition. Different stages of mites are found in colonies covered by white-silky webs on lower surface of leaves.

Management

- During egg stage and the resting stages, most acaricides are ineffective. At high temperatures, it may be necessary to apply these at an interval of two days.
- Acaricides like dicofol (0.05%) and wettable sulphur (0.3%) gives effective control of mites. Avoid use of synthetic pyrethroids for borer control.
- Collection and burning of severely infested plant parts reduces further multiplication of mites.
- Proper irrigation and clean cultivation are essential to reduce the damage caused by mites.

Tomato is a very popular constituent of Indian diet. It is devastated by an array of insect pests, however the major damage is caused by fruit borer, *Helicoverpa armigera*, serpentine leaf miner, *Lariomyza trifoli* and white flies, *Bemesia tabaci* (Shankar *et al.*, 2009b).

1. Tomato fruit borer, *Helicoverpa armigera* (Lepidoptera : Noctuidae)

Helicoverpa has been recorded on more than 200 hosts in India (Pawar, 1998) and is most damaging to cotton, pigeonpea, chickpea, groundnut, sorghum, pearl millet, tomato and many other crops of economic importance. Caterpillars are the immature stage of the insect which causes the considerable damage to the crops in various stages. In the pre-fruiting stage, the caterpillar feeds on the tender foliage including leaves, flowers, buds and even sometimes they make holes in the tender twigs which reflects the perforated look to the crops. After fruiting, the larvae bore large, clear, circular holes into fruits and feed on the pulp. Its name has been changed according to its host. Its larvae are stoutly built and robust. In tomato, it is responsible to destroy many fruits and make them unfit for the human consumption. Its larvae have peculiar feeding habit on host crop. The larvae thrusts its half of the body into the fruits and feeds the inner contents and rendering them unfit for the human consumption (Fig. 10). It causes damage up to the tune of 50 to 80 per cent if not intervened at right time.

Management

- Soil solarization in nursery beds as well as in the main field.
- Plant one row of marigold as trap crop for every 16 rows of tomato.
- Monitor and clip top three tender leaves for *Helicoverpa* eggs destruction.
- Early maturing varieties should be planted to escape the damage by borers.
- Installation of pheromone traps @ 5-7 per ha for early detection and 12-15 per ha for trapping and mass destruction for mating disruption (Fig. 11).
- Release *Trichogramma chilonis* @ 50,000 adults /ha per release (6 times) at weekly interval when eggs of *Helicoverpa* are noticed.
- Conservation of *Campoletus chloridae*, a potential parasitoid of *H.armigera.*
- Spraying of EPN *Steinernema carpocapsae* against *Helicoverpa* gave an excellent result in killing the larvae within 24 to 48 hrs (Fig. 12).
- Spray *Ha*-NPV 250 to 500 LE/ha (freshly prepared) 2-3 times at 10 days interval at evening hours (Sivaprakasam, 1998; Satpathy *et al.*, 1999; Satpathy and Rai, 2000).
- Spraying of bacterial formulation *Bt* @ 500 g/ha also may proved beneficial in regulating the *Helicoverpa* larvae in the field condition.

2. Serpentine leaf miner, *Liriomyza trifolii* (Diptera: Agromyzidae)

Leaf miners lay eggs below the epidermal layer of leaves and cause heavy leaf mining on upper or lower side of the leaves (Fig. 13). Mines starts from the margins of leaves and progress towards the base/centre of leaf (Fig. 13 a). Mines are initially narrow and gradually enlarge often twisting through leaf which causes stippling of young seedlings. Severe leaf mining slows down the plant growth and accelerates the leaf drop.

Management

- Nuvon/DDVP or dimethoate @ 2 ml per lit of water are found to be effective against leaf miner.
- Hymenoptera (Eulophidae) an endo-larval parasiotid of leafminer, *Hemiptarsenus varicornis* is found to be very effective.
- Neem seed kernel extract (NSKE) @ 5 per cent concentration found promising in controlling leaf miner followed by Karanj, Pongamia pinnata oil next in the order of efficacy.

Insect Pests of Chillies

1. Chilli mites, Polyphagotarsonemus latus (Acarina: Tarsonemidae)

Chilli crops are attacked by yellow mite, *Polyphagotarsonemus latus* which has become a serious problem in the entire country. Mite affected leaves of the chilli plant turn towards underside and look like an inverted boat. As young and adult mites attack the tender leaves, almost all new leaves curled downward and lower surface appears silvery whereas, the upper surface of leaves look more dark green as compared to healthy ones. Further the plant growth is checked and the internodes become shorter.

Management

- Since last few decades, farmers have normally been using different kinds of insecticides for the control of mites. Being insecticides these were not able to give satisfactory control of mites, due to which mites also developed resistance.
- Some insecticides have been reported to be effective for mite control (chilli pest complex) - triazophos 40 EC @ 2ml per lit of water or Dicofol 18.5 EC @ 2ml per lit of water or propargite 57 EC @ 3-4 ml per lit of water at the beginning of mite attack followed by another spray at an interval of 12 days.
- Seedling root dip treatment with imidacloprid 200 SL @ 1ml per lit of water before planting and 3 sprays with the same at 15 days interval.
- The following lines/varieties have been reported resistant/ tolerant to mites viz., Pusa Jwala.

2. Thrips, Scirtothrips dorsalis (Thysanoptera: Thripidae)

Thrips cause severe damage to the plants in the nursery and in transplanted crop and is the main constraint in the chilli and *Capcicum* cultivation. Both the nymphs and adults of chilli thrips lacerate the leaf and imbibe the oozing sap. Sometimes even buds and flowers are also attacked. Tender leaves and growing shoots are preferred for feeding, while the older leaves are rarely attacked. The infested leaves start curling and crumbling and ultimately shed while, buds become brittle and drop down. Main visible symptom are curling of leaves upward (normal boat shaped), small leaf size and shorter internodal space with stunting growth of plants, no fruiting or small deformed fruiting in infected plants takes place (Fig. 14 and 14 a).

Management

- Rouge out the infected plants and burn them.
- Grow resistant varieties.

- Seed treatment of imidacloprid 70 WS at 10, 20 and 30g per kg protected the seedlings in the nursery up to 45 days and foliar treatment of imidacloprid 200 SL at 250, 375 and 500 ml/ha reduced the thrips population significantly and also recorded higher yield of chilli fruits (Santharam *et al.*, 2003).
- Periodic spray of metasystox @ 2 ml per lit of water or imidacloprid @ 0.5 ml per lit water at 10 days interval for vector (*Bemesia tabaci*) Management.

Insect Pests of Cucurbits

1. Cutworms in cucurbits: Agrotis ipsilon (Lepidoptera: Noctuidae)

It is a polyphagous insect pest which feeds mainly in seedling stages of many vegetables, potato, maize and other crops. Caterpillar is the only damaging stage of this pest. The short stout larvae are highly destructive. At night, larvae cut the plant at ground level at an early stage of their growth (Fig. 15). A few caterpillars in the field kill many small plants during one night. They fell/cut more plants rather consume.

Management

- The population of cut worm larvae can be considerably suppressed by applying 1.3% lindane dust @ 20 kg/ha or drench the soil with chlorpyriphos @ 2.5 ml per lit of water or carbaryl @ 4 g per lit of water.

2. Red Pumpkin Beetle, Aulacophora foveicollis (Coleoptera: Galerucidae)

The beetles eat leaf lamina, cause perforation, irregular holes or complete defoliation and damage is severe, particularly at cotyledonary stage of crop (Fig. 16). The grubs feed on the underground stem and root portion of the host plant forming holes/galleries leading to dry up of branches or whole plant. They may also attack the fruits touching the ground surface by making holes.

Management

- Plant the crop in November to avoid the damage from this insect.
- After harvesting the cucurbits, the field should be ploughed to exposed and kill the grubs in soil.
- Spray Of the following insecticides are recommended against red pumpkin beetles- carbaryl @ 2g per lit of water or endosulfan @ 2ml per lit of water.

3. Fruit flies, Bactrocera cucurbitae (Diptera: Tephritidae)

Cucurbit fruit fly is a major pest of all cucurbitaceous vegetables. It is serious and destructive pest of cucurbits like melon, pumpkin, cucumber and gourds. Adult flies are reddish brown with lemon yellow markings on thorax. Female fly lacerates or puncture the fruits by its pointed ovipositor and laid eggs on soft fruits. Female secrets the gummy secretion which cements the punctured tissue and make the entrance water proof. The most damaging stage is the maggots. Maggots feed on the pulp of fruit and cause premature dropping of fruits and also make them unfit for the human consumption (Fig. 17). Sometimes secondary bacterial infection also takes place. Oozing of brown resinous fluid from the fruits and fruits become distorted and malformed. In severe condition or in rainy season 50-100 % damages are reported.

Management

- In endemic areas, change the sowing dates, as the fly population is low in hot dry conditions and its peak during rainy season.
- Collect infested fruits and dried leaves and burn it in deep pits. Expose the pupae by deep ploughing and turning over soil after harvest.
- Use ribbed gourd as trap crop and apply carbaryl (0.15%) on the congregating adult flies on the undersurface of leaves.
- Use poison baiting in severe infestation: Saturated sugar solution 5 ml+ Malathion 50EC 5ml+ gur or jaggary 100 ml are kept in the container and put it to the four corners of the field as a poison baits (Fig. 17 a).
- Use fly traps@ 50 traps/ha : Use 5 g wet fish meal in a polythin bag (20×15 cm in size and having 6 holes of 3 mm in diameter) and add 0.1 ml diclorovos (1 drop) in the cooton plug and put it in the bag. For better performance, dichlorovos is added every week and fishmeal should be renewed after 20 days.

Insect pests of Okra

1. Okra Fruit borer, Earias vittella - pest of humid areas and E. insulana - pest of dry regions (Lepidoptora: Noctuidae)

Eggs are laid singly on buds and flowers and occasionally on fruits. During the early stage of crop's growth, the eggs are laid on shoot tips. When the crop is only a few weeks old, the freshly hatched caterpillars bore into tender shoots and tunnel downwards. These shoots wither, droop and ultimately the growing points are destroyed, and side shoots may arise giving the plant a bushy appearance. With the formation of buds, flowers and fruits, the caterpillars bore inside these and feed on inner tissues (Fig. 18).

Management

- Use of resistant or tolerant varieties.
- Intallation of pheromone traps @ 25 per ha is quite effective.
- As soon as flowering starts give 3 sprays at fortnightly interval with any of the following insecticides carbaryl @ 2g or endosulfan@ 3ml or fenvalerate @1 ml per lit of water.

Caterpillar Pests of Cole Crops / Crucifer's Vegetables

The cruciferous crops (cole), which includes economically important crops like Cabbage, *Brassica oleracea* var. *capitata*; Cauliflower, *B. oleracea* var. *botrytis*; Knol khol, *B. caulorapa* (L.) throughout the world. However, poor productivity is mainly attributed to the incidence of diamondback moth (DBM), *Plutella xylostella*; cabbage caterpillar, *Pieris brassicae*; leaf webber, *Crocidolomia binotalis*; cabbage head borer, *Hellula undalis* and aphid *Brevicoryne brassicae* infesting the crop in various seasons (Shankar *et al.*, 2009a).

1. Cabbage butterfly, Pieris brassicae and *P. rapae)*

It is also known as imported cabbage worm. Two species of cabbage butterflies i.e. large butterfly, *Pieris brassicae* and small butterfly, *P. rapae* have been recorded and found damaging the cole crops in Jammu province. They are serious threats to the crops grown for seed purpose especially in hills as well as in plain areas of Jammu. Adults are small to large sized butterflies, pale white with black markings. Female butterflies have two black circular dots on the dorsal side of each forewing. After mating, female butterfly lay about 100-150 conical, yellowish eggs in clusters on lower or upper side of the leaf. The eggs hatch in 3-15 days depending upon the environmental temperature. Just after hatching, the caterpillars feed gregariously on leaves in early stages. They passes from five larval instars and become full grown in 2-3 weeks. Their incidence has been recorded in fields from October to April-May except in cool winter months during December-January. They completely skeletonize the leaves leaving only mid ribs (Fig. 19). Due to complete loss of photosynthetic area crops failed to produce the economic yields.

Management

- Hand picking and mechanical destruction of caterpillars during early stage insect pest attack is beneficial.
- *Cotesia glomeratus* are recorded as a potential parasitoids against cabbage caterpillar larva (Fig. 19 a).
- *Phryxe vulgaris* (dipteran fly) are also observed abundantly in cole crops fields to parasitize the caterpillars.

- Foliar spray of *Bt* formulations @ 500 g/ha along with sticker (0.5 ml/lit of water) is found promising to control all lepidopteran insect pests.
- Need based application of cypermethrin @ 1 ml per lit or malathion @ 2ml per lit or endosulfan @ 3ml per lit of water are beneficial in case of severe infestation.

2. Cabbage head borer - Hellula undalis (Lepidoptera: Pyraustidae) and *Helicoverpa armigera*

Cabbage head borer is seen sporadically but its attack coincides with the head/curd initiation stage and a single larva is responsible to damage the entire head/curd (Fig. 20). The early instar may feed on the growing point, and bore inside the head and damaging the head and excrements which prevents head formation. When fully grown, larvae pupate inside the head, or on stems, or in soil. *Helicoverpa armigera* larvae have also been recorded at several places in Jammu region as head borer of cabbage. It may be due to absence of preferred host in winter months or some environmental and behavioural change.

Management

- Monitoring and mechanical destruction of larvae at the time head/curd initiation may proved to be good.
- Bracon spp. are important larval parasitoids against head borer.
- Fungal formulations like *Beauvaria bassiana* and entomopathogenic nematodes (EPN) may be beneficial to suppress the attack of head borer.
- Spraying of synthetic pyrethroids like cypermethrin, fenvalerate, deltamethrin @ 1 ml per lit of water is effective to control head borer.

3. Diamondback moth, Plutella xylostella (Lepidoptera: Plutellidae)

Diamondback moth is the universal pest of crucifers' crops and remains one of the most serious problems for the intensive cole crops growing areas. Plants at all stages of growth may be attacked by and completely devastated by this pest. In Jammu province, diamondback moth is recorded only on early and on late season cole crops and on crucifers i.e., during September-November and from February onwards (Fig. 21).

Management

- Use of pheromone traps are useful in monitoring and detection of the DBM pest popualtion in early stages.
- Practice of trap or intercropping of mustard, coriander and marigold with cole crops may reduce the DBM pest infestation.

- Release of *Cotesia plutellae*, natural bio-agent of DBM larvae is effective to suppress its population in field condition (1000 adults per release every 2 week interval up to harvest).
- Use of resistant varieties are also effective against DBM.
- Some safer insecticides and neem products are recorded to be effective against DBM.
- Spray endosulfan@ 3ml, cypermethrin@ 1ml, cartap hydrochloride@ 2ml, spinosad@ 1ml and *Bt* formulation@ 500 g/ha may be promising in controlling the DBM pest population.

4. Tobacco catepillar : Spodoptera litura (Lepidoptera: Noctuidae)

Spodoptera litura has also wide host range and economic insect pest of cole crops and tomato but it is usually kept under damaging levels by management practices targeted to *Helicoverpa*. They are voracious feeders and scrape the leaves and later on it may completely defoliate the leaves. The insect is active during the entire year except severe cold months and females can lay up to 300-600 eggs each, usually in groups of about 100. Eggs are laid on the underside of lower leaves and are covered with fuzzy, white creamy scales. Under warm conditions, eggs hatch within three to four days. The neonate larvae feed from one to three weeks, in groups at younger stage and scattered on the plants when larger (Fig. 22).

Management

- Summer ploughing is beneficial to expose the hibernating pupae to natural enemies and for bird predation.
- Hand picking and mechanical destruction of egg masses, caterpillars and spraying of NSKE 5% during early stage may proved beneficial.
- Installation of pheromone traps @ 25 to 30/ha may be the useful tool for early detection and also for mass trapping and destruction of tobacco caterpillar.
- Spraying of *Splt* NPV @ 250 LE with gur or jaggary (10 g/lit) and sticker during evening hours may be the promising to control tobacco caterpillar (Monobrullah and Shankar, 2008).
- Foliar spray of Bt formulations @ 500 g/ha is the good for controlling the caterpillars.
- EPN may be quite promising to suppress the Spodoptera larval population in the field condition.
- Chemical sprays recommended for the DBM control will also work against this pest but in this case, spot application may be enough if the attack is not wide spread in the field.

Onion

Onion thrips, *Thrips tabaci* (Thysanoptera: Thripidae)

Besides onion, several other species of thrips infest tomato, brinjal, capsicum, and on cucurbits. They are tiny, slender insects that may vary in colour from yellow to dark brown or black. Both nymphs and adults (Fig. 23) rasp the epidermis of tender leaves and imbibe the oozing sap. As a result of which leaves develop numerous white specks. Heavily infested leaves curl up and get distorted and ultimately the top of attacked plants wither away. The symptoms develop by feeding on foliage may cause a silvering of foliage (Fig. 24).

Management

- In the case of onion, the only reliable method of controlling onion thrips is through the use of tolerant varieties. Advantages of pest-resistant crop varieties include ease of use; compatibility with other integrated pest management tactics; low cost; cumulative impact on the pest (each subsequent generation of the pest is further reduced); and reduced negative impact on the environment.
- The following lines/varieties have been reported resistant/ tolerant to onion thrips: White Persian, Guano, Sweet Spanish, Crystal Wax, Spanish White, Bombay White, Pusa Red Udaipur 103, N 780-1, Pbr-3, PBR-4, VL-1, No. 18, No. 19, Pusa Ratnar.
- Some insecticides and neem products have been reported to be effective against thrips.
- Biological control agents like minute pirate bugs, lacewing larvae and ladybirds prey on thrips.
- Application of systemic insecticides like imidacloprid or thimethoxam or dimethoate after 10 days of transplanting may help to suppress the thrip population.

Conclusion

Lack of knowledge about biocontrol agents and lack of availability of IPM devices and other inputs is the major constraints for farming communities to adopt the IPM practices on their farm. By enhancing the knowledge base of the farmers with modern technology and farmer's friendly policies, improving the skills of farmers, availability of inputs at various levels can bring a significant change in vegetable crop production. Extra attention is to be given in case of the chemical control of insect pests which leave toxic and hazardous residue on vegetables. As most of the vegetables are being eaten row, so, proper care and precautions are to be taken care of while recommending the use of pesticide. Sole reliance on chemical control should be avoided. The preference should be given to the other effective methods of control like cultural practices, host plant resistance, biocontrol, use of biopathogen/botanicals to minimize the use of pesticides. The present technology should embrace conservation of land and water resources, adoption of integrated nutrient and pest management, and application for hybridization and biotechnology towards food quality and high yield production for domestic and export market of vegetable crops.

References

Alam, S.N., Hossain, M.I., Rouf, F.M.A., Jhala, R.C., Patel, M.G., Rath, L.K., Sengupta, A., Baral, K., Shylesha, A.N., Satpathy, S., Shivalingaswamy, T.M., Cork, A., Talekar, N.S. 2006. Implementation and promotion of an IPM strategy for control of eggplant fruit and shoot borer in South Asia. Technical Bulletin No. 36. AVRDC publication number 06-672. AVRDC–The World Vegetable Center, Shanhua, Taiwan. 74 p.

Alam, S.N., Rashid, M.A., Rouf, F.M.A., Jhala, R.C., Patel, J.R., Satpathy, S., Shivalingaswamy, T.M., Rai, S., Wahundeniya, I., Cork, A., Ammaranan, C. and Talekar, N.S. 2003. Development of an Integrated pest management strategy for eggplant fruit and shoot borer in South Asia, Technical Bulletin TB 28, AVRDC-The World Vegetable Centre, Shanhua, Taiwan, PP. 66.

Ali, M.I., Ali, M.S. and Rahman, M.S.1980. Field observation of wilt disease and shoot & fruit borer attack on different cultivars of brinjal. Bangladesh. *J. Agril. Sci.*, 7: 193-194.

Anonymous, 2009. Indian Horticulture database. pp.2.

Balasubramanian, P. 2004. The Vegetable Industry in India: Critical IPM Needs. The Asian Regional IPM CRSP Workshop- Impact and lessons learned. June 29-30, 2004, Philippines.

BARI, 1995. Annual Research Report 193-94. Bangladesh Agricultural Research Institute, Joydebpur, Gazipur, Bangladesh.

Castanera, P. and Arroyo, M. 1979. An Del Inst. *Xlacional deivtigaciones. Agrar Serie Protection Vegetal*, 11. 93-102.

Cork, A., Alam, S.N., Das, A., Das, C.S., Ghosh, G.C., Farman, D.I., Hall, D.R., Maslen, N. R., Vedham, K., Phythiam, S.J., Rouf, F.M.A. and Srinivasan, K. 2001. Female sex pheromone of brinjal fruit and shoot borer, *Leucinodes orbonalis* blend optimization. *J. Chem. Ecol.*, 27: 1867-1877.

Cork, A., Alam, S.N., Rouf, F.M.A., Talekar, N.S. 2003. Female pheromone of brinjal fruit and shoot borer, *Leucinodes orbonalis*: trap optimization and preliminary mass trapping trials. *Bull. Entomol. Res.*, 93: 107-113.

Dhandapani, N., Shelkar, U.R. and Murugan, M. 2003. Bio-intensive pest management (BIPM) in major vegetable crops: An Indian perspective. *Food, Agriculture and Environment*, 1(2): 333-339.

Dhooria, M.S. and Sukhija, B.S. 1986. Susceptibility of different selections of muskmelon to vegetable mite, Tetranychus cinnabarinus under field conditions. Punjab *Vegetable Grower*, 21: 42-43.

DPPQS, Directorate of Plant Protection, Quarantine and Storage, 2003. *Integrated Pest Management Package*, Government of India. *(IPM Package No:18,21,2223)*.

Eigenbrode, S.D. and Trumble, J.T. 1994. Host Plant Resistance to Insects in Integrated Pest Management in Vegetable Crops. *J. Agric. Enlomol.*, 11(3): 201-224.

Food and Agriculture Organization, 1975. *Rep. FAO Panel of Experts on Integrated Pest Control, 5th, Oct. 15–25, 1974.* Rome, Italy: FAO-UN, Meeting Rep. 1975/M/2.41 pp.

Gill, C.K. and Chadha, M.L. 1979. Resistance in brinjal to shoot and fruit borer, *Leucinodes orbonalis. Indian Journal of Horticulture.* 36 (1) : 67-71.

Gill, C.K., Kanwar, J.S. and Nandpuri, K.S. 1980. A note on field evaluation of different tomato varieties for resistance to fruit borer (*Heliothis armigera* Hubner). *Punjab Vegetable Grower*, 1: 34-36.

Kale, P.B., Mohod, U.V., Dod, V.N. and Thakare, H.S. 1986. Screening of brinjal germplasm (*Solanum* spp.) for resistance to shoot and fruit borer (*Leucinodes orbonalis* Guen.) under field conditions. *Vegetable Science*, 13 (2): 376-382.

Kalloo, G. 1988. Solanaceous crops. In: Vegetable breeding, Vol. II. CRC Press, Boca Raton, Florida.

Kashyap, R.K., Butani, R.D. and Kalloo, G. 1982. Relative susceptibility and variability studies against fruit borer in tomato. *Punjab Horticulture Journal*, 12: 190-193.

Kashyap, R.K. and Verma, A.N. 1983. Relative susceptibility of okra to shoot and fruit borer. *Indian J. Ecol.*, 10: 303-309.

Kaur, C., Brar, T.S., Sukhija, B.S. and Saimbhi, M.S. 1978. Relative susceptibility of different varieties of onion (*Allium cepa* L.) to *Thrips tabaci* Lind. Thripidae: Thysanoptera. *Punjab Vegetable Grower*, 13: 1-4.

Landis, D.A., Wratten, S.D. and Gurr, G.M. 2000. Habitat management to conserve natural enemies of arthropod pests in agriculture. *Annu. Rev. Entomol.*, 45:175-201.

Luckmann, W.H. and Metcalf, R.L. 1982. The pest management concept. In : *Introduction to Insect Pest Management* (Metcalf, R. L. and Luckman, W. H. eds.), Second Edition, John Wiley & Sons, New York, 1-31 PP.

Mall, N.P., Pande, R.S., Singh, S.V., Singh, S.K. 1992. Seasonal incidence of insect pests and estimation of losses caused by shoot and fruit borer on brinjal. *Indian J. Entomol.* 54, 241-247.

Mehto, D.N. and Lall, B.S. 1981. Comparative susceptibility of different brinjal cultivars against fruit and shoot borer. *Indian J. Entomol.*, 453: 108-109.

Miller, S., Baltazar, A.M., Rapusas, H.R. 2003. Overview of the Southeast Asia Site in the Philippines. IPM CRSP Annual Highlights For Year 10 (2002-2003). Virginia Tech, Virginia, USA, pp. 42-50.

Monobrullah, Md. and Shankar, U. 2008. Sub-lethal effects of *Splt*MNPV infection on developmental stages of *Spodoptera litura* (Lepidoptera: Noctuidae). *Biocontrol Science and Technology*, 18(4):431-437.

Mote, U.N. 1978. Studies on the varietal resistance of okra (*Abelmoschus esculentus* L.) to fruit borer (*Earias fabia* Stoll.) under field conditions. *Vegetable Science*, 5: 47-49.

Nath, P. And Thakur, M.R. 1965. Evaluation for red pumpkin beetle resistance in gourds. *Indian Journal of Horticulture*, 22:330-331.

Nath, P., 1966. Varietal resistance of gourds to the fruit fly. *Indian J. Hort.*, 23 (1&2): 69-78.

Panda, N., Mahapatra, A. and Shoot, M. 1971. Field evaluation of some brinjal varieties for resistance to shoot and fruit borer, *Leucinodes orbonalis* (Guen.). *Indian Journal of Agril. Sci.*, 41: 597-601.

Patil, C.S., Moholkar, P.R. and Ajri, D.S. 1986. Development of fruit borer *Earias vittella* Stoll. on resistant and susceptible varieties of okra. *Vegetable Science*, 13: 385-388.

Pawar, C.S. 1998. Helicoverpa armigera a natural problem which needs a national policy and commitment for its management. *Pestology* 22 (7): 51-59.

Rahman, M.S., Alam, M.Z., Haq, M., Sultana, N., Islam, K.S. 2002. Effect of some integrated pest management (IPM) packages against brinjal shoot and fruit borer and its consequence on yield. *J. Biol. Sci.* 2, 489-491.

Raju, S.V.S., Bar, U.K., Shankar, U. and Kumar, S. 2007. Scenario of infestation and management of eggplant shoot and fruit borer, *Leucinodes orbonalis* Guen, in India. *Resistance Pest Management Newsletter*, 16 (2): 14-16.

Raut, U.M. and Sonone, H.N. 1980. Tolerance in brinjal varieties to shoot and fruit borer (*Leucinodes orbonalis* Guen.). *Vegetable Science*, 7 (1): 74-78.

Sandhu, G.S. *et al.*, 1974. Sources of resistance to jassid and whitefly in okra germplasm. *Crop Improvement*, 1: 77–81.

Santharam, G., Kumar, K., Chandrasekaran, S. and Kuttalam, S. 2003. Bioefficacy and residues of imidacloprid in chillies used against chilli thrips. *Madras Agric. J.* 90 (7-9): 395-399.

Sardana, H.R. 2001. Integrated pest management in vegetables. In: Training manual -2, Training on IPM for Zonal Agricultural Research Stations.May21- 26.105-118 pp.

Satpathy, S. and S. Rai. 2000. Preliminary field evaluation of NPV and insecticides against tomato fruit borer, Helicoverpa armigera (Hübner). *Insect Environment*, 6(1): 21-22.

Satpathy, S., S. Rai and M. Chattopadhyaya, 1999. Field evaluation of NPV and insecticides against tomato fruit borer, Helicoverpa armigera (Hübner). *Insect Environment*, 5(3): 117-118.

Satpathy, S., Shivalingaswamy, T.M., Akhilesh Kumar, Rai, A.B. and Mathura Rai, 2005. Biointensive management of eggplant shoot and fruit borer (Leucinodes orbonalis Guen.). *Vegetable Science*, 32(1): 103-104.

Saxena, R.C. 1970. Relative susceptibility of different varieties of onion (*Allium cepa* L.) to *Caliothrips indicus* Bagn, (Thripidae: Thysanoptera). *Indian Journal of Entomology*, 32(1): 98-100.

Shankar, U and Kaul, V. 2010. Entomopathogenic Nematodes: A Tool in IPM for Vegetable Crops. *In:Quality Seed Production of Vegetable Crops* , 2010, Sharma, J. P. (ed.), Kalyani Publishers, Ludhiana, India.pp. 207-225.

Shankar, U., Kumar, D. and Gupta, S. 2010. Integrated pest management in brinjal. Technical Bulletin No. 4, SKUAST-J, Jammu.

Shankar, U., Kumar, D., Singh, S.K. and Gupta, S. 2009a. Pest complex of cole crops and their management. *Technical Bulletin No. 1*, SKUAST-J, Jammu.

Shankar, U., Kumar, D., Singh, S.K. and Gupta, S. 2009b. Pest complex of tomato and their management. *Technical Bulletin No. 2*, SKUAST-J, Jammu.

Shennan, C. 2008. Biotic interactions, ecological knowledge and agriculture. *Phil. Trans. R. Soc. B*, 363: 717-739.

Shivalingswami, T.M., Satpathy, S. and Banergee, M.K. 2002. Estimation of crops losses due to Insect pests in vegetables Ed. by Sarath Babu, B.,Varaprasad, K.S. Anitha, K., Prasada Rao, D.V.J., Chakrabarthy S.K. and Chandurkar, P.S. *Resource management in plant protection*.1: 24-31.

Singh, D. 1991. Pests of solanaceous vegetables. In: A decade of research on pests of host-crops 1980-90. Group discussion of entomologists working in coordinated projects of Hort crops held at CIHNP, Lucknow Jan. 28-29.

Singh, Y.P. and Singh, P.P. 2002. Nature and extent of damage and avoidable yield losses due to shoot and fruit borer (*Leucinodes orbonalis*) to brinjal (*Solanum melongena*) at medium high altitude hills of Meghalaya. *Indian J. Ent.*, 64 (1): 12-16.

Sivaprakasam, N. 1998. Control of tomato fruit borer (Helicoverpa armigera) by nuclear polyhedrosis virus in comparison with other treatments. *Indian J. Agric. Sci.*, 68(12): 801-802.

Smith, C.M. 1989. Plant resistance to insects. Wiley, New York, 286 pp.

Srinivasan, K. 1993. Pests of vegetable crops and their control. In: Advances in Horticulture, 6: Vegetable crops Eds. K.L. Chadha and G. Kalloo. Malhotra Pub. House, New Delhi. p. 859-886.

Srinivasan, P.M. and Narayaswamy, P.S. 1961, Varietal susceptibility in bhindi (*Abelmoschus esculentus* (L.) to the shoot and fruit borer, *Earias fabia* Stoll. and *Earias insulana* Boisd. *Madras Agricultural Journal*, 48 (9): 345-346.

Talekar, N.S. 2002. Controlling Eggplant Fruit and Shoot Borer: A Simple, Safe and Economical Approach. International Cooperators' Guide, AVRDC Pub. #02-534. Asian Vegetable Research and Development Center, Shanhua, Taiwan, p. 4.

Teli, V.S. and Dalaya, V.P. 1981. Studies on varietal resistance in okra (*Abelmochus esculentus* (Linn.) Moench) to the shoot and fruit borer. *Earias vittella* (Fab.). *South Indian Horticulture*, 29 (1):54-60.

Van Emden, H.F. 1991. The integration of host plant resistance to insects with other control measures, pp. 939-948. *In* Proceedings of the Brighton Crop Protection Conference on Pests and Diseases -1990, Vol. 3. British Crop Protection Council, Surrey United Kingdom, 474 pp.

Fig. 2 : Life cycle of Brinjal shoot and fruit borer

Fig. 3 : Clipping of infested twigs and fruits

Fig. 4 : Pheromone trap catches in brinjal field

Fig. 4a : Water traps and pheromone traps installed in brinjal field

Fig. 5 : Jassids damage

Fig. 6 : Little leaf of brinjal

Fig. 7 : Mosaic disease

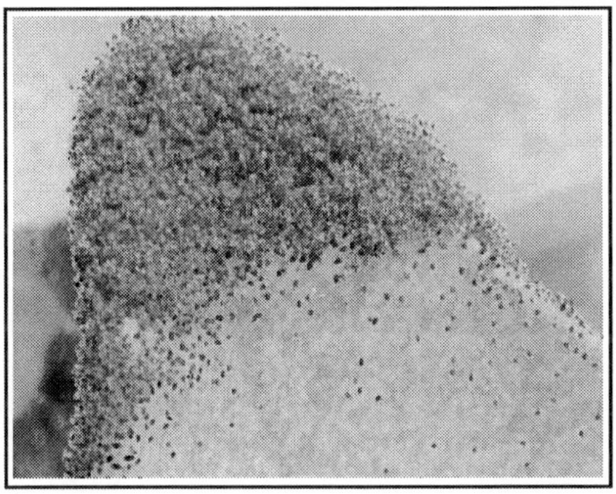

Fig. 8 : Red mite infested brinjal leaf

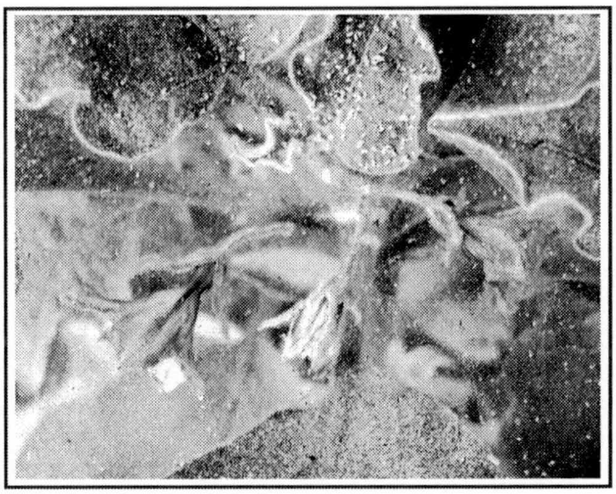

Fig. 9 : Flower buds infested with red mites

Fig. 10 : Life cycle of tomato fruit borer

Fig. 11 : Trap catches in tomato field

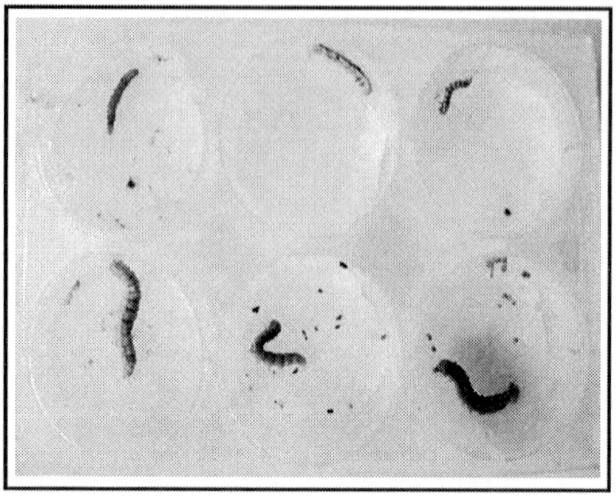

Fig. 12 : EPN infested Helicoverpa larvae

Fig. 13 : Leaf miner damage

Fig. 13a : Leaf miner ovipositing egg on tomato leaf

Fig. 14 : Chilli thrips

Fig. 14a : Thrips damage

Fig. 15 : Cut worm caterpillar

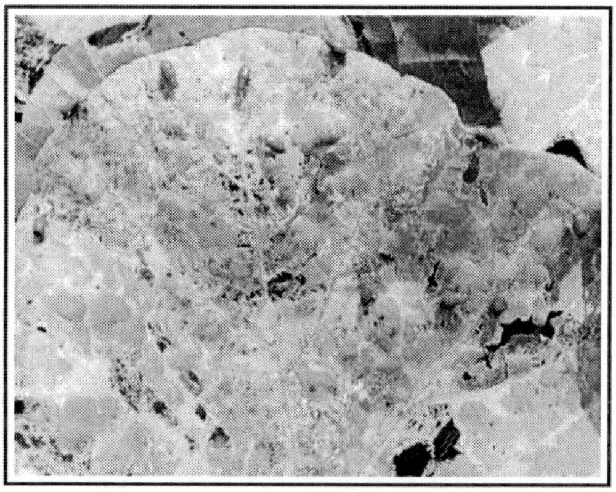

Fig. 16 : Red pumpkin beetle infestated leaf of cucurbits

Fig. 17 : Poison baiting

Fig. 17a : Fruit fly probing and oviposition on chappan kaddu

Fig. 18 : Larva of okra fruit borer

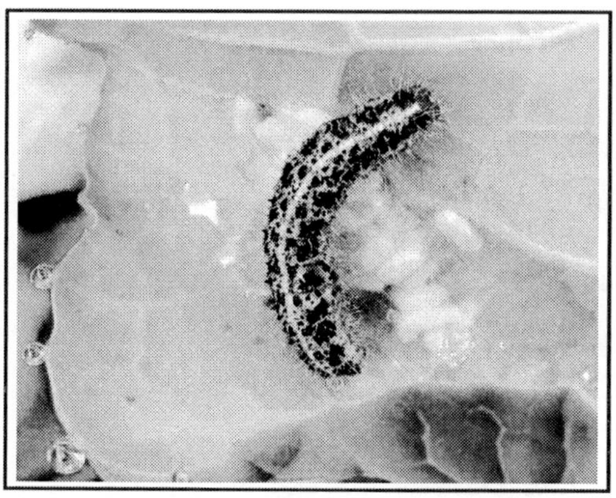

Fig. 19 : Parasitized larva of Pieris by Cotesia glomeratus

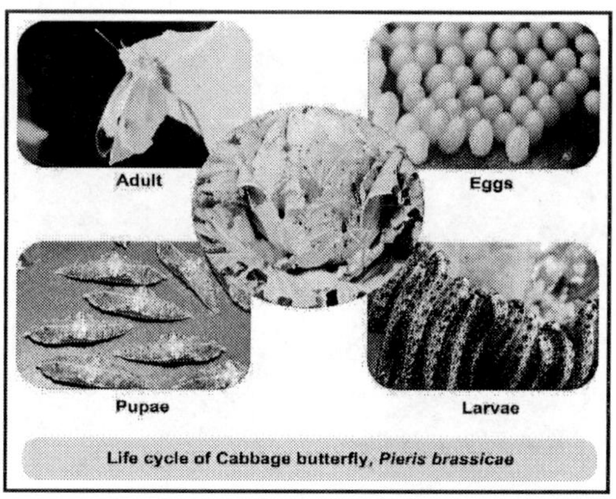

Fig. 19a : Life cycle of Cabbage butterfly

Fig. 20 : Cabbage head borer larva

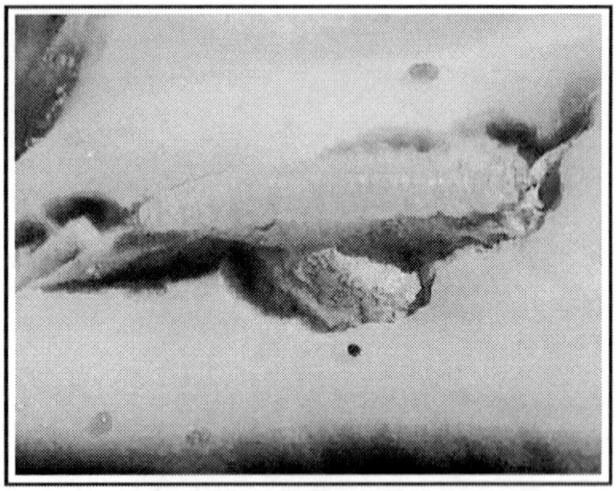

Fig. 21 : DBM larva

Fig. 21a : Diamondback moth adult

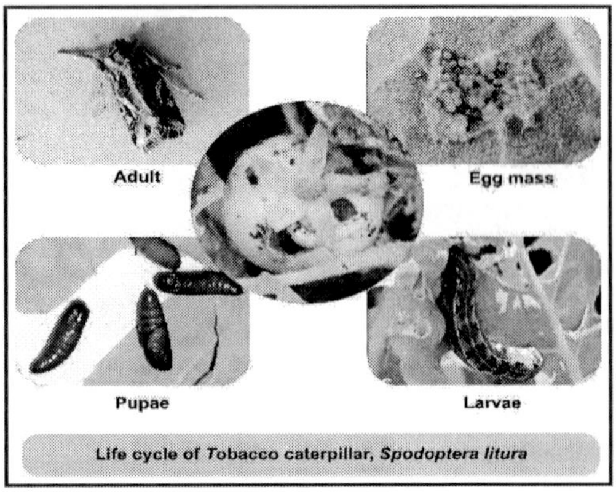

Fig. 22 : Life cycle of Spodoptera litura

Fig. 23 : Nymphs of onion thrips

Fig. 24 : Silvering of foliage

Ecologically Based Integrated Pest Management, 2012
© D.P. Abrol and Uma Shankar (eds.), pp. 651-663
New India Publishing Agency, New Delhi (India)
e-mail : info@nipabooks.com; website : www.nipabooks.com

33

Ecologically Based Integrated Disease Management for Vegetable Crops

MUKESH KUMAR PANDEY AND C.S. KALHA

Vegetables not only provide nutritional security but also income security to small and marginal land holders growing vegetables in the hinterlands of urban markets and dollar earners in global markets. India is endowed with favorable tropical, subtropical and temperate climate making it conducive for producing high quality vegetables round the year in one or other part of the county. As far as production is concerned, India is in number two in the world after China. But, as a matter of fact, in terms of productivity and subsequently for per capita availability, we are still lagging behind many countries of the world. Per capita availability of vegetables is 210-gm/capita/day against 300 gm/capita/day recommended by ICMR. There is urgent need to increase the productivity of vegetables in our country to feed the increasing population.

Vegetables being more succulent and rich in nutrient are prone to disease infection, Thereby incurring high yield losses during pre and post production period. Diseases are responsible for reducing 20-30% of the total yield of vegetables. Vegetable crops are threatened by various disease, insect and weed problems that may reduce the yield and quality of the produce and even destroy the plants. The impact of plant diseases on vegetables often appears less direct than that of insects and weeds. Disease pressure in vegetable crops from seedling stage to harvest caused by mainly fungi, bacteria and virus are the most important constraints for low production. Due to the easy accessibility and confirm action against pest, farmers prefer to use chemicals to control the pest population. We consume 13-14% of total pesticides to control pest in vegetables against 2.6% of cropped area of vegetables. Excessive and uncontrolled use of chemicals geared certain problems like pollution of natural resources (air, water and soil), residual toxicity, development of resistance / resurgence in pathogens and deleterious effects on non-target flora and fauna. Integrated disease management (IDM) is an approach which maximizes natural controls of disease populations, based upon knowledge of each pathogens, in its environment and its natural enemies, farming practices modified such as changes in the planting or sowing schedules, to affect the potential diseases adversely and to aid natural enemies of the pathogens. Thus, IDM is the optimization of pest control in an economically and ecologically sound manner. This is accompanied by the use of multiple tactics in a compatible manner to maintain pathogens damage below the economic injury level while providing protection against hazards to humans, animals, plants and the environment.

A successful disease control programme depends on a crop production system which is closely aligned with the goals of disease management. The five components of an IDM programme are prevention, monitoring, correct disease diagnosis, development and use of acceptable thresholds and optimum selection of management tools. The management strategies available include genetic control, cultural control, biological control and chemical control. For disease management, it is important to understand the potential of a pathogen to infest and spread in the crop. The three parameters of disease progress are 1) the initial amount of inoculum, 2) the rate of disease increase and 3) the time the crop is grown. These parameters may interact in different ways to produce a rapid pathogen population increase. The rate of disease increase over time is dependent upon the interactions of the pathogen, plant host and the environment. However, environmental conditions also play a critical role in determining the nature of plant disease epidemics.

To elaborate and stress up on the use of IDM strategies in vegetable diseases is given below.

Accurate Diagnosis of Diseases : Proper disease identification is critical for making correct disease management decisions. This will save time, money and the environment. Effective use of pesticides depends on correct identification of the problem. The accuracy of any diagnosis depends upon the information supplied, the specimen material selected and the condition of the specimen when it arrives at a clinic. Digital images of the fresh specimen with symptoms and field-view images of the problem might be useful in some cases. Descriptive and pictorial manuals are helpful for identification of diseases commonly found. It is important to know the common diseases of a given crop specific to your area.

Forecast or warning of diseases : The chance of disease increase when the susceptibility of the host and virulence of the pathogen are greater, as the environmental conditions approach the optimum level for pathogen growth, reproduction and spread, and as the duration of all favorable combination is prolonged or repeated. For example, development of late blights of potato and tomato epidemic can be predicted with reasonable accuracy, if the moisture and temperature condition in the field remain within certain ranges favorable to the fungus. When constant cool temperature more then 10 and 24^0C prevail, and the relative humidity remains over 75 per cent for at least 48 hours, or is at least 90 per cent for 10 hours each day for 8 days, infection will take place and a late blight outbreak can be expected from 2 to 3 weeks later. Forecasting models, based on weather conditions may helps us to alarm from these situations and suggest right time to management the disease. Computerized predictive systems also have been developed for epidemics of late blight and several other diseases; in some such systems for example, BLITECAST for late blight, FAST for forecasting *Alternaria solani* on tomatoes, TOMCAST for tomato early blight. The purpose of these forecast/warning system is to warn farmers of the impending onset of an infection period or to inform them that an infection period has already occurred, so that they can take immediate appropriate control measures to stop recent infection from developing or prevent further infections from occurring.

Prevention and Management Practices

Host Plant Resistance : The use of resistant varieties of crops aims to avoidance of diseases rather than their control. This is the most desirable method where in crop does not suffer from a disease, thus losses are not encountered. The use of resistant varieties is the cheapest, safest and most effective means of controlling vegetable diseases. Among the vegetables, spectacular success has been achieved in potato production. In last two decades varieties like Kufri, Jyoti, Kufri Alankar and Kufri Chandramukhi with high degree of resistance to late blight have been developed by the CPRI, Shimla.

Although it is preferable to use varieties that have both vertical and horizontal resistance, most of the resistant varieties carry only one or a few major genes of vertical resistance. If varieties are resistant only to some diseases as happens with *Phytophthora infestans*. Appearance of new races leads to breakdown of resistance of the popular, ruling genotype. As a result, varieties with vertical resistance need to be replaced at frequent intervals. With the development in biotechnology it is possible to introduce distant resistance genes in the plant by manipulation, genetic modification and multiplication of plants through techniques such as tissue culture and genetic engineering. Induction of Acquired resistance, in which the plant acquires localized or systemic resistance during through the effect of chemicals or microorganisms, using phosphates and carbonates as foliar therapy has been found to "turn on" the resistance genes. Rhizobacteria are also known to induce systemic acquired resistance in the foliar parts against many diseases simultaneously. Resistance to early blight has been identified in *Solanum* species such as *S. phureja* and *S. chacoense* that has been exploited for breeding resistant cultivars (Stewart *et al.*, 1994). Even biochemical resistance of non-genetic nature can be developed in plants by chemotherapy or host nutrition, but this type of resistance is temporary. Temporary physiological resistance in plants can be developed through chemotherapy, when systemic fungicides and antibiotics applied to the foliage or through the roots persist in the plants and while their toxic level is maintained the pathogen cannot invade the tissue. Nutrition cannot change a susceptible variety to a resistant variety, but making major- and micro-nutrients available there is strengthening of the tissue that can ward off invasion of the pathogen. It is very important to choose cultivars with multiple pathogen and nematode resistance whenever possible. Control of many diseases of vegetables (Fusarium wilt, Verticillium wilt, and gray leaf spot for tomato) is achieved primarily by this method. Recently, varieties resistant to *Tomato Yellow Leaf Curl Virus* (TYLCV) and *Tomato Spotted Wilt Virus* (TSWV) have been identified. Firm-fruited, crack-resistant tomato fruit can escape infection of some pathogens.

Cultural Methods

Although resistant varieties and chemical treatments are very important method of disease control, their efficiency can be improved, made more economical as well as environmentally acceptable by modification of cultural practices. The judicious use of fertilizers or nutrients, adjustment of sowing time, proper soil and water management, crop rotation, spacing, sanitation and proper weed management help in reducing the losses due to diseases, thereby resulting in better yields.

Minimum tillage : Minimum tillage are effective way to reduce the disease incidence in field. Jackson *et al.* (2004) suggested that farmers should alternate between conventional and minimum tillage, with frequent additions of organic matter to enhance several aspects of soil quality, and reduce disease and yield problems that can occur with continuous minimum tillage.

Irrigation Management : High soil moisture enhances the development of soil borne pathogens including *Phytophthora* sp., *Pythium* sp. and the bacterial wilt pathogens. Excess water damages roots by depriving them of oxygen and creates conditions that favour infection by certain soil borne pathogens. Irrigation management, based on plant needs, will help to create an environment unfavorable for pathogen survival and disease development. Pandey *et al.* (2007) suggested that improved practice like Soil Moisture-Based Seepage Irrigation Management reduced the incidence of plant disease compared to the conventional practice which resulted in crop failure in some fields. Thus, use of the improved practice reduced irrigation water use without affecting the crop yield.

Soil and Fertilizer Management : Plant nutrition and soil pH can play an important role in regulating incidence and severity of diseases in the field condition. Fertilizers with a higher proportion of nitrate nitrogen (NO_3) than ammoniacal nitrogen (NH_4) will help to reduce the incidence of Fusarium wilt on tomato. Increasing soil pH by liming is a good management strategy to reduce Fusarium wilt incidence as well as botrytis gray mold severity. Optimum calcium nutrition and higher soil pH may reduce the incidence of bacterial wilt in the field. Adequate calcium is necessary to minimize blossom end rot and to provide for overall healthy growth. Avoiding excessive nitrogen leads to less dense canopies, thus improving air movement in the canopy. There are encouraging reports stating that the application of inorganic fertilizers discouraged the survival of certain soil borne pathogens primarily because this promotes soil borne biocontrol agents that check the pathogen populations.

Avoidance of the pathogen : This can be done by a proper selection of the land or field, choice of time of sowing and selection of disease-free seed and planting stocks. The aim of these measures is to avoid the host coming in contact with the pathogens during vulnerable growth stage of the plant. For example, staking of plants helps in reducing the buck-eye rot of tomato. Correct time of sowing may help in avoidance of certain diseases. Adjustment of sowing time can help in reducing the damage due to white rot caused by Sclerotinia in the north western hill region. The November sown crop is healthier while the October planted is severely damaged by this disease. Certain varieties of crops escape damage because of their maturity period. For example powdery mildew of pea (*Erysiphe polygoni*) can be effectively managed by planting early maturing varieties that escape the disease.

Some of the diseases can be effectively kept under check by using healthy planting materials. For example, in foot rot and leaf rot of piper betel and rhizome rot of ginger, it is advised to obtain planting material from disease free sources. Tuber indexing has been resorted for obtaining virus free crop of potato.

Sanitation : Sanitation is another effective means of controlling soil and foliar diseases. Removal of crop debris, burning of the left over biomass (stubbles), and regulating the flow of irrigation water from field help in reducing the primary inoculum and the subsequent spread of the disease. Sanitation is of importance in reducing the extent of damage inflicted by soil-borne pathogens such as the *Fusarium, Verticillium, Rhizoctonia, Sclerotium* and *Sclerotina* spp. Flooding the field and impounding the irrigation for 15-30 days promote anaerobic respiration and decay of *Fusarium oxysporum* that causes wilt diseases.

Decoy crops : Decoy or trap crop are non-host crops, and are sown with the purpose of making soil borne pathogens waste their inoculum potential. This is achieved by stimulating the dormant pathogen propagules in the absence of the host. Bhat and Subbarao (2001) suggested that broccoli can be used as attractive rotation crop for the management of Verticillium wilt (*V. dahliae*) in many cropping systems of vegetable crops. Some of the pathogens controlling by this methods are given below in Table 1.

Table 1 : Decoy crops for the reduction of pathogen population

Crop	Pathogen	Decoy crop
Eggplant	*Meloidogyne incognita* *M. javanica*	*Tagetes patula* *Sesamum indicum*
Tomato	*Meloidogyne incognita* *Pratylenchus allent*	*T. patula*, casterbean, groundnut, Chrysanthemum

Crop rotation and mix cropping : Crop rotation is a very important practice, especially for soil borne disease control. For many soil borne diseases, at least a 3-year-rotation using a non-host crop reduces pathogen populations. 3-4 years crop rotation help in reducing the primary inoculum of early blight of potato and tomato caused by *Alternaria solani*. This practice is beneficial for Phytophthora blight of pepper and Fusarium wilt of watermelon. Land previously cropped to alternate and reservoir hosts should be avoided whenever possible.

Table 2 : Crop interval and rotation crop for diseases of vegetable crop management

Disease	Pathogen	Crop interval (Years)	Rotation crop
White rot of pea	*Sclerotinia sclerotiorum*	8-10	Non-host
Anthracnose of pea	*Colletotrichum pisi*	2+	Non-host
Golden nematode of potato	*Globodera rostrochinesis*	8	Beans, red clover
Potato scab	*Streptomyces sp.*	1+	Soybean
Soil rot of tomato	*Rhizoctonia solani*	1	Pangola grass

Eradicating alternate and collateral hosts : Weed control is important for the management of viral diseases. Weeds may be alternate hosts for several important vegetable viruses and their vectors. Elimination of weeds might reduce primary inoculum. Cover crops help to reduce weed populations that may harbor pathogens between seasons. For this purpose use cover crops that grow fast and provide maximum biomass. Non-host cover crops help to reduce weed populations and primary inoculum for soil borne pathogens.

Excessive handling of plants such as in thinning, pruning and tying may be involved in spread of pathogens, particularly bacteria. Because some pathogens can only enter the host through wounds, situations which promote plant injury should be avoided. During pruning process and harvest, workers should periodically clean their hands and tools with a disinfectant, such as isopropyl alcohol. If applicable, plants can be staked and tied for improved air movement in the foliar canopy. A more open canopy results in less wetness, discouraging growth of most pathogens.

Soil Solarization : Disease control by mulching soil with transparent plastic traps during summer is a relatively recent innovation. Soil solarization involves covering soils with clear, thin polyethylene sheets in mid summer, generally for 4-6 weeks, before the host crops. During soil solarization, the plastic sheets trap solar energy and prevent heat losses caused by evaporation and convection, thus creating a greenhouse effect (Mahrer, 1979). The principle effect of solarization is to raise soil temperature to levels which are lethal to soil borne pathogens.

Soil solarization has been found effective method to control soil borne pathogens in different crops and soil eco-systems (Katan, 1981). Raj *et al.* (1997) reported that soil solarization with transparent polyethylene mulch (25 μm) was found effective to control damping-off of different vegetable crops in the nursery. Nursery of ten different vegetable crops raised later in solarized bed recorded 18.3 to 42.0 per cent higher seed germination, lower incidence of post-emergence damping-off and had better seedling vigour. Soil solarization during May and June reduced the population (89.2-98.3%) of *Sclerotinia sclerotiorum* in soil. It also enhanced the biological control activities of bioagents (Sharma *et al.*, 2005) and weed control. Candido *et al.* (2007) reported that all the solarizing films (innovative films and traditional film (LDPE)) controlled weeds reducing their number by 72% and 97% in melon and lettuce crop, respectively.

In solarized soil, population of *Trichoderma* spp. increased significantly as compared to unsolarized soil (Abu-Blan and Abu-Gharbieh, 1994). Soil solarization has been found more effective against soil borne pathogens when integrated with soil amendments, limited use of fungicides and biological control agents (Gamliel and Stapleto, 1993; Stevens *et al.,* 2003). Bernal *et al.* (2005) reported that the soil solarization enhanced the beneficial organisms like fluorescent Pseudomonas and *Bacillus* spp. in treated plots with the incorporation of organic amendments buried chopped corn plant at the mature green corn stage. Several nonchemical management practices (Urban Plant Debris, solarized soil and Sunn Hemp treatments with plastic mulch) can reduce the impact of soil borne pathogens and its diseases like *Phytophthora* blight and *Pythium* root rot for fresh market bell pepper production in a humid, subtropical climate (Chellemi, 2006).

In solarized field, propagules of soil borne pathogens viz. *Fusaerium solani, Rhizoctonia spp. Pythium* sp. *and Pythium aphanidermatum* reduced greatly or eliminated between 0 and 25 cm depth. However, population densities of thermotolerent and thermophilic fungal organisms remained relatively higher following solarization. Population of *Talaromuces flavus, Aspergillus* spp. and *Penicillium* spp. also increased in solarized soils. Similarly, soil bacterial populations and actinomycetes are also reported to be affected by solarization. The density of green fluorescent *Pseudomonads* increased in the rhizosphere of sweet potato roots after solarization. *Bacillus* spp. and fluorescent *Pseudomonads* increased in the rhizosphere of lettuce plants in solarized soils. Solarization greatly reduced the populations of *Rhizobium* species. However, mycorrhizal fungus *Glomus fasiculatus* survived solar heating to the extent that colonization of host roots was not affected.

Populations of *Clavivecter michiganensis* sub sp. *michioganensis* were reduced drastically by solarization. Population of *Agrobacteria* in two naturally infested Italian nurseries decreased by 99 per cent and 92 per cent after solarization. Stapleton and DeVay (1984) reported a reduction of 72 per cent in populations of *Agrobacterium* species after solarization.

Populations of actinomycetes reduced by 45 to 58 per cent following solarization (Stapleton and Devay, 1984), however, Kaewruang *et al.* (1989) observed increased populations of actinomycetes antagonistic reduced the population of *F. oxysporum, F. solani* and *R. solani* in solarized soils as compared to non-solarized soil.

Hot water seed treatment : Hot water seed treatments of vegetable crops are highly effective to eliminate the fungi and bacteria from seeds. Heating of white cabbage and cauliflower seed by 30 minutes immersion in hot water maintained at 45°C or for 20 minutes at 50°C provided excellent control of *Alternaria brassicae* and improved seed germination by 13 per cent, besides elimination of other fungi. Ellis (1968a&b) reported that complete elimination of *Alternaria* infection from cruciferous seeds by hot water treatment for 25 minutes at 50°C. Sharma (1981) recommended hot water treatment at

52°C for 30 minutes followed by 30 minutes dip in Streptocycline (100 ppm) solution.

Botanicals : Higher plants are untapped reservoir of various valuable chemicals awaiting their intensive exploitation for integrated disease management. Kurucheve and Padmavathi (1997) reported that chilli seed treated with garlic bulb extract was most effective against damping-off (*Pythium aphanidermatum*) in nursery management. Prophylactic sprays of aqueous and ethanolic leaf extracts of *Reynoutria sachalinensis* has also been claimed to provide 40-50 per cent control of late blight of potato (Herger *et al.*, 1988). Application of plant extract like *Lantana camara, Gaussurea lappa*, garlic and neem extract has been also found effective in inhibiting the myceliogenic germination of sclerotia (Sharma and Kapoor, 1999 and Sharma *et al.*, 1999).

Biological Management

The successful control of soil borne diseases is likely to be related to the increased levels of soil microbial activity leading to increased competition and antagonism in the rhizosphere, the presence of beneficial root-colonizing bacteria and fungi, increased levels of vesicular-arbuscular mycorrhizal colonization of roots have all been identified as contributing factors; in the control of soil and root diseases. This is an unexplored area where native organisms provide protection against other harmful organisms. The composition and concentration of subterranean are important in improving the performance of the antagonist. The bioagents like *Trichoderma harzianum, T. viride, T. koningii and T. psedokoningii, Pythium oligandrum,* and bacterial bioagents such *as Pseudomonas fluorescens, Bacillus subtilis, Agrobacterium radiobacter* have been shown to be antagonistic to several plant pathogens and are most frequently studied fungi in relation to biological control. One of the fastest-growing areas in pesticide development, biopesticides present some of the greatest hope for control of highly destructive pathogens. *Trichoderma* spp. is the fungi that are present in substantial numbers in nearly all agricultural soils. *Trichoderma spp*. are effective against a wide range of plant pathogenic fungi including *Pythium* spp., *Rhizoctonia solani, Fusarium* spp., *Botrytis cinerea, Sclerotium rolfsii* and *Sclerotinia homoeocarpa* (Vora *et al.*, 2008).

Examples of commercially available biocontrol agents include the fungi *T. harzianum* and *T. virens*, an actinomycete *Streptomyces griseoviridis Paecilomyces lilacinus* and a bacterium *Bacillus subtilis, Pseudomonas fluorescens* has been commercially available in the market (Table 3). Soil inoculation with *T. harzianum, T. viride* and *Laetisaria arvalis* gave good control of *P. indicum* and there was manifold increase in seed germination in comparison to control (Krishnamoorthy and Bhaskaran, 1990). Soil application of these bioagents conferred greater protection than the seed dressings (Sreenivasaprasad and Manibhushanro, 1990).

Besides these fungi and bacteria, PGPR (Plant Growth-Promoting Rhizobacteria) may help compensate for reduced *Pseudomonas syringae* pv. *lachrymans* population and increased plant growth without methyl bromide fumigation in cucumber (Raupach and Kloepper, 2000). Vesicular-arbuscular mycorrhiza (VAM) fungi also proved their potential to check a number of soil borne diseases (Table 4). Inoculation of mycorrhizal fungus *Glomus fasciculatum* rather simultaneously or 2 weeks before inoculation with *P. aphanidermatum* reduced damping-off and increased plant height in compression to control (Hegde and Rai, 1984). VAM fungus *Gigaspora intaradices* with bacterial antagonist inoculated in onion seedlings reduced the root disease severity and enhanced the growth of crop (Linderman *et al.*, 2003).

Table 3 : Microbial bio-pesticides commercially available for use in disease management

S.No.	Name of agent/s	Commercial product
1.	*Trichoderma spp.*	BINAB-T, Trichodermin, Ecofit, Tricho-XP, Mnitor WP, Monitor-S, Trichomycil, T-2 Trichoderma spore powder and granules
2.	*Trichoderma harzianum*	F-Stop, TrichoDowels, Trichoject
3.	*T. viride*	Antagon TV, Sun-Derma, Trichominidowels, Trichoseal
4.	*T. virens*	Gliogard, Soilgard 12G
5.	*Aspergillus niger AN 27*	Kalisena
6.	*Pseudomonas fluorescence*	*P. fluorescence* wattable powder
7.	*Paecilomyces lilacinus*	Yorker, Tcicho X-P (Combi product)

Table 4 : Inhibitory effect of VAM fungi on soil borne fungal pathogens

Plant	Disease	Pathogen	Source
Onion	Wilt	*Fusarium oxysporum* f.sp. *cepi*	Becker, 1976; Dehne, 1982
Tomato	Wilt root rot	*Fusarium oxysporum* f.sp. *lycopersici*	Dehne and Schonbeck, 1979; Caron *et al.*, 1986; Caron *et al.*, 1966
Lettuce	Damping-off	*Ophiostoma brassicae*	Schonbeck and Dehne, 1979
Black pepper	Foot rot	*Phytophthora capsici*	Sivaprasad *et al.*, 1997
Chili	Wilt	*F. oxysporum* f.sp. *vasinfectum*	Newsham *et al.*, 1995

Chemical Control

Fungicides and bactericides are an important component of many disease management programmes. Since the 1960s the development of fungicides has been improved by the invention of systemic, more specific fungicides that have also offered the development of integrated disease management. It is important to remember that chemical use should be integrated with all other appropriate tactics. Information regarding physical mode of action of a fungicide is helpful in timing of fungicide applications. Physical modes of action of fungicides can be classified into four categories: protective, after infection, pre-symptom, and anti-sporulant (post-symptom). Protectant fungicides include the bulk of the foliar spray materials available. In order to be effective, protectant fungicides, such as copper compounds, mancozeb, etc., need to be on the leaf (or plant) surface prior to arrival of the pathogen. Systemic (therapeutic) fungicides, based on their level of systemically are active inside of the leaf (can penetrate at different rates through the cuticle). Systemic fungicides may stop an infection after it starts and prevent further disease development. If necessary, fungicides must be used based on recommended fungicide resistance management strategies. A new strategy to chemically manage plant diseases without direct interference with the pathogen is the triggering of plant defense reaction. Chemicals must be used at recommended rates and application frequencies.

Natural Product-Derived fungicides : Pesticides of microbial origin, so called agricultural antibiotics, are generally specific to target organisms and considered to be easily decomposable in the environment. Some antibacterial products viz. Streptomycin, Dihydrostreptomycin and Oxytetracycline used in vegetable crops against bacterial diseases may be incorporated in integrated management strategies.

References

Abu-Blan, H.A. and Abu-Gharbieh, W.I. 1994. Effect of soil solarization on winter planting of potato, cauliflower and cucumber in Central Jordan Valley. *Appl. Sci.* 21:203-213.

Becker, W.N. 1976. Quantification on onion Vesicular arbuscular mycorrhizae and their resistance to Pyrenochaeta terrestris, Ph.D. Dissertation, Uni., Illinois, Urbaria, pp.72.

Bernal, R.F.; Mendoza, Y. and Orihuela, C. 2005. Alternatives to Methyl Bromide in Uruguay. *Acta Horticulturae.* 15: 698-700.

Bhat, R.G. and Subbarao, K.V. 2001. Reaction of Broccoli to Isolates of Verticillium dahliae from Various. *Hosts Plant Disease.* 85(2) 141-146.

Candido, V.; Miccolis, V.; Castronuovo, D.and Manera, C. Eco-compatible Plastic Films for Crop Mulching and Soil Solarization in Greenhouse, *Acta Hort.* 761.

Caron, M.; Fortin, J.A. and Richard, C. 1966. Effect of inoculation sequence on the interaction between Glomus intaradices and Fusarium oxysporum f.sp. radices lycopersici tomatoes. *Can. J. Plant Pathol.* 8:12.

Caron, M.; Fortin, J.A. and Richard, C. 1986. Effect of phosphorus concentration and Glomus intaradices on Fusarium crown and root rot of tomatoes. *Phytopathol.* 76:942-946.

Chellemi, D.O. 2006. Effect of Urban Plant Debris and Soil Management Practices on Plant Parasitic Nematodes, Phytophthora Blight and Pythium Root Rot of Bell Pepper. *Crop Protection*. 25(10) 1109-1116.

Dehne, H.W. and Schonbeck, F. 1979. Influence of endotrophic mycorrhiza on Fusarial wilt of tomato. *Z. Pflkrankh.* 82:630.

Dehne, H.W. 1982. Interaction between vesicular arbuscular mycorrhizae fungi and plant pathogens. *Phytopathol*. 72: 1115.

Ellis, M.B. 1968a. Alternaria brassicae.Commonwealth Mycological Institute. Description of pathogenic fungi and bacteria. No. 162. CMI, Kew, Surrey, England.

Ellis M.B. 1968b. Alternaria brassicae.Commonwealth Mycological Institute. Description of pathogenic fungi and bacteria. No. 163. CMI, Kew, Surrey, England.

Gamliel, A. and Stapleton, J.J. 1976. Characterization of antifungal volatile compounds evolved from solarized soil amended with cabbage residues. *Phytopathol.* 83: 899-905.

Hegde, S.V. and Rai, P.V. 1984. Influence of *Glomus fasciculatum* on damping-off of tomato. *Curr. Sci.* 53 (11): 588-589.

Herger, G.; Klingauf, F.; Mangold, D.; Pommer, E.H. and Scherer, M. 1988. Efficacy of extract of Reynoutria sachalinensis Nakai against fungal diseases, especially powdery mildew. *Nachrichtenbl. Disch Pflanzenschutzdienst.* 40: 56-60.

Jackson, L.E.; Ramirez, I.; Yokota, R.; Fennimore, S.A.; Koike, S.T.; Henderson, D.M.; Chaney, W.E.; Calderón, F.J. and Klonsky, K. 2004. On-farm Assessment of Organic Matter and Tillage Management on Vegetable Yield, Soil, Weeds, Pests, and Economics in California Agriculture. *Ecosystems and Environment.* 103(3):443-463.

Katan, J. 1981. Solar heating (solarization) of soil for control of soil borne pests. *Ann. Rev.Phytopathol*. 19: 211-236.

Krishnamoorthy, A.S. and Bhaskaran, R. 1990. Biological control of damping-off diseases of tomato caused by Pythium indicum Balakrishnan. *J. Biol. Control* 4(1): 323-324.

Kurucheve, V. and Padmavathi, R. 1997. Fungitoxicity of selected plant products against *Pythium aphanidermatum. Indian Phytopath.* 50(4): 535-539.

Linderman, R.G.; Marlow, J.L. and Davis E.A. 2003. Contribution of microbial associates of VA mycorrhizae to mycorrhiza effects on plant growth and health. In: "Int. Con. on Mycorrhiza", Montreal University, Canada, (Abs.).

Newsham, K.K.; Fitter, A.H. and Watkinson, A.A. 1995. Arbuscular mycorrhizas protect annual grass from root pathogenic fungi in the field. *J. Ecology*. 83: 991-1000.

Pandey, C.; Shukla, S. Obreza; T.A. 2007. Development and Evaluation of Soil Moisture-Based Seepage Irrigation Management for Water Use and Quality. *Journal of Irrigation and Drainage Engineering*. 133 (5) 435-443.

Rai, H.; Bhardwaj, M.L. and Sharma, N.K. 1997. Soil Solarization for the control of damping-off of different vegetable crops in the nursery. *Indian Phytopath*. 50(4): 524-528.

Schonbeck, F. and Dehne, H.W. 1979. Unterschungen zum Pilzliche spores parasiten, Olpidium brassicae, TMV, Z. pflanzenrankh. *Pflanzenschutz*. 86: 103.

Sharma, B.K.; basandrai, A.K. and Singh, A.K. 1999. Management of white rot pea. *Pl. Dis. Res*. 14(2): 161-164.

Sharma, C.L. and Kapoor, A.S. 1999. In vitro evaluation of herbicides and plants extract against Sclerotinia sclerotiorum causing white rot in peas. *Pl. Dis. Res*. 14(10): 79-81.

Sharma, S.L. 1981. Control of black rot of cauliflower by hot water seed treatment and field sprays with Streptocycline. *Indian J. Mycol. Pl. Pathol*. 11: 17-20.

Sivaprasad, P.; Joseph, P. and Balakrishnan, S. 1997. Management of foot rot of black pepper with arbuscular mycorrhizal fungi. Int. Con. on integrated disease management for sustainable agriculture. IARI, New Delhi, pp. 191-226.

Sreenivasaprasad, S. and Manibhushanro, K. 1990. Biological potential of fungal antagonists Gliocladium virens and Trichoderma longibrachiatum. *Z. Pflanzenkr. Pflanzenschutz* 97(6): 570-579.

Stevens, C.; Khan, V.A.; Kabana, R.; Ploper, L.D.; Bakkan, P.A.; Collins, D.J.; Brown, J.E.; Wilson, M.A. and Igmedge, E.C.K. 2003. Integration of soil solarization with chemical, biological and cultural control for management of soil borne diseases of vegetables. *PI Soil* 253: 493-496.

Stewart, H.E.; Bradshaw, J.E.; Wastie, R.L., Mackay, G.R.; Erlich, O.; Livescu, L. and Nachmias, A. 1994. Assessing progenies of potato for resistance to early blight. *Potato Res*. 37:237-269.

□□□

Ecologically Based Integrated Pest Management, 2012
© D.P. Abrol and Uma Shankar (eds.), pp. 665-679
New India Publishing Agency, New Delhi (India)
e-mail : info@nipabooks.com; website : www.nipabooks.com

34

Mites Injurious to Economic Plants and their Management

D.P. ABROL AND A.K. BHOWMICK

Introduction

Mites and ticks belong to acarology an offshoot of animal science Zoology a common man mites and insects look alike but former can be easily differentiated by absence wings and antennae and presence of four pairs of legs in most farms. Parasitic mites and ticks live on domestic animals, pets and poultry (Baker and Pritchard, 1960; Baker and Tuttle, 1994; Krantz and Walter, 2009). They are responsible for inducing paralysis, host intoxication, irritation carriers of pathogens and other health disorders to them. The severe bites of ticks cause tick toxicosis producing deep wounds, which bleed and are open to secondary infection. Parasitic mites are among the serious pests of the bees the world over. At times they cause mass mortality of brood and put a heavy loss to apiculture industry. Both solitary and social bees that are important pollinators are parasitized by Messostigmatid mites reducing their life span and foraging efficiency. They parasitize silkworm larvae, pupae and adults. By their toxication silkworms die in few hours. Man himself is not free from the attack of mites. Human itch mite, scabies mite and pyroglyphid mites causing allergic asthma are notorious examples of parasitic mites on man. Dancing on the head of a pin' is definitely impossible for most animals. However, a large number of mites, the smallest Arachnida could do it. The size of these creatures ranges from a fraction of a millimetre, invisible to the naked eye, to about one centimetre in ticks.

Many species of mites and ticks are severe pests of crops and stored products or ectoparasites of man, livestock and pets. Furthermore, many species are potential carriers of diseases. A large number of plant-parasitic mites considerably reduce yield of crops and might eventually kill their host plant. Particularly in agricultural systems, spider mites became severe secondary pests. Due to their stylet-like piercing-sucking mouthparts, the mites are able to pierce and suck juices from individual plant cells. Thus, mites can avoid most of the defence mechanisms of the plant. In addition, short generation times of one to three weeks paired with a high reproductive potential easily allow the animals to over-exploit their hosts. The piercing-sucking action of mites causes typical signs of damage on the affected plant tissues.

Pests of crops are spider mites, red spiders or twospotted mites (Tetranychidae) and gall, rust and blister mites (Eriophyidae). Some species of the latter family cause an abnormal development of the epidermal plant cells forming galls or other deformations of the leaf surface. The latter kind of damage is referred to as erineum and can be confused with fungous growth. The piercing-sucking action of mites can also result in grey or brown mottling of the foliage. In PNG, one Eriophyidae species reportedly damaged *Terminalia brassii*. Important agents for biological control belong to the predacious families Phytoseiidae and Stigmaeidae. Mites like *Typhlodromus* and *Phytoseiulus* are reared commercially in large quantities. After their release in an infested area, these predators effectively suppress plant-parasitic mites and other small insect pests (Helle and Sabelis, 1985; Bolland *et al.*, 1998; Beard *et al.*, 2003).

Plant Mites

Among the non-insect pests, mites form an important group that inflicts damages to almost all agricultural and horticultural crops (Table 1). More than 6 000 species of plant feeding (phytophagous) mites are known worldwide. The majority of plant feeding species belong to the obligate plant parasitic Eriophyoidea (e.g. gall mites, erinose mites, bud mites, rust mites) and Tetranychoidea (e.g. spider mites, false spider mites), while a number of species belong to other lineages (e.g. Eupodoidea, Tarsonemidae, and single oribatid mites). Plant feeding mites form an integral and important part of the natural ecosystem. Some species, especially eriophyoid mites, can be utilised for the biological control of weeds. Many plant feeding mites are of economic importance as pests of crop plants (Gupta, 1985; Sadana, 1985;

Rather, 1989, 1993). All eriophyoid mites are plant feeding. They are extremely tiny, the majority are less than 300 micron long, and essentially invisible to the unaided eye. They have elongated, worm-like bodies, with only two pairs of legs. Eriophyoid mites are commonly known as gall mites, bud mites, rust mites, erineum mites, witches' broom mites, blister mites and so on, referring to the symptoms caused by a particular species. The feeding of almost half of eriophyoid species known, though, does not cause visible damage to their plant hosts. The eriophyoid mites belong to three families: Phytoptidae, Eriophyidae and Diptilomiopidae. About 3 400 species are known, but these probably represent only as little as 5% or less of extant eriophyoid species. Most woody and many herbaceous flowering plants, and gymnosperms and ferns most likely host these minute mites. Most eriophyoid species are highly host specific, restricted to one or single closely related plant species (Jeppson *et al.*, 1975; Hislop and Jeppson, 1976; Ochoa *et al.*, 1994; Tseng, 1990; Zhang, 2003).

The small size of mites is responsible for the lack of information available on them. They have gained increasing importance from last two decades due to sudden upsurgance and have caused great concern to both farmers and agricultural scientists all round the world. In some commercial crops losses due to these new pests have been estimated as 50-80% in mango, 20-25% in paddy , 20-30% in sugarcane, 15-30% in red gram, 20-30% in cotton, 10-15% in vegetables, 5-11% in tea and 10-15% in mushrooms (Gupta, 1991).

Eriophyid Mites

Mites are among the most common organisms in nature. More closely related to spiders and ticks than to insects, these tiny members of the arachnid class are unique in their physical structure and habits. Most mites characteristically have eight legs and lack both antennae and body segments. Many mites are common plant pests, and feed by inserting their mouthparts into their host plants and drawing out the plant's juices. Certain mites cause damage to the plant in the form of small dots (stippling) on the leaf tissue, and some cause growth distortions of the plant such as galls. Their feeding activities can lead to a general decline in the plants health, cause occasional leaf drop, or rarely, cause plant death. Among the many different types of mites are the eriophyid mites, including the gall, blister, bud and rust mites, each of which distorts plant tissue in the manner they are named for. All eriophyid mites are very tiny; a hand lens or a microscope is needed to see them. Generally what most people see is the distol1ion or alteration of the host plant's tissues that has resulted from the mite's feeding activities. The relationship between eriophyid mites and their host plants is unique; in response to the mite's activities, host plants may exhibit an array of different responses from harmless tissue alteration such as galls, to plant death. Eriophyids are selective in their choice of host plants, and each species prefers a narrow range of host plants. Most familiar are the gall-inducing eriophyids, though other eriophyids move freely about on the leaf surface, seeking shelter in natural irregularities on the plant's surface. Still other eriophyid mites bury themselves under bud scales or leaf bases to obtain food and cover. In Alaska, eriophyid mite activity is commonly observed as galls, or fuzzy mats on the foliage of both ornamental plantings and native hosts such as birch, mountain ash, and willow, usually with little or no discernible ill effects to the plant. Certain insects also induce galls on plants, and their activities are also easily observed.

Description

Eriophyid mites are among the most specialized of plant feeding organisms, maintaining a unique relationship with their host plant. In response to the mite's attack on the host plant, a localized growth reaction occurs, creating the visible gall or other plant abnormality. The mites are very tiny and observation requires a hand lens or microscope. Unlike the familiar spider-like forms of other mites, eriophyid mites are generally elongated, worm-like, and have only two pair of legs instead of the usual four pairs. Because of the mite's small size, the galls or other plant abnormalities such as russeting, leaf folding, or blistering are the most observable clues that eriophyid mites are, or were, present.

Life History

The life cycle of many eriophyid mites is fairly simple; within a spring-summer season the mites develop through four growth stages: egg, first nymph, second nymph, and adult, which is also the overwintering stage. Certain eriophyid mites have a more complex life cycle which involves the alternation of a female-only generation with a male-female generation. This alternation of generations is more common in eriophyids that feed on deciduous plants, and seems to be an adaptation based on the seasonal changes of the host material. Overwintering, female-only mites are called deutogynes, and females associated with female-male generations are called protogynes. Deutogynes and protogynes within the same species can be entirely different in shape and size, and are consequently difficult to identify. Eriophyid mites travel between hosts by air currents, or by hitching rides on insects or birds.

Plant Abnormalities Caused by Eriophyid Mites

Each species of eriophyid mite has a particular association with its host plant, and their feeding activities are concentrated on the plant's fleshy tissues. Plant abnormalities such as galls occur when the mites inject their saliva, which contains chemicals that act as plant growth regulators, into the host plants tissue's. The galls serve as special sites that mites develop for themselves to provide protection for their brood and to supply themselves with additional food -made available through the extra tissue in the gall. Eriophyid induced galls can be described in many different ways, and all occur on the soft parts of the host plant. Bladder, bead, finger, and pouch galls are examples, and each gall, regardless of type, is an individual gall, complete with an escape hole on the underside. Many galls become covered with hairs, or erinea, which are mite-induced plant hair growths. Erinea may also be induced by non-gall making mites which use the fuzzy growth much like the gall-making species use galls; the erinea is induced for food and shelter. Plants hosting erineainducing eriophyids have characteristic hairy or fuzzy pads on their leaf surfaces or other plant parts, and the erinea can often be colorful.

Eriophyid mites can induce twig elongation, excessive bud production, or other forms of brooming as a result of their growth regulating salivary constituents. Eriophyids may also feed in developing individual buds, causing bud swelling referred to as "big bud", or may induce the plant to produce several bud galls along a twig. Eriophyid mites can further alter a plant's normal appearance by inducing leaf edgerolling, or causing the leaf surface to blister in response to their feeding activities.

Management of Eriophid Mites

Managing eriophyid mites in ornamental plantings is usually quite easy; removing and destroying the affected plant parts (generally leaves with galls or erinea) eliminates the immediate mite population as well as the visual signs of the mites. Early season examination of plant material can provide quick detection and allow for removal of newly infested parts. Some species overwintering as adults on deciduous trees may be controlled with an ultrafine horticultural oil spray applied in the fall, prior to leaf drop. Beneficial predaceous mites, however, may also be eliminated. Spraying the plant will not remove the galls nor erinea, but a late season application may control those exposed adult mites seeking overwintering sites, and consequently reduce the possibility of mite-induced galls or erinea appearing on the next year's new foliage. Eriophyid mites occurring on native Alaskan plants are generally not considered serious pests, and often provide the observer with a chance to examine a unique relationship between a host plant and an inhabitant.

Spider Mites

Mites are common pests in landscapes and gardens and can be found feeding on many fruit trees, vines, berries, vegetables, and ornamental plants (Mignon and Flechtmann, 2004). Although related to insects, mites are not insects but members of the arachnid class along with spiders and ticks. The spider mites, also called webspinning mites, are the most common mite pests and among the most ubiquitous of all pests in the garden and farm. The web spinning spider mites include Pacific spider mite, twospotted spider mite 81K], strawberry spider mite, and several other species. Most common ones are closely related species in the *Tetranychus* genus and cannot be reliably distinguished in the field. However, there is little need to do so since their damage, biology, and management are virtually the same.

Identification

To the naked eye, spider mites look like tiny moving dots; however, you can see them easily with a 10X hand lens. Adult females, the largest forms, are less than 1/20 inch long. Spider mites live in colonies , mostly on the under-surfaces of leaves; a single colony may contain hundreds of individuals. The names "spider mite" and "webspinning mite" come from the silk webbing most species produce on infested leaves. The presence of webbing is an easy way to distinguish them from all other types of mites. Adults have eight legs and an oval body, with two red eyespots near the head end of the body. Females usually have a large, dark blotch on each side of the body and numerous bristles covering the legs and body. Immatures resemble adults, except the newly hatched larvae have only six legs. Eggs are spherical and translucent, like tiny droplets, becoming cream colored before hatching.

Life Cycle

spider mites may feed and reproduce all year on plants that retain their green leaves through the winter. In colder areas and on deciduous trees that drop their leaves, web spinning mites overwinter as red or orange mated females under rough bark scales and in ground litter and trash. They begin feeding and laying eggs when warm weather returns in spring. Spider mites reproduce rapidly in hot weather and commonly become numerous in June through September. If temperature and food supplies are favorable, a generation can be completed in less than a week. Spider mites are generally favored by hot, dusty conditions and are usually found first on trees or plants adjacent to dusty roadways or at margins of gardens. Plants under water stress are also highly susceptible. As foliage quality declines on heavily infested plants, female mites catch wind currents and disperse to other plants. High mite populations may undergo a rapid decline in late summer when predators overtake them, host plant conditions become unfavorable, the weather turns cooler, or following rain.

Damage

Mites cause damage by sucking cell contents from leaves. A small number of mites is not usually reason for concern, but very high populations—levels high enough to show visible damage to leaves—can be damaging to plants, especially herbaceous ones. At first, the damage shows up as a stippling of light dots on the leaves; sometimes the leaves take on a bronze color. As feeding continues, the leaves turn yellow and drop off. Often leaves, twigs, and fruit are covered with large amounts of webbing. Damage is usually worse when compounded by water stress.

Loss of leaves will not cause yield losses in fruit trees during the year of infestation unless it occurs in spring or very early summer, but it may impact next year's crop. On annual vegetable crops, such as squash, melons, and watermelons, loss of leaves can have a significant impact on yield and lead to sunburning. On crops such as sugar peas and beans, where pods are attacked, spider mites can cause direct damage. On ornamentals, mites are primarily an aesthetic concern, but can kill plants if populations become very high on annual plants. Spider mites are also important pests of field-grown roses.

Management

Spider mites have many natural enemies that often limit populations. Adequate irrigation is important because water-stressed plants are most likely to be damaged. Broad-spectrum insecticide treatments for other pests frequently cause mite outbreaks, so avoid these when possible. Sprays of water, insecticidal oils, or soaps can be used for management. Always monitor before treatment.

Monitoring

Mites are tiny and difficult to detect. Usually plant damage—stippling or yellowing of leaves—will be noticed before you spot the mites themselves. Check the undersides of leaves for mites, their eggs, and webbing; you will need a hand lens to identify them. To observe them more closely, shake a few off the leaf surface onto a white sheet of paper. Once disturbed, they will move around rapidly. Be sure mites are present before you treat. Sometimes the mites will be gone by the time you notice the damage; plants will often recover after mites have left.

Biological Control

Spider mites have many natural enemies, which limit their numbers in many landscapes and gardens, especially when undisturbed by pesticide sprays. Some of the most important are the predatory mites, including the western predatory mite, *Galendromus (=Metaseiulus) occidentalis,* and *Phytoseiulus* species. Predatory mites are about the same size as plant-feeding mites but have longer legs and are more active. Various insects are also important predators—the sixspotted thrips *(Scolothrips sexmaculatus),* the larvae and adults of the spider mite destroyer lady beetle *(Stethorus picipes),* the larvae of certain flies including the cecidomyid *Feltiella acarivora*, and various general predators such as minute pirate bugs, bigeyed bugs, and lacewing. The purchase and release of predatory mites can be useful in establishing populations in large plantings or orchards, but the best results are obtained by creating favorable conditions for naturally occurring predators—for instance, by avoiding dusty conditions and pesticide sprays.

The major predator mites commercially available for release are the western predatory mite and *Phytoseiulus*. The western predatory mite is more effective under hot, dry conditions. These predators do not feed on foliage or become pests; thus if pest mites are not available when predatory mites are released, the predators starve or migrate elsewhere. If you wish to establish predators in a heavily infested orchard or garden that has few predators, use a soap spray or selective miticide to bring pest mites to a lower level and then release predatory mites. A good guideline is that one predator is needed for every ten spider mites to provide control. More than one application of predatory mites may be required if you want to reduce pest populations rapidly. Concentrate releases in hot spots where spider mite numbers are highest. Once established on perennials, predatory mites may reproduce and provide biological control indefinitely without further augmentation unless nonselective insecticides are applied that kill the predators.

Cultural Control

Cultural practices can have a significant impact on spider mites. Dusty conditions often lead to mite outbreaks. Apply water to pathways and other dusty areas at regular intervals. Water-stressed trees and plants are less tolerant of spider mite damage. Be sure to provide adequate irrigation. Mid-season washing of trees and vines with water to remove dust may help prevent serious late-season mite infestations.

In gardens and on small fruit trees, regular, forceful spraying of plants with water will often reduce spider mite numbers adequately. Be sure to get good coverage, especially on the undersides of leaves. If more control is required, use an insecticidal soap or oil in your spray (as described below), but test the product on one or two plants to be sure it is not damaging to plants.

Chemicals

Spider mites frequently become a problem after the application of insecticides. Such outbreaks are commonly a result of the insecticide killing off the natural enemies of the mites, but also occur when certain insecticides stimulate mite reproduction. For example, spider mites exposed to carbaryl (Sevin) in the laboratory have been shown to reproduce faster than untreated populations. Carbaryl, some organophosphates, and some pyrethroids apparently also favor spider mites by increasing the level of nitrogen in leaves. Insecticides applied during hot weather usually appear to have the greatest effect on mites, causing dramatic outbreaks within a few days.

If a treatment for mites is necessary, use selective materials, preferably insecticidal soap or insecticidal oil. Petroleum-based horticultural oils or neem oils are both acceptable. Do not use soaps or oils on water-stressed plants or when temperatures exceed 90°F. These materials may be phytotoxic to some plants, so check labels and/or test them out on a portion of the foliage several days before applying a full treatment. Oils and soaps must contact mites to kill them so excellent coverage, especially on the undersides of leaves, is essential and repeat applications may be required. Sulfur dust or spray can be used on some vegetables, but will burn cucurbits. Do not use sulfur dust if temperatures exceed 90°F and do not apply sulfur within 30 days of an oil spray. Sulfur dusts are skin irritants and eye and respiratory hazards. Always wear appropriate protective clothing.

Causes of Upsurgence

The random and indiscriminate use of broad spectrum insecticides especially chlorinated hydrocarbons against various pests eliminated natural enemies of mites. Organo-phosphorus compounds chiefly in sub-lethal doses developed resistance in mites. DDT and synthetic pyrethroides increased reproductive potential of mites as post sprayer head. Besides cultivation of high yielding varieties, improved agricultural practices and application of higher doses nitrogenous fertilizers provided better nourishment to mites and induced mite multiplication. It can be concluded that in the absence of insects for competition for food, coupled with elimination of natural enemies, introduction of susceptible varieties, improved agriculture practices, development of resistance in mites by some chemicals there population increased enormously in the agro-ecosystem and pose great problem.

Symptoms and Injury Caused by Plant Mites

Direct Damage

Spider and false spider mites penetrate the leaf tissue through their needle like chelicerae and suck the exuding fluid. They destroy chlorophyll (causing reduction in photosynthesis) and loss of water through the punctures produced by them. Leaves are covered with dense webs which entangle dust particles and hinder the normal physiological activities of the leaves. Yellow or white spots, curling and crinkling appear on leaves. They become rusty due to thousands of mites present on them. Plants suffer premature defoliation. Growth is stunted; quality and quantity of produce are affected. Problem becomes severe in summer and early autumn, seasons.

Eriphyds, some tenuipalpids and tarsonemid mites injects toxins through saliva while feeding on plant tissues causing phytotoxaemia and gall formation. The symptoms induced by eriophyid mites are generally confused with other pathogens since they remain concealed inside the plant tissues. They produce malformation (mango), gall and erineum formation (walnut, grapevine, litchi), witches broom effect (mango, chrysanthemum), blisters (pears, walnut, salix), russetting (citrus), malfunctioning (mango, sophora) etc. Their feeding inside the buds cause bud blast, fall of blossoms and deformation of fruits. In severe infestation flower and fruit formation is prevented as in chilli. When plant products research to storage house they are infested by stored product mites. Grains lose their vigour. Heavily infested commodities become unfit for consumption.

Indirect Damage

Some of them transmit viral and fungal diseases to crops like pigeon pea sterility mosaic, fig mosaic, wheat streak mosaic. Sugarcane streak mosaic, peach mosaic, dolichosenation mosaic, Stewart's bud rot, bulb diseases etc.

Beneficial Mites

However, it is not that all agricultural mites are our foes as there are many who our friends are. They help us by predating upon injurious mite species as well as some insect pests like mealy bugs, aphid, scales etc and suppress their populations. Some of them parasitize various insect pests and help in their biological control. For instance parasitic mites on locusts decrease their reproductive potential and are utilized for their control.

Soil mites outnumber all other micro-organisms in forests, pastures, compost heaps and arable soils. They play a significant role in nutrient cycling, energy flow, bio-indication of soil conditions, decomposition of plant matter etc., thus contribute to soil fertility, water management and soil erosion. Some of them prey upon nematodes and help in their control.

Water mites live in salt and fresh water. Besides other activities some of them feed upn weeds and help in weed control. Orthogalumna terebrantis is utilized in many countries for the control of water hyacinth.

Dispersal of Mites

They spread from one field to another rapidly through a unique phenomenon called "ballooning" in which they hang themselves from leaves by fine silken threads produced by their body and come down on the soil or else remain hanging for carrying away by the wind. They are also carried by insects, birds and man himself from on e field to another.

Causes for Mites

Stress Encourages Pest Problems

Mites are opportunistic pests, usually attacking plants that are already struggling from other pest problems and/or drought or lack of light. All pest insects target weakened plants that lack vigor to resist their attack. Researchers are finding that glutathione, a chemical produced by stressed plants, enhances the reproduction, growth, pesticide resistance of pest insects. This may explain why they are attracted to already suffering plants. It also suggests that the best mite control strategy is to treat the immediate mite problem first, then determine, if possible, the underlying problem that is causing the plant stress and correct that.

Some Causes of Plant Stress

Some causes of indoor or outdoor plant stress include: Too much or too little light or water. Inappropriate plants for the climate or location. Overuse of nitrogen-rich fertilizer, which encourages too much leafy plant growth. Drastic pruning of trees or shrubs, which stimulates growth of succulent suckers. Use of broad-spectrum pesticides which also kill the mites' natural predators. Transplant shock which makes plants vulnerable to mites for a few days to a week.

Control of Mite Pests

All means of control are integrated to develop management practices of mites.

1. Cultural Control

a) Burn the damage leaves another infested plant materials to destroy eggs and hibernating females. This will also help in controlling fungal and viral diseases as well.

b) Remove alternate hosts as weeds which provide source of reinfestation to next crop.

c) Grow resistant varieties or varieties less resistant to mite attack.

d) Avoid monocropping instead follow intercropping by using non host crops.

e) Adjust sowing dates. Since sowing has been found effective against several vegetable mites and viral diseases.

f) Apply moderate doses of nitrogenous fertilizers

2. Biological Control

a) Predatory mites especially phytoseiids have shown promise as efficient predators and can be mass multiplied.

b) Arthropod predators such as coccinellids beetles, thrips, bugs, spiders etc prey upon mites.

c) Some pathogenic diseases have also been reported from mites which give control.

For utilization of these natural enemies of mites in the field judicious use of acaricides/pesticides is highly desirable.

3. Chemical Control

a) Spray acaricides when pest population is at low level to achieve better kill and avoid heavy populations latter.

b) Use selective acaricides

c) Spray fluid should give good coverage of the crop to reach concealed niches like auxiliary buds, galls etc.

d) Acaricides like Docofol18.5 EC at 1.5 ml/litre, Tetradifon 8 E C 2ml/litre, Ethion 50 EC 1 ml/litre, Amitraz 20 EC 2ml/litre, Dimethoate 30 EC 1.5 ml/litre or neem seed kernel extract at 5% have provided good control of mite pests in general.

e) Acaricides that are chemically unrelated should be applied in rotation to avoid mite resistance.

Table 1 : Distribution of mites attacking plants

S.No	Mite	Pest status	Distribution	Crops attacked
1	Two spotted spider mite	Major in Jammu Kashmir and ladakh	All over India	Bhindi, brinjal, spinach, tomato, clover, cabbage, cucurbits, sunflower, castor, cotton, alfalfa, tea, fruits and ornamental plants
2	Red vegetable mite	Major in Jammu Division	All over India	Bhindi, brinjal, spinach, tomato, clover, cabbage, cucurbits, sunflower, castor, cotton, alfalfa, tea, fruits and ornamental plants
3	Orange mite	Major in Kashmir valley	Jammu & Kashmir	Cucurbits, fodders, bean, Frenchbean, mulberry, walnut, almond, apple, cherry, grape, citrus, peach, soyabean

Contd...

Table 1. Contd...

4	carmine mite	-	All over India	Dahlias, potato, watermelon, tomato, pyrethrum, sunflower
5	Oriental mite	Minor in Jammu Kashmir	All over India	Citrus, maize, papaya, sunflower, ber, plum, peach, cotton, ornamental and medicinal plants
6	Dark red spider mite	Minor in Jammu Kashmir	All over India	Fruits especially mango and cashew
7	Tea crimson mite	-	N.E&South India	Tea, jute
8	Paddy white mite	-	South India	Cereals especially paddy
9	Brown wheat mite	-	Rajasthan, Punjab, Uttar Pradesh	Cereals and spices, Wheat, barley, triticale, oats, cotton and grasses. Crops are at risk during warm, dry periods.
10	White patchy mite	-	South and North India	cereals and sugarcane
11	European red mite	Major in Kashmir valley	Jammu Kashmir and Himachal Pradesh	Apple, plum, peach, almondvines, mulberry, blackberry tomato
12	Rust mite	Major in Jammu Kashmir and Ladakh	Jammu Kashmir, Himachal Pradesh and Western Uttar Pradesh	Apple, quince, loquat, hawthorn
13	Citrus red mite	Major in Jammu Division	Jammu and Punjab	Citrus, strawberry
14	Raddish brown spider mite	Major in Jammu Division	Jammu, Central and South India	Bhindi, gourd, pumpkin, cucumber, beans
15	Greenish yellow spider mite	Major in Jammu Kashmir and Ladakh	All over India	Malva, ficus, grape, rose, apple, apricot, hops, shrubs ornamental plants.

Contd...

Table 1. Contd...

16	Yellow broad mite	-	All over India	Chilli , pulses, cotton, tea, gingelly
17	Scarlet mite	Minor in Jammu Kashmir	All over India	Fruits, ornamentals, citrus, grapevine, brinjal, pomegranate, jack fruit and other plantation crops
18	Ornament al flat mite	Major in Jammu Division	Jammu, Kashmir and Tamilnadu	Walnut, tea
19	Mango bud mite	Minor in Jammu Division	All over India	mango
20	Cotton creamy	Minor in Jammu Division	Gujarat	Fibre crops
21	Vine bud mite	Major in Jammu	Jammu Kashmir, Andhra Pradesh, Gujarat, West Bengal	grape
22	Walnut gall mite	Major in Jammu	Jammu Kashmir and Himachal Pradesh, Uttar Pradesh	walnut
23	Pear blister mite	Major in Kashmir valley	Kashmir and Himachal Pradesh	Pear, apple

References

Baker EW and Pritchard A.E. 1960. The tetranychoid mites of Africa. *Hilgardia* 29(11): 455-574.

Baker E.W. and Tuttle D.M. 1994. *A guide to the spider mites (Tetranychidae) of the United States.* Indira Pub. House, West Bloomfield, MI: 347 pp.

Beard J.J., Walter, D.E. and Allsopp P. 2003. Spider Mites of Sugarcane in Australia: A Review of Grass-feeding *Oligonychus* Berlese (Acari: Prostigmata: Tetranychidae) with the description of four new species. *Australian Journal of Entomology,* 42: 51-78.

Bolland, H.R., Gutierrez J.and Flechtmann, C.H.W. 1998. World Catalogue of the Spider Mite Family (Acari: Tetranychidae). E.J. Brill: Leiden.

Gupta, S.K. 1985. *Hand book - Plant mites of India.* Zoological Survey of India, Calcutta, pp. 520.

Gupta, S.K. 1991. The mites of agricultural importance in India with remarks on their economic status (P. 502-522). In : Dusbabek, F. and Bukva, V. (eds.), *Modern Acarology* Vol I. Academia, Prague and SPB Academic (Publ.).

Helle W. and Sabelis M.W. (eds.) 1985. Spider Mites, Their Biology, Natural Enemies, and Control, Vol. 1A. Elsevier: New York.

Hislop, R.G. and Jeppson L.R. 1976. Morphology of the mouthparts of several species of phytophagous mites. *Annals of the Entomological Society of America* 69: 1125–1135.

Jeppson L.R., Keifer H.H. and.Baker E.W. 1975. Mites Injurious to Economic Plants, University of California Press: Berkeley.

Krantz G.W. and Walter, D.E. 2009. *A Manual of Acarology*. IIIrd Edition, Texas Tech. University, Press, Texas, USA.

Mignon A and Flechtmann, C.H.W. 2004. First additions and corrections to the World Catalogue of the Spider Mite Family (Acari: Tetranychidae). *Intern. J. Acarol.* 30: 143-152.

Ochoa, R., Aguilar, H. and Vargas, C. 1994. Phytophagous Mites of Central America: An Illustrated Guide CATIE, Turrialba, Costa Rica.

Rather, A.Q. 1989. Studies on mites (Acari) associated with stone fruits in subtropical, temperate and cold-arid zones of Jammu and Kashmir (P. 183-189). In : ChannaBasavanna, G. P. and Viraktamath, C.A. (eds.), *Progress in Acarology,* Vol. 2, Oxford & IBH (Publ.), pp. 483.

Rather, A.Q. 1993. Mites injurious to viticulture in Jammu AND Kashmir and their control (P. 211-217). In : Mukherjee, A.B., Somchoudhury, A.K. and Sarkar, P.K. (eds.)., *Contributions to Acarological Researches in India,* Bidan Chandra Krishi Vidyala, West Bengal, pp. 445.

Sadana, G.L. 1985. *Plant feeding mites of India,* Kalyani Publishers, New Delhi, P.99.

Tseng Y.H. 1990. *A monograph of the mite family Tetranychidae (Acarina: Trombidiformes) from Taiwan.* Taiwan Museum Special Publication series 9. 224 pp.

Zhang Z.Q. 2003. Mites of greenhouses: Identification, Biology and Control. CABI Publishing, Wallingford: 244 pp.

❑❑❑

Ecologically Based Integrated Pest Management, 2012
© D.P. Abrol and Uma Shankar (eds.), pp. 681-710
New India Publishing Agency, New Delhi (India)
e-mail : info@nipabooks.com; website : www.nipabooks.com

35

Integrated Pest Management in Dry Land Horticultural Crops

SANJEEV RAI AND UMA SHANKAR

Introduction

India is bestowed with a wide variety of agro-climatic conditions and has unique comparative advantage for growing almost all horticultural crops (fruits and vegetables) in the country. Horticulture has remained as a major sub-sector of agriculture and forms an integral part of food and nutritional security, employment generation and enhancing the farm income. The growing economic importance of these crops could be attributed to increasing demands in domestic as well as in overseas market. The total area under fruit is 6.10 million hectare and total production is 68.47 million tonnes which accounts for 11 per cent of the world total fruit production (Anonymous, 2009).

Since, there is a limited scope to enhance the horticultural production in traditionally growing areas, emphasis has to be laid on harnessing the potential of vast wasteland and dry land spread over more than half of the geographical area of the country. Dry and rainfed area constitutes a major share of wasteland and untapped land resource where several indigenous horticultural crop species can be commercially exploited to strengthen the agro-based industries and providing the livelihood to the masses.

The arid regions are spread over about 38.7 million ha including both hot and cold arid regions mainly in the states of Rajasthan, Gujarat, Andhra Pradesh, Punjab, Haryana, Jammu and Kashmir and Himachal Pradesh (More and Singh, 2008). The Indian arid zone is characterized by high temperature/ low temperature and ranging from low to variable precipitation which limits the scope for high productivity, diversification and choice of crops in horticultural crops. Dry land area largely favours the development of several fruits like mango, citrus, guava, litchi, aonla, pomegranate, phalsa, ber, beal, karounda, lasora and peach.

In view of increasing costs of labour and inputs, farming has become less remunerative. Today, farmers are in search of new alternatives, especially when several incentives under Horticulture Mission and export promotion are provided by the Government. Farmers in the dry land are resource poor and mostly concentrated upon subsistence farming for their livelihood. Their primary requirements are food, fodder, fuel, fibre, timber, etc. in these areas. Therefore, proper cropping or farming systems should be developed to meet their requirements from the available land and water resources. Although, several advanced technologies have been developed, the extent of their diffusion and rate of adoption among farming communities are vary slow. Technology such as high density planting, micropropagation technique of planting materials, drip irrigation to enhance the water use efficiency and special technologies for problems areas have not been adopted by the growers. Thus, there is a need to strengthen the means of technology transfer and improve the lab to land programme.

Integrated Pest Management

A comprehensive IPM program has been worked out in many countries to test the knowledge of the grower to identify the important pests and natural enemies in the agri-horti-eco-system. Although, several advanced technologies have been developed but the extent of their diffusion and rate of adoption among farming communities are very slow. Technology such as high density planting, meadow plantings, micro-propagation technique of planting materials, drip irrigation to enhance the water use efficiency and special technologies for problems areas have not been adopted by the masses. Thus, there is a need to strengthen the means of technology transfer and improve the lab to land programme. IPM involves a combination of various measures to ensure effective pest management without disturbing the ecosystem, reducing environmental pollution and eliminate direct and indirect health hazards to human beings. These methods can be broadly divided into three categories: Observation methods, Preventive methods and Intervention methods.

Observation methods include identification of pests and beneficial insects, surveillance of insect pests in prevailing climatic condition, various types of trap catches observations for forecasting the possible pest or disease appearance, scouting the pests and beneficial insects in farm unit, Proper diagnosis and Economic threshold level determination. ETL for all major pests should be standardized and this scientific information should be disseminated among farmers. But, due to more complexity in determining the ET level, it has been replaced with Agro-Ecosystem Analysis in modern agriculture. Preventive methods comprises cultural practices like summer and pre monsoon ploughing, suitable crop rotation, incorporation of intercrops, use of resistant /tolerant varieties, proper water management by judicious irrigation while Intervention methods mainly focused on mechanical, biological and chemical control.

IPM developed as an umbrella to provide an effective, all encompassing, low-risk approach to protect resources and people from pests. Coordination of the use of pest biology, environmental information and available technology to prevent unacceptble levels of pest damage by the most economical means, while posing the least possible risk to people, property, resources and the environment became imminent. Thus, *Integrated Pest Management* (IPM) *is a dynamic and constantly evolving system in which all suitable control tactics and available surveillance and forecasting information are synthesized into a holistic management programme delivered to farmers at intervals as part of sustainable crop management technique".*

Current status of fruits

Agriculture is the most important and diversified industry of India which contributes about 18.5 per cent of the national income and gives employment to 52 per cent of its population (Anonymous, 2007). The horticulture sector is most vibrant sector for economic development which cover 18.98 million hectares i.e. 13.27 per cent of total cultivated area and contribute 28 per cent of the gross domestic product in agriculture (Mittal, 2007).

The major fruit crops grown in the Jammu & Kashmir state are apple, mango, walnut, almond, pear, ber, apricot, peach, plum, citrus, grapes, olive etc. The per unit land profits from horticultural crops is more than the cereals and these crops are labour intensive and help in generating additional income and employment opportunities through on-farm packing, processing and marketing of value added products. With increased environmental degradation and low water availability, the diversification to horticultural plantations is on increase. It is therefore essential to increase the area as well as the productivity of the fruit crops for providing nutritive and protective food in addition to improving farm income, enhancing the beauty of through landscape and conserving the ecosystem.

Changing pest scenario in fruits

In recent past, many new cultivars have been introduced to increase the production growth rate of fruits. These intensive, high yielding hybrids with good agronomic traits and yield attributes have gradually changed the pest situation. Consequently, a shift in pest status has been realized in time and space. Many pests have adapted new hosts, developed resistance to pesticide and often there are secondary out breaks. New pests like mango stone weevil showed its severity in mango in southern India are well documented. Earlier shoot gall maker and *Deudorix isocrates* of aonla known to be minor pest is gradually becoming a regular problem in aonla, whereas, fruit fly severity of occurrence and infestations have been intensified due to climatic change.

Current status of pest control in fruits

Among the various methods of pest control, chemical method still enjoys first choice because of its quick, certain and efficacious action. As the fruits are harvested at short intervals, chemical method of plant protection has become risky and hazardous and also contains pesticide residue. The persistent residue of major groups of insecticides has been detectable to considerable extent which hampers the export potential of the fruit market. India being the home of maximum fruit crops is a large reservoir of the natural enemies of pests and pollinators involved with them. But careful studies are required to use the predators/parasitoids/pllinators as a tool of pest management and increasing the fruit yield. Release of *Trichogramma* spp. (egg parasitoid) (Fig. 1), *Chrysoperla carnea* (predator) (Fig. 2), syrphid fly (Fig. 3) and coccinellid beetle grub and adults (Fig. 4 and 4a) has shown immense potential which has to be adopted through framing certain mandatory policy by the existing government. The feasibility of biocontrol agents in fruit is comparatively less because of the annual nature of crop with short growing period. However, use of bioagents is limited to some areas only. The components for successful Integrated Pest Management consists mainly of insect pests monitoring, integration of various methods, decision-making and implementation.

Non chemical approaches

Cultural practices

This component of Integrated Pest Management requires no additional budget for the farmers. Rather, these aspects are more sustainable and eco-safe. But it needs careful consideration with sufficient planning in advance. Crop husbandry of fruit crops is radically different to that of field crops and under many situations the pest problems are permanent nature. Several cultural practices if performed regularly in orchard ecosystem would be helpful in either elimination or atleast reducing the carryover of pest population. Routine agronomic practices like selection of varieties, intercultural operation like sanitation, tillage, pruning and thining and raking of the soil etc., trap cropping/intercropping should be suitably modified to make uncongenial condition for pests and to enhance the activity of natural enemies.

Keeping in mind the diversity and intensity of pests in particular place, selection of resistant/less susceptible varieties, holds well in pest management. Plant resistance provides a built in ability to allow less pest and cut off the extra load of insecticides in the crop. Being completely safe, host plant resistance fits well with all other components. Unlike cereals, pulses, oilseeds and vegetables in fruits very less number of resistant/less susceptible varieties have been developed Table 1.

Table 1 : Resistant/tolerant varities of dryland fruit crops against different insect and mite pests in India

Fruit crop	Resistant/tolerant stock	Insect/mite pest
Citrus	Redblush, Foster, Marsh seedless, Fallglo, Nova mandarin, Star ruby	Whitefly, *Dialeurodes citri* Ashmed
	Cleopatra, Rubidoux, Orange Michal, Coorgcitron, Deshndo, Nagpur mandarin, Gal-gal, Kagzi lime	Psylla, *Diaphorina citri* Kuwayama
	Jatti Khatti, Citrumelo, Carrizo, Troyer, Orange-Michal, Coorg citron, Deshndo, Eureka lemon, Savage, Sacaton, Rangatra L.J.	Leaf-miner, *Phyllonistis citrella* Stainton
	Citron, Redblush, Marsh seedless	Lemon caterpillar, *Papilio demoleus* L.
	Nagpur mandarin, Ikeda Unshiu, Long sportvalencia, Campbel valencia, Washington Naval	Aphids, (Black) *Toxoptera aurantii* Boyer de Fonscombe (Green) *Aphis gossypii* Glover
	Washington Naval, Hazara, Jenru tenga	Scales, *Aonidiella aurantii* Maskell *Coccus hesperidium* L.
	Chakotra, Blood red, Kagzi lime, Malta, Mosambi, Kinnow, Galgal, Sweet lime, Etrong citron, Coorg, Wilking, Sylhet, Valencia champman	Mites, *Eutetranychus orientalis* Klein *Panonychus citri* Mc Greger

Guava	Nasik, China Surkha, Behat Coconut, Pear shaped, Red flesh, Smooth Green hybrid	Fruit fly, *Bactrocera. zonatus* Saunders, *Bactocera dorsalis* Hendel
	Red flush	Shoot borer, *Indarbela tetraonis Moore*
	Bangalore Round, Bapatala, AC 10, Seedless	Tea mosquito bug, *Helopeltis antonii* Signoret
Mango	Pulhora, Kala Hapus, Keshar Basti, Annanas, Baneshan Bangalora, Chinnarasam and Khander	Leaf Hoppers, *Amritodes atkinsoni* Lethierry
	Annanas, Anain, Delicious, Gulabkhas, KO7, KO11, Maharaja of Mysore, Salem, Banglora, Vellakachi	Gall insect, *Procontarinia matteriana* Kieff. & Cecec
	Toranjo, Monteiro, Manjurad	Fruit fly, *Bactocera dorsalis* Hendel
	Deshi Malgoba, Alam Baneshan	Leaf gall midge, *Dasyneura mangiferae* Felt.
	Mohandas, Malgoa, Pulliadi	Mites, *Aceria mangiferae* Sayed

(*Source :* Sharma, 2006)

Tillage

Many insects like mango mealy bug, gall makers, leaf-eating weevil, litchi nut borer, fruit fly of guava, pear and peach, *Bactrocera dorsalis, B. zonatus* and fruit fly of ber, *Carpomyia vesuviana* are found to be controlled by ploughing the soil during May-June as the gorged female lay eggs in the soil. The eggs are exposed to either intense heat or for predation/parasitization of birds and natural enemies. Similarly, the pre- pupal and pupal populations of *Bactrocera dorsalis* occurred in the soil and shallow ploughing with cultivator.

Pruning

Pruning the citrus plants in December-January will effectively check the incidence of citrus leaf minor, *Phyllocnistis citrella.* Pruning of branches in mango reduces the prevalance of leaf webber and gall insects.

Sanitation (collection and destruction)

These practices generally involve the removal and destruction of pests mostly by using manual labour. At small scale, it is very effective and profitable especially homestead fruit growers. Collection and destruction of insect stages is a predominating and effective method in checking the out break of many pest e.g. hand picking of lemon butterfly caterpillar, *Papilio demolius* Linn., brushing of scales of mango, dropped guava, ber, pear, peach etc., fruits infested by fruit fly maggot *Bactrocera* spp., dropped mango leaves mined by grubs of leaf cutting weevil, mummies of litchi for litchi nut borer. Removal and destruction of infestedplant parts is also effective in regulating the pest populations of citrus leaf minor, mango stem borer, *Bactrocera rufomaculata* and guava (*Microcolona technographa* Meyrick). Regular removal and burying of fallen fruits at leastat 60 cm depth has been proved to be very effective against fruit fly, *Bactrocera dorsalis* in guava, peach and pear orchards.

Mechanical practices

Exclusion through screens and barriers

Application of fluppy band of 15cm wide or a band of a folded slippery sheet like alkathene around the trunk of mango against mealy bug (Fig. 5). Wrapping individual fruits of pomegrannate and citrus with butter paper envelops to save them from the attack of anar butterfly and fruit sucking moths. Covering the fruits with polythene bags for controlling fruit flies in case of guava and mango, squirrels in guava.

Traps

Different types of traps is also recommended for the monitoring, reducing the pest population and will also help in application of in the orchard. Bait trap Banana or other over ripe waste fruit pulp (1kg) + yeast (10g) + citric acid (5g) @ 100g/plastic made flat tray (Fig. 6). Monitoring may also be made with methyl eugenol (2ml) + malathion (2ml) + water (1lit) by keeping in a jar @ 500ml per jar and No mate® cu-lure block (Fig. 7) for the fruit fly. A circular whole should be made in the jar to facilitate the entry of fruit fly. Pheromone trap and light traps is also effective for the many lepidopterous insects.

Preparation procedure for bait traps

The plastic tray should be covered with another inverted tray of the same type. The gap between the rims of the both the trays should be kept 4 cm to avoid the problem of bird poisoning and deposition of dust and the rain water.

Jarring

Placing cut pieces of banana pseudostem to attract banana pseudostem weevil, *Odoiporus longicollis*. Installing jam jars suitably baited with sugar, yeast and fruit juice and sunk to the rim in soil to trap crickets or hung on tree to trap fruit sucking moths.

Natural enemies

Fruit ecosystem is endowed with a large number of natural enemies consisting of predators, parasites and parasitoids causing damage to different stages of the insect to varying extent. Till now except a few, most of these natural enemies are under exploitation for pest management in vegetables.

Among the egg parasites *Trichogramma chilonis* Nagraja has been utilized to check the anar butter fly @ 2,50,000 parasitised eggs/ha at 10 days interval reduce the fruit borer infestation. Release of *Cryptolaemus montrouzeri* @ 20-30/tree gives excellent control of *Ferrisia virgata* on guava. *Verticillium lecanii* @ 1x10^8 spore/ml against mango leaf hopper. In citrus release of *Cotesia nigita* @ 10-15/tree and *Aphytis melinus* De Bach @ 2000/tree for red scale. *Verticillium lecanii* Zimmemlan @ 1x10^8 spore/ml for green scale. Another predator, which needs immediate attention, is Coccinellds, syrphid fly and *Chrysoperla carnea*. These are effective predators for the controlling of soft bodied insects i.e. aphid, jassid and eggs as well as early instar larvae of some lepidopterous borers. Beneficial spiders comprise the majority of species in healthy ecosystems. They are the generalists predatprs which feeds on a variety of insects ranging from small to large (Fig. 8). Due to the perennial and undisturbed nature of fruit crops, many species of spiders are commonly found predating upon the pest species in fruit crops. Mass rearing of these predator needs to be streamlined for economising its use as biocontrol agent. Development of efficient and economic rearing technique of these bioagents will be the first step towards successful biological control. Besides, improvement of field release technique to maintain optimum potential is most important. The strains having potency to survive in toxic environment will fit well into the Integrated Pest Management system. Careful consideration of control methods to maintain and enhance the resident natural enemies is the most useful way, if the area is endowed with sufficient population of particular natural enemy.

Table 2 : Natural enemies of insect pests of some important fruit crops

Bioagents	Pest species
Predators	
Rodolia cardinalis	Cottony cushion scale (*Icerya purchasi* Maskell) Citrus mealy bug (*Planococcus citri* Risso)
Cryptolaemus montrouzieri	Mealy bug (*Ferrisia virgata* Cockerell), *Planococcus citri, Maconellicocus hirsutus* Green)
Mallada boninensis	Citrus blackfly (*Aleurocanthus woglumi* Ashby), Mango hopper (*Amritodes atkinsoni* Lethierry)
Cryptolaemus nigritus	California red scale (*Aonidiella aurantii* Maskell)
Parasitoids	
Tetrastichus radiatus	Citrus psylla (*Diaphorina citri*)
Leptomastix dactylopi	Mealy bugs (*Planococcus citri*)
Fungus	
Verticillium laccani	Mango hoppers (*Amritodes atkinsoni*)

(*Source :* Singh *et al.*, 2001)

Microbial control

Through this method, entomopathogenic bacteria, fungi and viruses are utilized for control of insect pests. These disease-causing microorganisms are produced in the laboratory and sprayed in the field to spread the disease in the host insect. Microbial pesticides as a component of Integrated Pest Management are cheaper than pesticides. Besides, they are ecofriendly and compatible with natural enemies existing in the ecosystem. Fruit pest control through microbial intervention is so far limited to few pests only.

Considering the wide spread use of chemicals and consequent residue problem in fruits, biological control of fruit pests became more viable. Large-scale cultural operations, short cropping period and chemical intervention did not allow the natural enemies to grow to their potential. Thus conservation and periodical release of natural enemies in fruit ecosystem should be taken up. Besides host specific insect viruses and other entomopathogenic microorganisms when available in epizootic form or used through artificial application to the target site, provides good control of several pests of fruits. Some important parasitoids, microbial agents recommended against fruit pests are given in Table 2 (Singh *et al.,* 2001).

Judicious selection of chemicals

Due to the easy accessibility and sure action against the target organism, chemical control still remains first choice of farmers. There is large number of reports of successful chemical control of fruit pests. But various ill effecrs by the insecticides like, insecticide resistance, secondary pest outbreak, pest resurgence and killing of non target organisms etc. Have now threatens the widespread use of synthetic insecticides. However, these insecticides can be judiciously used considering certain aspect like waiting period of chemical, economic threshold level and initiation of control measures through bait trap, pheromone/light trap catches.

Choice of chemical

Residue of major group of insecticides has been detected to considerable extent in fruit ecosystem. To avoid the persistent residues, choice of chemical plays a crucial role. Considering the picking interval and waiting period, the insecticides should be chosen. Fruit with less picking interval should be sprayed with insecticide having less waiting period. Chemicals like plant origin insecticides, growth regulators attractants and even synthetic pyrethroids can fit well in the Integrated Pest Management. Botanicals being non persistent and safe to mammals possess in depth promise in pest management. Neem products have shown efficacy against insect like jassids, aphids and leaf miners. Application of crude neem seed kernel extract (4%) controls leaf hopper and mealy bug in mango.

Economic threshold level (ETL)

Instead of going for routine schedule application of insecticide, need based spraying on the basis of economic threshold level will help to reduce pesticide consumption and environmental abuse. ETL based application schedule helps maintaining minimum pest residue for survival of natural enemies. The information in literature showed that little efforts are made on determining the ETL on fruit crops and there is need to develop the strategies to avoid the insecticides residues on fruits. Moreover, to obtain high efficacy of insecticides on fruit trees the following factors must be taken into consideration:

- Right stage, site and activity period of the pest
- Proper selection of insecticide, dosage and application equipment.
- Coverage of the tree canopy.

Table 3 : Economic Threshold Level of different insect & mite pest

Insect/mites	Economic threshold levels
Green citrus aphid, *Aphis spiraecola* Patch	5-10% infested shoots
Melon aphid, *Aphis gossypii* Glover	25% infested shoots
Black citrus aphid, *Toxoptera aurantii* Boyer de Fonscolombe	25% infested shoots
Citrus whiteflies, *Dialeurodes citri* Ashmed	5-10 nymphs/leaf on mandarin
Citrus blackflies, *Aleurocanthus woglumi* Ashby	First colonies occurrence
Citrus mealy bug, *Planococcus citri* Risso	5-10% infested fruits
California red scale, *Aonidiella aurantii* Maskell	3-5 nymphs/leaf
Citrus flower moth, *Prays citri* Millière	50% infested flowers
Mediterranean fruit fly, *Ceratitis capitata* Wiedemann	20 adults/trap/week
Citrus bud mite, *Eriophyes sheldoni* Ewing	50-70% infested buds
Pink citrus rust mite, *Aculops pelekassi* Keifer	2-3% infested fruits
Citrus red mite, *Panonychus citri* Mc Gregor	3 specimens/leaf or 50% infested leaves
Two-spotted spider mite, *Tetranychus urticae* Koch	2% infested fruits or 10% infested leaves

(*Source :* DPPQS, 2001)

Selection of safe chemicals

Chemical insecticides cannot be out rightly rejected from fruit pest control schedule. But selection of insecticides, which are comparatively safe to the insect natural enemies, should be taken into consideration. Application of these components will maintain the insect residue, which will keep the pest population below economic threshold level.

Mango (*Mangifera indica*), is the important fruit crops of India and grown almost every state. Mango is a tropical fruit and can be grown on a wide variety of soils and can withstand drought admirably under varied climatic conditions. Although, India is the largest mango producing country, accounting about 60 per cent of world production, the export of fresh fruit is limited to Alphonso and Dashahari varieties. India's share in the world mango market is about 15 per cent. A comprehensive list of pests of mango has also been documented by Tandon and Verghese (1985) and Veeresh (1989). Lists of mango pests with special reference to details of life histories and control of mango pests are also discussed by Golez, 1991 and Murrey, 1991. Since last three years, bark aeting caterpillar, mango hoppers, mango mealy bug and fruit fly pest complex are the common phenomenon on fruit crops in dry land of Jammu province and inflicting significant economical loss.

Mango hoppers

Fifteen species have been reported on mango from Asia. However, only 3 species are serious pests such as *Amritodus atkinsoni, Idioscopus clypealis* and *Idioscopus niveosparsus.* The mango leaf hoppers (Fig. 9) are small wedge shaped insects which continuously suck the saps from the inflorescence thereby reducing the vigour of the plants and particularly destroying the inflorescence and causing fruit drop. Heavy puncturing and continuous draining of the sap cause curling and drying of the infested tissue. They also damage the crop by secreting a sweet sticky substance which encourages the development of sooty mould *Maliola mangiferae* and *Capnodium mangiferae* on leaves and inflorescence and thereby adversely affecting the photosynthetic activities of the leaves. (Kaul *et al.*, 2009). A low population of hoppers has been recorded in mango orchards throughout the year but it shoots up during February-April and June-August. High humidity and rainfall conditions during blossoming are congenial for their rapid development and multiplication. The extent of damage depends upon the critical crop stage and hopper population. Cloudy weathers and precipitations are conducive conditions which favours the pest population build up at the time of blooming. Temperature and relative humidity constitute important environmental factors regulating the population of the leaf hoppers (Tandon *et al.*, 1983).

Management

1. Conserve the natural enemies like coccinellids (Fig. 9a), chrysopids and spiders as they are potential predators of nymphs of hoppers .
2. Fungus pathogen, *Beauveria bassiana* are the useful bioagents to control this pest.
3. In senile and neglected orchards, dense tree canopy should be pruned heavily to have better light penetration.
4. Three sprays of insecticides at critical crop stages, namely, starting of bud burst stage, panicle emergence and after fruit set (pea stage) are important to avoid hopper infestation. Spray cypermethrin or fenvalerate (1ml per lit of water) followed by imidacloprid (0.5 ml per lit of water) at fortnightly interval at evening hours. Do not spray when trees are in full bloom to avoid killing of pollinators.

Mango fruit fly

Fifty two species are placed in the *Bactrocera dorsalis* complex in Asia and eight of which are considered of economic importance (Drew and Hancock, 1994). Fruit flies are polyphagus in nature and their extent of damage depends upon type, size and condition of fruit, availability of different hosts and population density. Most of the mango orchards are located in fruit fly infested areas, and growers suffer significant direct and indirect economic losses resulting from fruit fly damage (Singh, 1991; Fig. 10). Seasonal incidence of fruit fly has also been documented on subtropical fruits like mango, guava (Fig. 11), citrus (Fig. 12), phalsa and ber under rainfed conditions of Jammu (Rai *et al.*, 2008). Female fruit fly lays eggs below the fruit epidermis (1-4mm) deep. On hatching, the maggots feed on fruit pulp and the infested fruit starts rotting due to further bacterial infection. Attacked fruits usually show signs of oviposition punctures and ripe fruits with high sugar content exude a sugary liquid. The maggots come out of the affected fruit to pupate in soil. Due to climate change and adoptability of insects the unripe and immature fruits are also claimed to be damaged by the fruit fly. On the average, 36-40% fruits of mango have been observed to be damaged by *B. dorsalis* and *B. zonata* (Mohyuddin and Mahmood, 1993). Observations recorded on varietal resistance revealed that most of the commercial varieties are susceptible, however, infestation is quite low in Langra, Dashehari and Bombay Green varieties (Jothi *et al.*, 1994).

Management

1. Fourty five days prior to harvest, deep plough the soil around trees to expose and destroy the fallen and pupating fly populations.
2. Although trap devices are not sufficient enough to suppress the fruit fly population, install sixteen methyl eugenol bottle traps (0.1%) per ha or

one Flig-T trap (Fig. 13) per tree (PCI Pvt. Ltd.) for catching the male fly to disrupt the mass mating for further multiplication of pfly opulation.

3. One months before harvest, spray decamethrin (0.5 ml/lit) or malathion (2ml/lit) along with gur solution and sticker.
4. Traet the harvested fruit with hot water for 1 hour at 48°C.
5. Early harvesting of mature fruits to avoid fruit fly infestation and collect fruit fly infested and dropped fruits and destroy them.

Mango mealy bug

Mango mealy bugs (*Drosicha mangiferae, Drosicha stebbingi* and *Rastrococcus iceryoides*), a polyphagous pests in India (Fig. 14) are recorded as serious pests from Asia on several host crops (Tandon and Verghese, 1985). Nymphs and adult female bugs are flat, oval and covered with waxy white powder which makes them hard to control. Generally eggs are laid in silken pouch in ending May to June, which hatches out during ending December to starting January. The newly hatched nymphs start crawling on trees and settle on inflorescence causing flower drop and affecting fruit set.

Management

1. Deep ploughing of orchard immediately after harvest or during summer months to expose eggs and pupae of mealy bugs to natural enemies and sun heat. Heavy irrigation of orchard in October also helps in destruction of eggs of mealy bugs.
2. Raking of soil around the tree trunk and mixing with methyl parathion 2% dust @ 250 g per tree or drench the tree basin with imidacloprid @ 0.5 ml per lit of water for controlling early instar nymphs of mealy bugs in the month of November–December.
3. Release *Cryptolaemus montrouzieri* at 10 beetles/plant (Tandon, 1995). Mealybug destructors are observed devouring the mealybugs in Jammu conditions (Fig. 15).
4. *Menochilus sexmaculatus, Rodolia fumida* and *Sumnius renardi* are important predators in controlling the nymphs. The entomogenous fungus *Beauveria bassiana* is found to be an effective bioagent in controlling the nymphs of the mealy bug.
5. After mud plastering 25 cm wide, 400 gauge alkathene (polythene) sheet should be fastened to the tree trunk about 30 cm above the ground level (See Fig. 5) to prevent migration of freshly hatched nymphs of mealy bugs in the month of December-January (Kaul *et al.*, 2009).
6. Put a heap of fine sand around the tree trunk so that the mealybug crawlers cannot climb on the trees (Fig. 16).

Bark eating caterpillar

This pest is found damaging a variety of plants including a number of fruit trees, forest trees and ornamentals throuout the country. This is a serious problem in fruit orchards like mango, citrus, litchi, guava, aonla, jackfruit and ber. The old and shady orchards are more prone to attack by this pest. Larvae of this moth feed on the bark and bore into the main trunk which eventually weakening the tree. The moth is light grey in colour with dark brown dots. Eggs are laid by female in batches on cuts and crevices in the bark. On hatching the larvae bore into the bark or main stem and remain within the bored holes during day and come out at night to feed on bark. They feed and shelter under the silken galleries/tunnels full of frass and faecal matters on the bark surface. The tunnels are made by the larvae over the bark, with pieces of bark, excrements and silken thread. The larvae regularly maintain the tunnels (i.e. active tunnels) with fresh materials and when they die or remain inactive during the pupal stage, the tunnels get detached from the tree trunk, due to lack of maintenance. The absence of fresh bark feeding, followed by detachment of the larval tunnels are indications that the larva has pupated. The pupation will be followed by moth emergence, during which the pupa wriggles out to the mouth of the borer hole. When the moth emerges out, the "pupal skin" remains at the mouth of the borer holes and the presence of projecting "pupal skin" (Fig. 17) is considered as evidence for moth emergence (Sasidharan and Varma, 2008). This problem is often noticed in neglected and senile orchards where yield and quality of fruits have been adversely affected.

Management

1. Orchard sanitation should be maintained.
2. As soon as noticing the infestation, kill them mechanically by inserting the iron spike into hole.
3. In case of severe infestation, insert the cotton swab soaked in dichlorovos @ 2ml/lit or emulsion of Monocrotophos (0.05%) in each hole and plug the hole with mud plaster.
4. Cleaning of the infested area and plugging holes with petrol, nuvacron or formaline soaked material is advocated for the control of bark eating caterpillar (Singh and Singh, 2002).

Citrus

The production of sweet orange, lime and lemon is largely favoured by dry, semi-arid to subtropical conditions. Kinnow appears to be very exacting in its climatic requirements and requires warm cool temperature with a chilling temperature during winter for good cropping and high quality fruits. Citrus crops are influenced by wide array of insect and mite pests, which can cause serious quantitative and qualitative losses. More than 300 species of insects and mites have been recorded on different *Citrus* spp. from Asia and other countries. However, the key pests are leafminer, psylla, mealybugs, scales, blackfly, whiteflies, aphids, fruitfly, fruit sucking moths, mites and thrips. In Jammu condition, six insect pest viz., citrus psylla, leaf miner, citrus caterpillar, whiteflies, mealybug, and aphids have been recorded as the major pests of citrus (Kumar, 2008).

Citrus Psylla

The citrus psylla is widely distributed in Asia that's why it is also being called Asian psyllids. Adults are grey coloured actively flying insects and while at rest, they raise their body upward. The nymphs are orange yellow in colour, flattened and circular in shape. The damage is caused by the nymphs and adults who suck sap from buds and leaves (Fig. 18 and 18a). The affected leaves get curled and shoots become dry. The psyllid also acts as a vector of greening disease. There are no systematic data available on extent of damage, however, citrus psylla has been reported causing loss to mandarin to the tune of Rs. 40 million (about US $ 1.04 million) alone in Vidharva region of Maharashtra in India.

Management

1. Several species of predators and syrphids have been reported feeding on eggs and nymphs of citrus psylla.
2. At the initiation of new flush, spray monocrotophos (1.5 ml per lit) or dimethoate (3ml per lit) or Imidacloprid (0.5 ml per lit). If required, repeat the spray at 15 days interval, once or twice (Tandon, 1991).

Citrus Leafminer

Citrus leafminer, *Phyllocnistis citrella* Stainton adult is a tiny silvery white moth with black eyes and narrow fringed white hind wings. Eggs are minute, rounded and yellowish green. The caterpillars are legless and pale yellow in colour with brownish head. The larvae feed on the epidermis of tender leaves making serpentine mines of silvery colour (Fig. 19). Severely infested leaves become distorted, crumpled and finally fall off. Attack of leafminer encourages the incidence of canker during rainy season (Fig. 20).

Management

1. For the effective management of citrus leafminer, clipping of infested leaves and their pruning is advised.
2. Commencement of new flushes should be sprayed with fenvalerate (0.05%) and neem seed extract (2%), alternatively, at 10-12 days interval (Jothi *et al.*, 1990a, b). Release of parasitoids *C. quadristriatus* and *T. phyllocnistoides* is also recommended.

Citrus caterpillar

Two species of citrus caterpillar butterfly namely *Papilio demoleus* and *P. Polytes* have been recorded in Jammu causing complete defoliation of citrus (Kumar, 2008). The pests are active throughout the year except winter months. The infestation is more pronounced in nurseries and on young trees (Fig. 21). Peak pest activity coincides with the new flushes in citrus. This pest has also been found defoliating the beal fruit (personnel observation). The list of citrus insect pest succession with special reference of management of citrus caterpillar in Jammu region have been portrayed by (Kumar, 2008).

Management

1. Collect and destroy the larvae which look like bird excreta in early stage.
2. Spray endosulphan @ 2ml per lit or carbaryl @ 2g per lit of water.

Guava

Guava is very hardy in nature and can grown successfully in tropical and subtropical regions up to 1,500m above MSL. Out of 80 species of insect pests recorded on guava, only few of them are considered as pest of regular occurrence and causing serious damage to the guava trees. These are bark eating caterpillar (*Indarbela* spp.), fruit fly (*Bactrocera* spp.), fruit borer, guava aphids, mealy bug and scale insect (*Chloropulvinaria psidii*). Intensive surveys of guava growing regions of Indian subcontinent revealed that guava fruit borer, *Deudorix isocrates* (Fabricius) incidence is rapidly increasing (2.5–22.5%) with crop loss range from 5.00 to 35.00 per cent. Common occurrence of another fruit borer, *Dichocrocis punctiferalis* (Fig. 22) in rainy season guava was also noticed throughout the guava growing belts in the country. Four other species of fruit borers, viz., *Cryptophleba illepida, Rapala varuna, Deudorix isocrates* (Fig. 23) and *D. epijarbas* have been reported damaging the Guava fruits (Kaul and Kesar, 2003). Guava fruit flies have been a major limiting factor in production of rainy season guava. Infestation of fruit fly ranged from 20.00 to 46.00 per cent with crop loss of 16.00-40.00 per cent, which is a matter of serious concern.

Management

1. Lucknow 49 (round big) is the most resistant cultivars to fruit damage by *Bactrocera spp.*
2. Release of *Cryptolaemus montrouzieri* @ 20 adult beetles/tree is found effective against *C. psidii* in farmer's field (Mani and Krishnamoorthy, 1990). *Cryptolaemus montrouzieri* has controlled the pest effectively within 30 days of its release (Mani *et al.*, 1989).

3. For increasing the productivity in existing guava orchards it is required to adopt proper canopy and ground management practices, adequate nutrient application, good irrigation management, use of suitable rootstocks and effective and timely control of pest and diseases through eco-friendly approaches (Kumar *et al.*, 2007).

Litchi

Growth and production of litchi is hampered by the attack of 50 species of insect and mite pests. These include bark eating caterpillar, fruitfly, fruit borers, eriophyid mite, hoppers, leaf rollers and scales. It is a delicious fruit and generally considered to be free from serious insect pests and does not even require the regular control measures. However, it is being attacked by litchi stalk borer/litchi fruit borer (Fig.) and birds in Jammu condition. If nor regulated well in time they will inflict economical damage to the fruits.

Litchi Fruit Borers

Four species of fruit borers, viz., *Cryptophleba illepida, Rapala varuna, Deudorix isocrates* and *D. epijarbas* have been reported damaging the fruits. *C. illepida* causes extensive damage ranging from 40-60%. The caterpillars bore into the developing fruits and feed on seeds. The bore hole gets filled with the excreta of the caterpillar and infested fruits start rotting.

Management

1. Collect all fallen infested fruits and destroy them.
2. At early stage of fruiting which coincides with egg laying, spray monocrotophos (2 ml per lit) or cypermethrin (1ml/lit of water). Repeat twice at 10-12 days interval.
3. Bark eating caterpillar can be managed as suggested for mango.

Birds menace

Birds menace is quite high in fruit tree cultivation. Several birds were found infesting the fruits of litchi crop such as parrots, bulbul and crows (Fig. 24).

Management

1. Birds damage can be reduced by using local made sound devices (Fig. 24 a) or using some ribbon reflectors or cassette tapes reel.
2. Bird scarer and reflecting ribbons can be used to keep away the bird's damage.

Aonla

Aonla or Indian goose berry (*Emblica officinalis* Gaertn.), king of arid fruits, can be best grown in various kinds of wastelands, dry lands and survives well in adverse agro-climatic conditions. It is a rich source of vitamin C and has gained prominence as a crop of economic importance because of its therapeutic properties. Its cultivation is increasing fast and the crop is becoming popular in Jammu and Kashmir where the moisture availability is very low. Increasing health consciousness among people as well as growing popularity of alternate medicine and herbal products is also enhancing the requirement of aonla both in domestic and international markets.

The major constraint in aonla fruit production is ravages caused by the insects, diseases and some physiological problems throughout the world. Among the insect pests, the aonla shoot gall maker, leaf rolling caterpillar, bark eating caterpillar, mealy bug have been reported to be of major importance (Chadha, 2003). Shankar *et al.* (2007) also recorded shoot gall maker, *Betousa stylophora* Swinhoe (Fig. 25); bark eating caterpillar, *Indarbela tetraonis* Moore; leaf rolling caterpillar, *Garcillaria acidula* Forster (Fig. 26); aphid, *Cerciaphis emblica* Patel and Kulkarni (Fig. 27) and anar fruit borer, *Deudorix isocrates* Fabricius in Jammu province. Certain new insect pests have also been encountered on aonla during 2009-10 such as mealy bug, stick insects, aonla looper etc. which has yet to be identified.

Management

1. Clean culture, ploughing and regular elimination of weeds around the orchard creates unfavourable condition for insect pest development. Weeds nearby the orchard and on bunds should be cleaned regularly.
2. An eulophidae parasitoid are found parasitizing the larval-pupal stages of leaf rolling caterpillar (Fig. 28).
3. Use calendar based spray after dormancy in aonla: Alternat use of systematic insecticide (Imidacloprid, dimethoate, monocrophos, thiomethoxam, metasystox etc.) followed after 15 days by contact and stomach insecticides (endosulpfan, cypermethrin, fenvalerate etc.) to control pest complexes on aonla.

Pomegranate

Pomegranate (*Punica granatum*) is a favourite table fruit in tropical and sub-tropical regions of the world and grows well under semi-arid conditions. It thrives best under hot dry summer and cold winter provided irrigation facilities are available. The tree requires hot and dry climate during fruit development and ripening.

Pomegranate Butterfly, *Deudorix isocrates Fabricius*

The larvae bore into the fruit can destroy up to the 50 % of the fruits. The female butterfly lay egg singly on the calyx of flowers or small fruits. On hatching, the larvae bore inside the developing fruits and feeds on aril and seeds (Fig. 29). The entry hole allows secondary infection of fungi and bacteria to cause rotting of the fruits and rendering them unfit for consumption. The affected fruits ultimately fall down.

Management

1. Removal and destruction of fallen and affected fruits with exit hole.
2. Remove flowering weeds in and around the anar crops.
3. Release of *Trichogramma embryophagam* @ 2.5 lakhs per ha four times at 10 days interval.
4. Bagging of fruits with butter paper gives good protection.
5. Spray decamethrin @ 1ml per lit of water at the time of 50 % fruit set. Repeat after 15 days with fenvalerate 0.5 ml per lit of water during non-rainy season.

Pomegranate Thrips

Thrips are minute insects having piercing and sucking type mouth parts. They suck the sap from tender leaves and flowers causing leaf tip or margin curl and shedding of lowers. When they feed on tender fruit causing scab like symptoms and reduce the market value of fruits.

Management

1. Sparying of dimethoate @ 2ml per lit or imidacloprid @ 0.3 ml per lit of water before flowering. If incidence is high enough, Sparying of dimethoate @ 2ml per lit or imidacloprid @ 0.3 ml per lit after fruit set.

Peach

Peach crops have been found infesting by peach leaf curl aphid and peach fruit fly as major insect pests in Jammu.

Peach leaf curl aphid

It is a serious pest of peach which indirectly reduces the yield and quality of produce. Nymphs and adults suck the sap from all tender parts especially leaves of the plant. Affected leaves turn pale and curl up, blossom withers and fruits do not develop and drop prematurely (Fig. 30 and 30a).

Management

1. Conserve and encourage the coccinellids beetle to suppress the aphid populations.
2. Spray the trees with metasystox @ 1ml per lit or imidacloprid @ 0.3 ml per lit of water at the time of leaf emergence.

Peach fruit fly

Adult females lay eggs in small batches of 2-10 inside the ripening fruits by making punctures with their ovipositors (egg laying apparatus) (Fig. 31). On hatching, the maggots feed on the pulp and fruit becomes soft, ferments and drops. The attack is more serious on late maturing varieties. Full-grown maggots come out of the infested fruits and jump into soil for pupation.

Management

1. Fallen fruits should be collected and burnt to maintain sanitation in peach orchard. Summer ploughing just after harvest is essential to expose the hibernating pupae to sun and for parasitisation.
2. Methyl eugenol or Bait traps should be used to suppress the fly pest population (Fig. 32).
3. Flooding and drenching the orchard with fipronil through irrigation water is also beneficial to check the pupating population of fruit fly.

Ber

Ber, one of the ancient and common fruits of India is often called poor man's fruit. It grows under varying conditions of climate all over India and can withstand extremely hot conditions but is susceptible to frost. It is extremely drought hardy owing to its deep root system and other xerophytic characters. The tree prefers atmospheric dryness for development of good quality fruits. The fruit fly (*Carpomyia vesuviana*) is a major pest of ber and causes serious damage to ber fruits.

Management

1. Collection and destruction of fallen fruit and digging the soil under the tree canopy also decrease their infestation.
2. A potential biocontrol agents, *Dichasmimorpha* sp. have been identified for the first time from Jammu and Kashmir in suppressing the fruit fly maggots population in the field condition (Fig. 33).
3. The chemical spray schedule consists of first spray with 0.03% monocrotophos at pea stage, second spray with 0.05% fenthion after 15 days and third spray with 0.1% carbaryl after another 15 days may be beneficial in checking the fruit fly infestation.

Phalsa

Phalsa (*Grewia subinaequalis*), is a hardy plant and can be grown and withstand drought. It requires distinct winter and summer for better yield and fruit quality. Fruit fly and mealybugs are the major insect pest recorded in dry land of Jammu province.

Management

A suggested in mango fruit fly and mealybug control.

Beal

Owing to hardy nature, beal (*Aegle marmelos*) tree has a wide adaptability to adverse soil and climatic conditions. It requires subtropical climate where summer is hot and dry, and winter is mild. Beal is also free from economic insect pets infestation except bark eating caterpillar (*Inderbela tetraonis*) and leaf defoliator (*Papilio demolius* and *P. polystes*). They are causing enormous damage in economical production of beal.

Management

As suggested in aonla.

Jamun *(Syzygium cuminii)*

Since Jamun is a hardy fruit crop, it can thrives well under both tropical and subtropical climates and requires dry weather at the time of flowering and fruit setting. Dry temperate areas suitable for apple cultivation in Himachal Pradesh and Jammu and Kashmir are most suitable for production of high-quality fruits having intense colour development, high sugar content and longer shelf-life. Comparatively, Jamun is free from serious fruit pests however bark eating caterpillar, thrips and whiteflies are found infesting the Jamun trees.

Management

As suggested in aonla.

Acknowledgement

This chapter is dedicated to **Late Dr. Sanjeev Rai** who conceived this concept but could not see the real impressions in the publication.

References

Anonymous, 2007. *Economic Survey 2007-2008*. II Edition. Directorate of Economics and Statistics, Planning and Development Department, Jammu and Kashmir Government.

Anonymous, 2009. Indian Horticulture Database- 2009. Ministry of Agriculture, Government of India, 85, Institutional Area, Sector-18, Gurgaon-122 015, India. *www.nhb.gov.in*

Chadha, K.L. 2003. Handbook of Horticulture, ICAR Publication, New Delhi, pp. 747.

DPPQS, 2001. *Integrated Pest Management Package,* Directorate of Plant Protection, Quarantine and Storage, Government of India. *(IPM Package No: 28, 31, 32, 34*)

Drew, R.A.I. and Hancock, D.L. 1994. The *Bactrocera dorsalis* complex of fruitflies (Diptera:Tephritidae:Dacinae) in Asia. *Bulletin of Entomological Research,* Supplement Series No. 2:68.

Golez, H.G. 1991. Bionomics and control of the mango seed borer, *Noorda albizonalis* Hampson (Pyralidae: Lepidoptera). *Acta Hortic.* 291:418-424.

Jothi, D.B., Tandon, P.L. and Verghese, A. 1994. Hot water immersion as a quarantine treatment for Indian mangoes infested with the oriental fruitfly, *Bactrocera dorsalis* (Hendel) (Diptera:Tephritidae). *FAO Plant Protection Bulletin* 42(3): 158–159.

Jothi, D.B., Tandon, P.L. and Verghese, A. 1990a. Evaluation of different plant oils and extracts against citrus leafminer, *Phyllocrustis citrella* Stainton. In *Proceedings of National Symposium on Problems and Prospects of Botanical Pesticides in IPM,* Rajamundry (A.P.), India, 21-22 January 1990.

Jothi, D.B., Tandon, P.L. and Verghese, A. 1990b. Evaluation of different plant oils and extracts against citrus aphid, *Toxoptera citricidus* (Kirk.). *Indian J. Plant Protection* 18:251-254.

Kaul, V. and Kesar, Y.K. 2003. Incidence and management of lepidopteran fruit borers of guava (*Psidium guajava* L.) in Jammu, India. *Journal of Asia Pacific Entomology* 6(2): 201-205.

Kaul, V., Shankar, U. and Khushu, M.K. 2009. Bio-intensive Integrated Pest Management in fruit crop ecosystem In: *Integrated Pest Management: Innovation Development Process Vol-1,* Peshin, R. and Dhawan, A.K. (Eds.) pp. 631-666. Springer Verlag.

Kumar, M. 2008. Succession of major insect pests of citrus with special reference to the management of citrus caterpillar, *Papilio demoleus* (L.). M.Sc. thesis. SKUAST-J, Jammu.pp. 70.

Kumar, N., Suresh, J. and Anbu, S. 2007. Problems and prospects of guava cultivation under Tamil Nadu conditions. *Acta Hort.* 735:329-334.

Mani, M. and Krishnamoorthy, A. 1990. Evaluation of the exotic predator *Cryptolaemus montrouzieri* Muls. (Coccinellidae, Coleoptera) in the suppression of green shield scale *Chloropulvinaria psidii* (Maskell) (Coccidae, Homoptera) on guava. *Entomon.* 15: 45-48.

Mani, M., Krishnamoorthy, A. and Singh, S.P. 1989. The impact of the predator, *Cryptolaemus montrouzieri* Muls. on the pesticide resistant populations of the striped mealybug *Ferrisia virgata* (Ckll.) on guava in India. *Insect Science and its Application.* 11: 167-170.

Mittal, S. 2007. Can horticulture be a success Story for india? *Indian Council for Research on International Economic Relations,* Working Paper No. 197.

Mohyuddin, A.I. and Mahmood, R. 1993. Integrated control of mango pests in Pakistan. *Acta Hortic.* 341:467-483.

More, T.A. And Singh, R.S. 2008. Conserving biodiversity in different areas. Survey of Indian Agriculture. 50-54.

Murray, R.C. 1991. Arthropod pests of mango in the Caribbean Islands. *Rev. Crop Prot. Univ. West Indies, Kingston.*

Rai, S., Shankar, U., Bhagat, R.M. and Gupta, S.P. 2008. Population dynamics and succession of fruit fly on sub-tropical fruits under rainfed condition in Jammu region. *Indian Journal of Entomology,* 70(1): 12-15.

Sasidharan, K.R. and Varma, R.V. 2008. Seasonal population variations of the bark eating caterpillar (*Indarbela quadrinotata*) in *Casuarina* plantations of Tamil Nadu. *Tropical Ecology,* 49(1): 79-83.

Shankar, U., Kaul, V. and Rai, S. 2007. New record of anar fruit borer, *Deudorix isocrates* (Fabricius) on aonla. *Journal of Research, SKUAST-J.* 6(2): 292-295.

Sharma, D.R. 2006. Management of insect pests of tropical and subtropical fruit crops. In: Proceedings of advance training course on Emerging Trends in Economic Entomology, Feb. 21–13, *March, 2006, Hissar, India, pp. 123–135.*

Singh, G. 1991. Loss assessment, ecology and management of mango fruitfly, *Dacus* sp. *Acta Hortic.* 291:425-436.

Singh, H.P. and Singh, B. 2002. Lychee production in India. In: Lychee production in asia-Pacific Region. FAO of United Nations, Regional Office for the Asia and The Pacific, Bankok, Thialand, March 2002.

Singh, S.P., Rao, N.S., Ramani, S. and Poorani, J. (*eds*) (2001) Annual Report 2000–2001, *Project Directorate of Biological Control, Bangalore. Bangalore, India; PDBC, 218 pp.*

Tandon, P.L. 1991. Problems and prospects of insect pest management in fruit trees. In Trends in Agricultural Pest management (Dhaliwal, G.S. and Arora, R. eds.). Commonwealth Publishers, New Delhi.

Tandon, P.L. and Verghese, A. 1985. World list of insect, mite and other pests of mango. *Technical Documentation Indian Institut Horticulture Research* 5.

Tandon, P.L., Lal, B. and Rao, G.S.P. 1983. Prediction of mangohopper, *Indioscopus clypealis* population in relation to physical environmental factors. *Entomon,* 8:251-261.

Tandon, P.L. 1995. Integrated insect pest management in mango. *In Proceedings National Symposium on Integrated Pest Management and Environment* organised by Plant Protection Association of India and FAO at Madras 2-4 February, 1995.

Veeresh, G.K. 1989. Pest problems in mango – world situation. *Acta Hortic.,* 231:551-565.

Fig. 1 : Egg parasitoid *Trichogramma* spp.

Fig. 2 : *Chrysoperla carnea*

Fig. 3 : Syrphid fly

Fig. 4 : Coccinellid beetle adult

Fig. 4a : Coccinellid grub feeding on hoppers

Fig. 5 : Band of alkathene sheet against mano mealybugs

Fig. 6 : Plastic made flat tray

Fig. 7 : Cu-lure block trap

Fig. 8 : Spider predating upon beetle

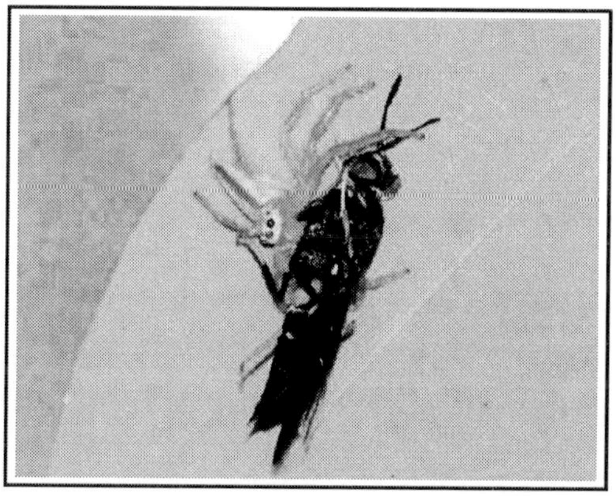

Fig. 8a : Oxyopes predating on fly

Fig. 9 : Mango hoppers infesting leaves and inflorescence

Fig. 9a : Coccinellid feeding on hoppers

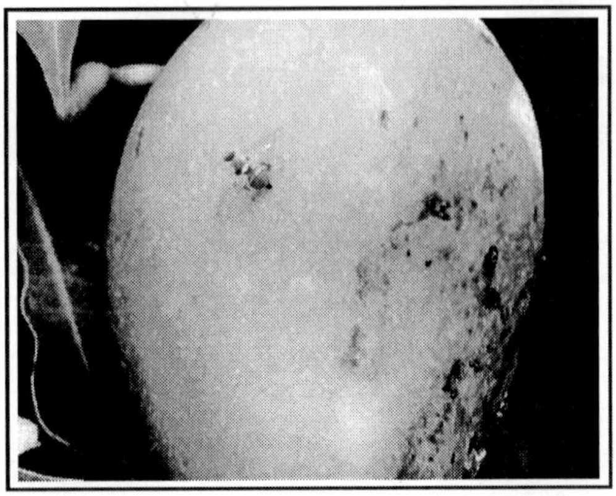

Fig. 10 : Fruit fly ovipositing on mango

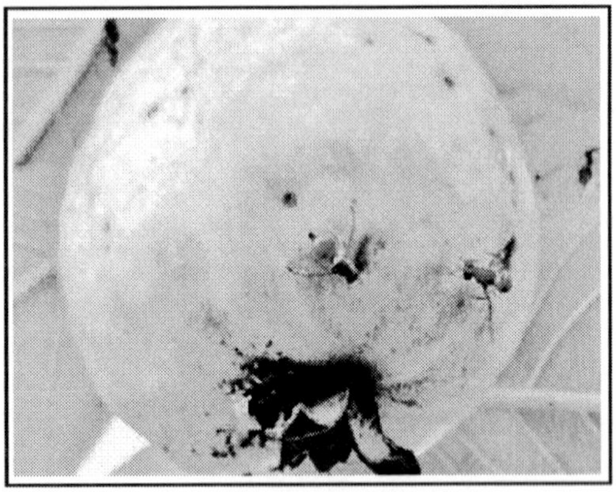

Fig. 11 : Fruit flies infesting guava fruit

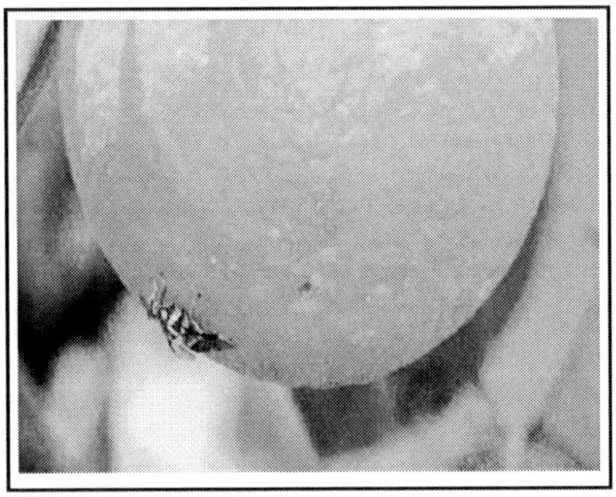

Fig. 12 : Fruit fly infesting citrus fruit

Fig. 13 : Fligh-T trap for catching fruit flies

Fig. 14 : Mealybugs infesting fruit and shoot of mango

Fig. 15 : Mealybug destructor

Fig. 16 : Sand heap around the trunk of mango to detain mealy bug crowlers

Fig. 17 : Life cycle of Inderbela in Jammu

Fig. 18 : Citrus psylla nymphs

Fig. 18a : Eggs and adult of citrus psylla

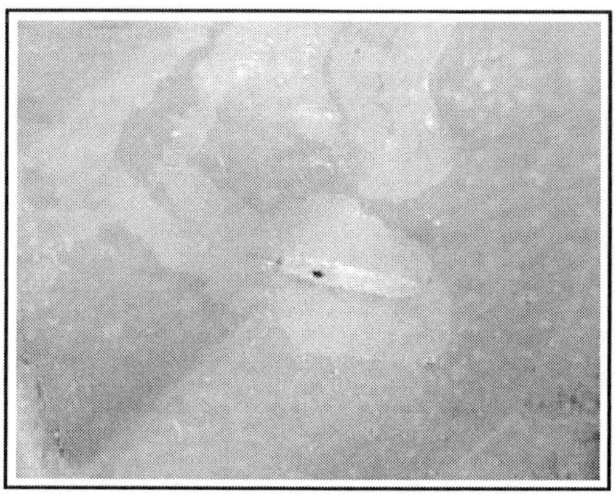

Fig. 19 : Citrus leafminer

Fig. 20 : Leafminer induced canker

Fig. 21 : *Papilio demoleus* larva

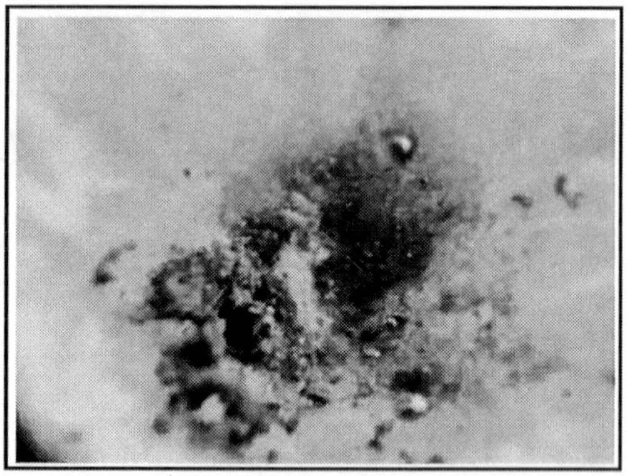

Fig. 22 : *Dichocrocis punctiferalis*

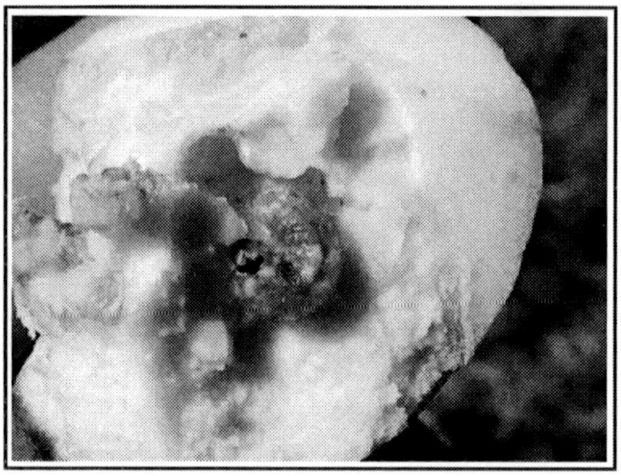

Fig. 23 : *Deudorix isocrates*

Fig. 24 : Locally made sound device to keep away the birds

Fig. 24a : Parrot feeding on guava fruit

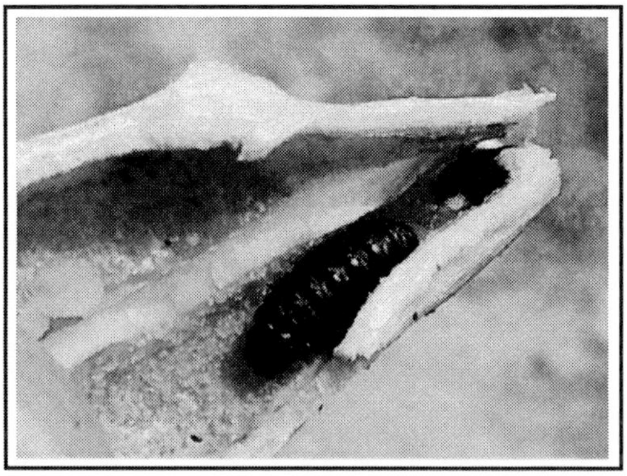

Fig. 25 : Shoot gall maker larva

Fig. 26 : Damage of leaf roller caterpillar on Aonla

Fig. 27 : Aonla aphids

Fig. 28 : Eulophidae parasitoid

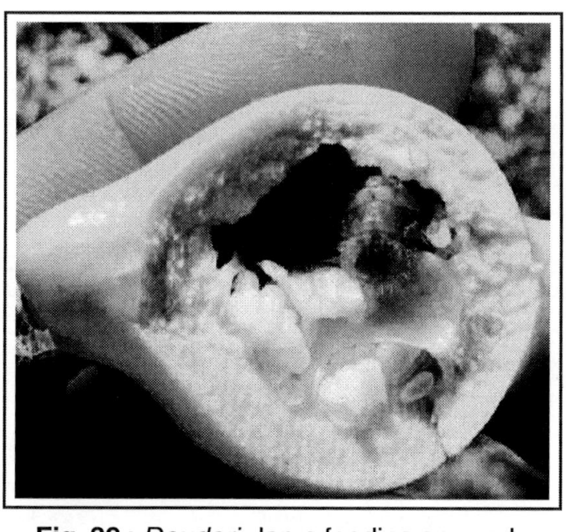

Fig. 29 : *Deudorix* larva feeding on seeds

Fig. 30 : Peach leaf curl

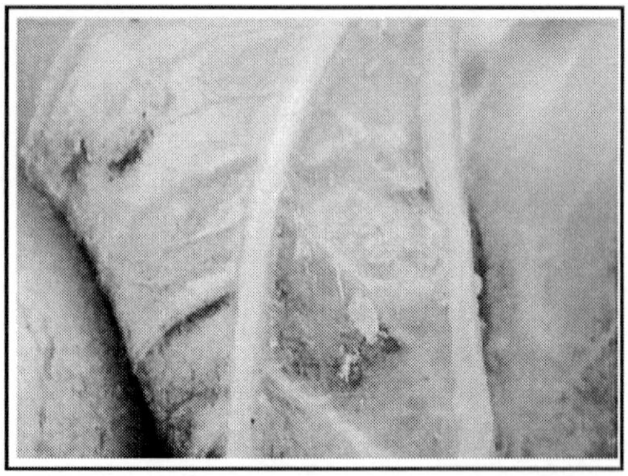

Fig. 30a : Peach aphid

Fig. 31 : Fruit flies ovipositing on peach fruit

Fig. 32 : Fruit fly bait in peach

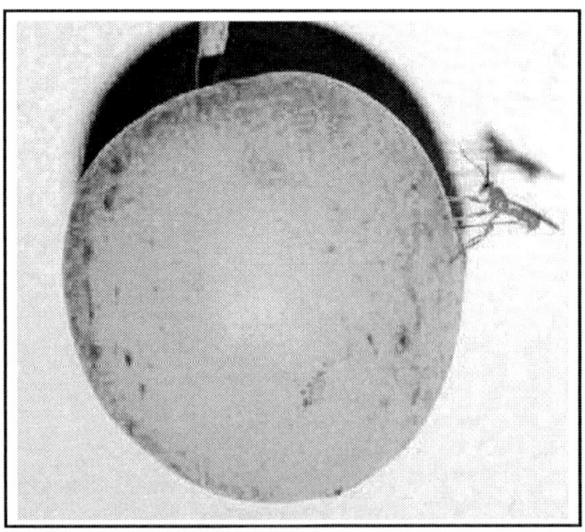

Fig. 33 : Parasitoid *Dichasmimorpha* sp. on ber fruit fly

Ecologically Based Integrated Pest Management, 2012
© D.P. Abrol and Uma Shankar (eds.), pp. 711-727
New India Publishing Agency, New Delhi (India)
e-mail : info@nipabooks.com; website : www.nipabooks.com

36

Insect Pests and Diseases of Temperate Fruit Crops and their Management

FAROOQ A. ZAKI

India is the second largest producer of fruits and vegetables. The area under fruits alone is 32.05 lakh hectares with a production of 329.55 lakh tones. Jammu and Kashmir is the most important horticultural state with unique temperate fruit and European type of vegetable crop production. The total area during 2005-06 under the fruit crops in the state was 2.83 lakh hectares with a production of 1504.01 thousand metric tones (Table 1). Almost 45 per cent of economic returns under agriculture sector are contributed by horticulture, thereby adding 7 per cent to the Gross State Domestic Product (Anonymous, 2007). Among the fresh and dry fruits, apple and walnut, respectively, are the most important fruit crops in the Jammu and Kashmir. However, there is a wide gap between the present yield and the potential yield of fruits. One of the limiting factors for this gap in fruit production is severe attack of arthropod pests and diseases affecting yield and quality. Unlike agricultural crops, horticultural crops are grown as perennial monocultures, therefore, harbor the insect pest and disease problems that are entirely different and complex in nature. These two factors impose huge losses upon horticulture in terms of alleviated production, reduced quality and cost of management.

Table 1 : Details of major horticultural crops in Jammu and Kashmir during 2005-2007

	2005-06	2006-07
Area (lakh ha)	2.68	2.83
Production (thousand m t)	1419.16	1504.01
Yield (m t/ha)	5.26	5.31
Revenue (Crores)	20,00	21,12

Pests of Economic Importance in Temperate Fruits

A large number of pests have been observed on different horticultural crops with the effects ranging from mild damages to serious losses in production and quality. The arthropod pest spectrum of various crops in Kashmir (Table 2, 3, 4) is very wide.

Table 2 : Major arthropod pests under Kashmir agro-climatic conditions

Pest	Scientific name	Active Period	Remarks
San Jose scale	*Quadraspidiotus perniciosus*	March-Oct.	Serious key pest
Wooly apple aphid	*Eriosoma lanigerum*	Feb.-Nov.	Perennial pest
Gypsy moth	*Lymantria obfuscata*	April-June	Attaining a serious status
Apple green aphid	*Aphis pomi*	May-August	Perennial pest
European red mite	*Panonychus ulmi*	April-Nov.	Key and serious pest
Two spotted spider mite	*Tetranychus urticae*	April-Nov.	Key and serious pest
Short hole borer	*Scolytoplatipus* sp.	April-Oct.	Moderately serious pest on apple but most serious on walnut
Stem borer	*Aeolesthus sarta*	April-Nov.	Serious pest
Pear psylla	*Psylla pyricola*	May-Sep.	Minor pest
Leaf minor	*Lyonetia prunifoliella*	April-June	Moderately serious
Chaffer beetles	*Protacta neglecta*, *Adoretus* spp.	Grub attacks throughout the year. Adult from June-July	Serious, and moderately serious, respectively
Tent caterpillar	*Malacosoma kashmiriensis*	April-Nov.	Moderately serious
Walnut weevil	*Alcides porrectirostris*	April-Sep.	Most serious

Contd...

Table 2. Contd...

Pin hole borer	*Scolytoplatipus* sp.	April-Oct.	Most serious
Cabbage butterfly	*Pieris brassicae*	March-Oct.	Serious
Mustard saw-fly	*Athalia proxima*	Feb.-May	Sporadic pest
Aphid	*Myzus persicae, Lipaphis erysimae*	Throughout the year	Most serious
Diamond backmoth	*Plutella xylostella*	June-Oct.	Moderately serious
Bark beetle	*Scolytus nitidus*	April-Oct.	Moderately serious
Cut worm	*Agrotis ipsilon*	May-June	Serious
Army worm	*Mythimna separata*	March-Sep.	Sporadic pest
Cabbage aphid	*Brevicoryne brassicae*	Throughout the year	Serious

Table 3 : Arthropod insect pests attacking apple in Kashmir

S. No.	Name of the pest	Scientific name	Order	Family
1	San Jose Scale	*Quadraspidiotus perniciosus*	Homoptera	Diaspididae
2	Woolly apple aphid	*Eriosoma lanigerum*	Homoptera	Aphididae
3	Leaf blister mite	*Phytoptus pyri*	Prostigmata	Eriophyidae
4	Gypsy moth	*Lymantria obfuscata*	Lepidoptera	Lymantriidae
5	European red mite	*Panonychus ulmi*	Prostigmata	Tetranychidae
6	Two spotted spider mite	*Tetranychus urticae*	Prostigmata	Tetranychidae
7	Shot hole borer	*Scolytoplatipus* sp.	Coleoptera	Scolytidae
8	Bark beetle	*Scolytus niditus*	Coleoptera	Scolytidae
9	Leaf miner	*Lyonetia prunifoliella*	Coleoptera	Phyllocnistidae
10	Codling moth	*Cydia pomonella*	Lepidoptera	Tortricidae
11	Stem borer	*Aeolesthes sarta*	Coleoptera	Cerambycidae
12	Tent caterpillar	*Malacosoma kashmirensis*	Lepidoptera	Lasiocampidae

San Jose scale, *Quadraspidiotus perniciosus* is one of the most serious and key pests on apple. Its first peak commences in the second or third week of May and second peak during second or third week of August and the 3rd generation overwinters. The pest is a sucking type, feeding on the young and old branches and twigs, reducing the vigour and vitality of trees, the surfaces of the branches appear ash grey, fruit is severe damaged rendering it unmarketable. **Woolly apple aphid** (*Eriosoma lanigerum*), a serious pest on apple commences during 2nd week of March continuing up to September. The pest migrates from roots to aerial portion and *vice versa* during spring and fall, with maximum dispersal during day between 4-6 pm. These form colonies on the bark of the tree trunk and the branches characterized by white cottony cushion and suck out the plant sap, weakening the tree and make it prone to the attack of other diseases. **Apple green aphid** (*Aphis pomi*) drains sap from the leaves, flowers, buds and terminals of apple trees reducing their vitalities. It remains active from 2nd week of April to last week of May. About 12 phytophagous mite species have been identified on apple in Kashmir (Rather and Masoodi, 1988) of which **European red mite** (*Panonychus ulmi*) and **two spotted spider mite** (*Tetranychus urticae*) are the most important. European Red Mite has emerged as a limiting factor in apple production in Kashmir. European red mite completes six generations with high biotic potential, starting its activity in mid spring and continuing till late autumn. The infestation of the mites causes yellowing and bronzing of the foliage, affecting the photosynthetic efficiency of the leaves and results in the premature leaf drop. The situation is worsened by the presence of *Alternaria* causing complete leaf blotch rendering the apple tree denuded (Zaki *et al.*, 2004). **Stem borer** (*Aeolesthes sarta*) and **bark beetles** (*Scolytus* sp.), the pests of both fruit as well as forest trees, have emerged as a serious problem for apple and walnut. **Codling moth** (*Cydia pomonella*) is a pest of international importance. However, in Jammu and Kashmir its incidence is confined to Leh and Kargil. This pest attacks many fruit trees e.g. apricot, pear, walnut, etc. besides all cultivars and varieties of apple. The pest inflicts direct damage to the apple and apricot rendering the fruit unfit for the human consumption and market. The extent of losses caused to apple crop alone is the tune of 50 per cent in Ladakh. **Indian gypsy moth** (*Lymantria obfuscata*), is a voracious polyphagous pest attacking many fruit and tree plantations including apple and walnut. **Chaffer beetles** (*Adoretus simplex, Brahmina coriacea, Holotrichia longipennis*) damage fruit crops especially High-density plantation of apple and walnut. Among 70 species of leaf feeding beetles recorded in temperate regions of India, scarabaeids, cetoniids, curculionids and chrysomelids in that order cause significant damage to pome and stone fruits. **Walnut weevil** (*Alcides porrectirostris*) prefers thin shelled varieties to the less valuable thick shelled walnut types and is restricted in its distribution to Tangdar area of Kupwara district. The kernel of the nut is damaged and the attacked fruit usually drops or is unmarketable. The dropping of the fruit starts from 2nd week of May till end of August. Another important pest associated with walnut is **grey weevil**

(*Myllocerus* spp.). It feeds on foliage by biting holes in the lamina or by peripheral feeding on margins. The maximum damage is caused from mid May to mid June. Among **leaf rollers**, folders *Archips pomivora* is the most important. It causes maximum damage in June. It attacks foliage but injury on fruits and their rotting cause economic losses. An important pest of Red cherry and Black cherry is **cherry fruit worm**. The extent of damage if caused by this pest is predominantly visible when the cherry starts to ripen. The cherry is also infested by Cherry fruit fly if the fruit harvest is delayed in certain localities. **Pomegranate fruit borer** (*Deudorix epijarbas* Moor) a widely distributed and predominant pest of pomegranate limiting its cultivation and causing considerable loss to the crop. **Almond mealy bug** (*Drosicha dalbergia*) is a serious pest on almond which feeds on both aerial and underground parts of the plant, colonizing at the collar region of the stem, in the crevices and the wounded sites. Infested plants exhibit sickly appearance, thereby, reducing the quality and quantity of the fruit.

Table 4 : Insect pests attacking walnut in Kashmir

S. No.	Name of the pest	Scientific name	Order	Family
1	Walnut weevil	*Alcides porrectirostris* (Marshall)	Coleoptera	Curculionidae
2	Stem borer	*Aeolesthes sarta* (Solsky)	Coleoptera	Cerambycidae
3	Pin hole borer or shot hole borer	*Scolytoplatipus* spp.	Coleoptera	Scolytidae
4	Chaffer beetle	*Protecta neglecta* *Adoretus simplex*	Coleoptera	Scarabaidae
5	Indian Gypsy moth or hairy caterpillar	*Lymantria obfuscata* (Walker)	Lepidoptera	Lymantriidae
6	Walnut leaf defoliater	*Chaetoprocta odata* (Hewitson)	Lepidoptera	Lycaenidae

IPM for Major Insect Pests on Fruit Crops in Kashmir

A) Mineral oil sprays

Horticulture mineral oils that are used for the management of important fruit pests are physical toxicants and result in the death of insect pests due to asphyxiation. They are eco-friendly and bio-safe as they come from earth and are formed as a result of transformation of plankton due to the action of high pressure and high temperature over millions of years. Comparative bioefficacy of different summer spray oils against San Jose scale and European red mite (ERM) in Kashmir during 2004-2008 resulted in a maximum mortality per cent of about 87.71 and 94.15, respectively (Table 5). A number of such mineral oils, viz., ATSO spray oil, HP spray oil, Arbofine spray oil, Mak all season spray oil, etc. at 2% and 0.75% concentrations during winter and summer, respectively, have been found effective against the scale and mites (Sofi *et al.*, 2008, Zaki *et al.*, 2008a, Zaki *et al.*, 2008b).

Table 5 : Comparative bioefficacy of different Summer spray oils against European red mite (ERM) in Kashmir

Treatment	Concentration (%)	Per cent Mortality			
		Arabal (Srinagar) 2004	Hanjiwara (Baramulla) 2003	Hanjiwara (Baramulla) 2004	Pooled Mean
PD Spray oil	0.75	83.53	92.42	82.00	86.83
	1.0	92.99	95.67	91.44	94.15
ATSO Spray oil	0.75	83.21	90.45	84.05	83.34
	1.0	89.20	96.74	88.18	92.45
Orchex-796 Spray oil	0.75	83.01	92.12	82.75	85.96
	1.0	93.35	94.69	87.09	91.71
Arbofine Spray oil	0.75	84.53	92.94	82.64	87.75
	1.0	92.35	95.24	91.08	94.01
HP Spray oil E	0.75	81.43		87.63	84.53
	1.0	90.18		88.40	89.29
HP Spray oil	0.75	77.98		81.21	79.59
	1.0	88.18		87.38	87.78
Sparrow 888 plus Spray oil	0.75	76.88		71.89	74.38
	1.0	88.69		80.20	84.44
Control		21.61		25.87	23.74

Each figure is mean of 15 replicates and each replicate is mean of 12 observations

B) Chemical control

The most critical issue on the use of chemical control is the precise timing of insecticidal application, which is important for achieving the desirable control of important insect pests. Every year SKUAST-K, Shalimar prepares a spray schedule (Annexure I) where the proper timing and application of particular pesticides is provided to farmers. By and large, two summer sprays of an acaricide/insecticide or insecticide-cum-acaricide is advocated during the third to fourth week of May and first to second week of July, respectively, to control and manage most of the arthropod pest of apple crop in Temperate Horticulture. Normally no acaricide/insecticide is recommended beyond that period, thereby ensuring that no acaricide/insecticide residues are in the fruit at the harvest. Need based sprays are recommended only if the mite and /or San Jose scale populations exceed economic threshold level during August or after harvesting.

Table 6 : Biological control agents reported from Jammu and Kashmir

Pest	Pest status	Biological Control Agents
San Jose scale (*Quadraspidiotus perniciosus*)	Serious key horticulture pest	*Aphytis proclia, A. paramculicornis, Azotia kashmirensis, Maristta camesi, Encarsia perniciosi* (Parasitoids); *Chilocorus bijugus, Pharoscymnus flexibilis, Cybocephalus* sp., *Sticholitis marqinalia, Adalia tetraspilota, Melachius* sp., *Harmonia* sp., *Coccinella septempunctata, Chrysoperla carnae* (Predators).
Apple aphid (*Aphis pomi*)	Perennial pest	*Coccinella septempunctata* Linnus, *Adalia tetraspilots* Hoe, *Hippodamia variegate* Goeze, *Harmonia breiti* Madar, *Pharoscymnus* sp., *Symnus grsoilis, Exochomus uropygialia, Plantynaspis* sp. Mulsant, *Propyles japonica* Thunberg, *Dropephylla almorensis* Cham, *Chrasoperis carnae* Stephen, *C. prestes* Banks, *Tomnoatethus (Ectemnus) paradoxus* Hetchinson, *Lygus* sp., *Episyrphus balteatus* DeGreer, *Eumerus elbirfrons, Metasyrphus confrator* Weid, *Eristalis tenax* Linnaeus, *E. arbustorum* Linnaeus, *E. correalis* Fabricius, *Scaeva pyrastri* Linnaeus, *Sphaerospheria ecripta* Linnaeus, *Aphidoletes aphidomyza* Rondani, *Pachyneuron* sp.
Wooly apple aphid (*Eriosoma lanigerum*)	Perennial pest	*C. septempunctata, C. bijugus* (Predator); *Aphelinus mali* (Parasitoid)
European Red mite (*Panonychus ulmi*)	Key and Serious pest of apple	*Anystis baccurum* Linnaeus, *Eusius vignus* Rishi and Rather, *Phytoseius domesticus* Rather, *Euseius insanus* Khan and Chaudhri, *Amblyseius finlandicus* Oudemans, *Phytoseius mixtus* Chaudhri, *Euseius* sp., *Cheyletus* sp., *Agistemus* sp. (Acarine predators); *Hippodamia* sp., *Stethorus punctillum, Stethorus punctum, Chrysoperla carnea, Oligota* sp., *Anthocoris* sp., *Scolothrips* sp. (Arthropod predators)

Contd...

Table 6. Contd...

Mealy plum aphid (*Hyalopterus arundinis*)	Moderately serious	*Coccinella undecimpunctata, Adalia tetraspilota, Hippodamia variegata, Scymenus gracilis, Episyrphus balteatus, Eumerus nepalensis, Sphaerophora indica, Syrphus* sp., *Anthocoris* sp., *Chrysoperla carnia, Trioxys rishii.*
Peach curl aphid (*Anuraphis helichrysii)*	Serious pest	*Scymenus gracilis*
Walnut aphid (*Callaphis juglandis)*	Serious pest	*Metasyrphus corollae*
Oystershell scale (*Lepidosaphes kirgisica*)	minor	*Aphytis paramaculicornis*

Annexure I

S.K. University of Agricultural Sciences and Technology of Kashmir
Directorate of Research
Shalimar, Srinagar-191121, J&K (INDIA)

Spray	Tree Stage	Fungicide/100 lit of water	Insecticide/Acaricide/ HMO per 100 lit of water
I	Green tip (Delayed Dormancy) (Mid March-Mid April)	-	Horticultural Mineral Oil (HMO) @ 2 litres (2%)
II	Pink bud (Mid April-End April)	Myclobutanil 10 WP (30 g) or Dodine 65 WP (60g) or Bitertanol 25 WP (50g) or Fenarimol 12 EC (40 ml)	*** 3-4 days after II fungicidal spray in case spray I is missed. Chlorpyriphos 20 EC (100 ml) or Endosulfan 35 EC (140 ml) or Summer Spray Oil (750ml) or Fenazaquin 10 EC (40ml) or Herbal (200 ml)
III	Petal fall (80 to 100 per cent)	14-18 days after II spray Hexaconazole 5 EC (30ml) or	3-4 days after III fungicidal spray Quinalphos 25 EC

Contd...

Annexure I. Contd...

		Difenaconazole 25 EC (30ml) or Triadimefon 25 WP (50g) or Diniconazole 25 WP (40g)	(100ml) or Methyl-o -Demeton 25 EC (80ml) or Phosalone 35 EC (140ml) or Summer Spray Oil (750ml) or Fenazaquin 10 EC (40ml)
IV	Fruit let (Pea size)	14-18 days after III spray Ziram 80 WP (200g) or Mancozeb 75 WP (300g) or Ziram 27 W/V (600ml) or Captan 50 WP (300g) or Propineb 75 (300g) or Mancozeb Flowable 35SL (300ml) or Zineb 75 WP (300g)	***3-4 days after IV fungicide spray Chlorpyriphos 20 EC (100ml) or Methyl-o- Demeton 25 EC (80ml) **or** Phosalone 35 EC (140 ml) Or Dicofal 18.5 EC (108ml) or Summer Spray Oil (750 ml) or Fenazaquin 10 EC (40 ml)
V	Fruit development I	14-18 days after IV spray Bitertanol 25 WP (50g) or Dithio non 75 WP (75g) or Penconazole 10 EC (50ml) or Fenarimol 12 EC (40ml)	3-4 days after V fungicide spray Chlorpyriphos 20 EC (100ml) or Methyl-o-Demeton 25EC(80ml) or Dimethoate 30 EC(100ml) or Docofol 18.5 EC (108ml) or Summer Spray Oil (750ml) or Fenazaquin 10 EC (40ml) or Abamectin 1.8 EC (55.5ml)
VI	Fruit development II	14-18 days after V spray Hexaconazole 5 EC (30ml) or Myclobutanil 10 WP (30g)	-

Contd...

		or Triadimefon 25 WP (50g) or Diniconazole 25 WP (40g)	
VII	Fruit development III	14-18 days after VI spray Difenoconazole 25 EC (30ml) or Dodine 65 WP (60g) or Dithionon 75 WP (75g) or Penconazole 10 EC (50g)	3-4 days after VI fungicidal spray Chlorpyriphos 20EC (100ml) or Dimethoate 30EC (100ml) or Endosulfan 35EC (140ml) or Methyl-o-Demeton 75 EC (80ml) **or Phosalone 35 EC (140ml) or** Fenazaquin 10EC (40ml) or Abamectin 1.8 EC (55.5ml)
VIII	Fruit development IV	*14-18 days after VII spray Hexaconazole 5 EC(30ml) or Bitertanol 25 WP (50g) or Fenarimol 12 EC (40ml)	-
IX	Pre-harvest	** 25 days before harvest Ziram 80 WP (200g) or Mancozeb 75 WP (300g) or Ziram 27 W/V (600ml) or Captan 50 WP (300g) or Propineb 75 WP (300g) or Mancozeb Flowable 35 SL (300ml) or Zineb 75 WP 300 mg	-
X	Post harvest	-	***** Ethion 50 EC(100ml) or Fenazaquin 10 EC (40ml) or HMO (75ml) or Chlorpyriphos 20 EC (100ml)

* Need based spray if leaf spotting incidence is more than20%

** Need based spray for long term storage of fruits.

*** Need based spray when 5-8 mites/leaf are observed.

**** Need based spray when 10 or more mites/leaf are observed.

***** Need based spray when 20 or more mites/leaf are observed during late August and spray at Fruit Development III (during August) has been missed.

C) Biological control

A large number of natural enemies of the horticultural and vegetable pests have been recognized in Kashmir. The biological control agents associated with the important pests of horticultural crops are presented in Table 6. The important biocontrol agents include *Aphytis proclia, Encarsia pernisiosi, Chilocorus bijugus, Adalia tetraspilota, Coccinella septumpunctata* and *Chrysoperla carnae*, against San Jose scale (Anonymous, 1988); *C. septumpunctata, A. tetraspilota* and *Symnus grsoilis* (against Apple aphid); *Aphelinus mali, C. septumpunctata, C. bijugus, Amblyseius finlandicus, Phytoseius mixtus* and *Stethorus* sp. (against European red mite and Two spotted spider mite); *Scymenus gracilis* (against Peach leaf curl aphid). During 1989-90 studies on the population record of predators associated with woolly apple aphid in Kashmir revealed presence of *Aphelinus mali* an eulophid endoparasite and as many as 7 predators. During the year 2000, observations in respect of predators were same. However, during 2002 only three predators including *Chilocorus* sp., *Coccinella* sp. and *Phasoscymnus balteatus* were found. The prevalence of pests has been observed to directly influence the population status of its natural enemies. Appearance and population built-up as well as disappearance of coccinellid beetle and syrphid fly was found to follow the population built-up pattern of their host aphid.

D) Cultural control

i. Burlapping

Burlapping is a technique of placement and wrapping of a gunny bag band or corrugated cardboard band around the tree trunk at 4-6 feet above ground level to provide artificial hibernating/nesting sites. This practice is followed for the management of codling moth in Ladakh where the larvae are collected and destroyed at an appropriate time which reduces the primary infestation for the next season or the next generation. Gunny bag burlaps have been found to be more effective and economic because these can be re-used. Burlapping in integration with the insecticidal spray at Kargil under moderate infestation showed that the number of larvae trapped in corrugated cardboard in different insecticidal spray ranged between 2.3 larvae in fenvalerate treatment to 12.3 larvae in endosulfan as compared to 17.3 larvae in unsprayed control. The results indicated that even without insecticidal application the efficacy of this practice is quite satisfactory. Gunny bag burlaps have been found to be more effective and economic because these can be re-used. The practice of burlapping have also been found effective in the trapping, collection and destruction of caterpillars and eggs of Indian gypsy moth, which otherwise take shelter under the bark of the trees. Bark scrapping for the control of the gypsy moth and San Jose scale reduces the pest infestation.

ii. Fruit destruction

Destruction of infested fruits is one of the important cultural practices for the management of various insect pests. This method can bring up the effective control of fruit borers eg. Cherry fruit fly, Pomegranate borer, codling moth, etc. Departments of Horticulture at Leh and Kargil had adopted a practice of destroying or burying the apple fruits infested with codling moth in farmers' fields and paying remuneration to farmers. The practice was carried out with an objective to destroy the larvae believed to be infesting the fruit and to reduce the larval population and subsequent incidence for the next year. However, during 1992-93 on the recommendation of the author the practice was stopped by the District Development Commissioner, Kargil inspite of the fact that 80 per cent of fallen fruits were found affected by the codling moth. This shall theoretically eliminate 35% of primary infestation. However, our observation indicated that by the time the department initiated the process of fruit collection and destruction in the farmers' field the full fed larvae are already out of the infested fruit to take shelter and undergo hibernation beneath tree barks, brickbats, stone bunds, adjacent houses, etc. The destruction of affected but non infested fruits from which the larvae exit had remained of no practical value for larval destruction over a period of many years, therefore, rendered redundant (Personal communication).

iii. Collection and destruction of wood

Wood is more susceptible to the infestation of stem borers, pin hole borers and short hole borers. Therefore, placing some logs of dry hard wood in the orchards may considerably prevent these orchards from infestation of such insect pests. The wood thus placed act as a trap strategy to these pests, which need to be collected and destroyed.

iv. Training and pruning

Proper training and pruning of fruit trees is very important cultural practice as it helps in the management of economically important insect and non-insect pests eg. Aphids, San Jose scale, European red mite, bark beetle, etc. Off season biology of the important pests is carried over on the tree branches and trunk/bark, therefore, pruning of infested parts reduces the initial pest load in the fruit orchards. The borer and bark beetle infested branches and twigs should be removed.

v. Fertilizer management

It has been found that the use of fertilizers help in improving the fertility of soil, however, the balanced dosage of fertilizers is equally important. When the nitrogen is applied in higher amounts, it increases the quantity of sap but at the same time increases the infestation of sap sucking pests eg. Green aphids, black aphids, wooly aphid, European red mite, peach leaf curl aphid, San Jose scale, two spotted mite, almond mealy bug, etc. Balanced doses of fertilizers, micronutrients and higher doses of potassium were found to reduce the infestation of ERM and aphids.

vi. Habitat management

The ground cover management around fruit trees is another important cultural practice as it provides habitation to various predators eg. *Amblyseius* sp. during application of pesticides and dormancy. It also harbours the prey for such predators during the off season. Therefore, in order to enhance the efficiency of biological control of European red mite, it is advised to allow some flora around the apple tree trunks.

vii. Scrapping

Scrapping of San Jose scale infested twigs with khurpi is another technique, yet uncommon among the farmers but can help in reducing the infestation and further increment of scale population considerably.

E) Quarantine control

The distribution of codling moth in India is restricted to Ladakh only. In order to prevent its accidental entry to other localities and particularly the temperate fruit growing states of Himachal Pradesh, Uttranchal, Arunchal Pradesh and Uttar Pradesh, the pest has been brought under domestic quarantine through an SRO promulgated by the Government of Jammu & Kashmir under " Destructive Insect Pest and Weeds Act, 1914 amended 1968" Clause-II of Government of India. The export of fresh fruit from Ladakh to other places is banned which is regulated by the Jammu and Kashmir State Department of Horticulture, at the Sonamarg post. The domestic quarantine has, therefore, checked its entry form Ladakh to other places. However, with the operating of Leh-Manali road and no quarantine check post on the highway there are all apprehensions of this pest getting entry into Himachal Pradesh through the fresh fruit transport by the tourists and exchange of plant materials. The issue needs to be reviewed seriously by the Government of Himachal Pradesh and the Government of India for Northern and N. E. states. Import of plant materials from other countries in the past without following proper regulations have introduced a number of pests and diseases not reported earlier. Root rot of apple in epidemic form in Kashmir and the report of dreaded blackline virus in walnut nursery at Zainpora, Kashmir which prompted the Department of Horticulture to destroy the plantations, are two examples of many unwanted troubles brought in due to poor quarantine. Fire blight caused by *Erwinia amylovora* reported infecting roses by Indian Institute of Horticulture, Bangalore (Personal communication), has probably been introduced in the country along with the rose plants imported for the new varietal introductions. The apple growing states of India should be vigilant on the introduction of rose plants infected with the bacteria through AICRP on Ornamental Crops, which may destroy apple crop on which the disease is not reported so far. The domestic quarantine measures need to be enforced on the movement of the rose plants in the apple growing states.

Constraints in Management of Insect Pests of Fruit Crops in Kashmir

Despite spectacular progress made in the area expansion and increase in the fruit production, the pace of development has started stumbling down due to devastating emerging challenges in management of insect pests and diseases causing poor productivity and quality in apple/walnut orchards in Kashmir, thereby, putting enormous pressure on the sustainability of horticulture and export of fruits to international markets.

European red mite (*Panonychus ulmi*), stem borer and bark beetles on fruit plantation have emerged as most serious pest problems, due to drought like conditions in the valley during last few years. European red mite has become one of the limiting factors in apple orchards in recent past, posing a serious challenge in its management. San Jose scale and wooly aphid have also upsurged in recent past. The following factors have been attributed to the control failures against these pests:

a) Mineral oil sprays

Dormant spray oils which were earlier used regularly against San Jose scale and would also kill over wintering eggs of European red mite, had been partially abandoned due to ease of application of synthetic insecticides, high input and labour cost, hostile weather conditions during dormancy, shift in priorities from pests to apple scab disease control when it appeared in epiphytotic form, replacement of diesel oil by summer pesticide spray. Such an omission in some orchards has resulted in population explosion of the ERM to damaging proportions.

b) Disruptive use of pesticides

Out break of ERM, San Jose scale and wooly aphid has also been induced by intensive and indiscriminate use of broad-spectrum pesticides resulting in the destruction of natural enemies, which acted as strong population stabilizing factors. Besides dormant oil replaced by summer sprays for the control of San Jose scale, resulted in the decimation of predatory mites and lady bird beetles. Consequently in the absence of natural controlling factors the mite population and San Jose scale increased to alarming proportions. The predator activity in the pesticide managed and unmanaged orchards showed higher activity of predators with lower mite infestation in the latter in comparison to the former (Zaki and Chan, 2003).

c) Insect resistance to insecticides

Development of resistance of mites to pesticides is a major challenge influencing the changing pattern of chemical pest control strategies and is, therefore, the most cogent reason for an integrated approach to pest control in apple orchards. ERM has developed resistance to almost the entire-spectrum of acaricides and insecticides-cum-acaricides used in tree fruit production (Zaki *et al.*, 2005) in Kashmir. European red mite has developed 7-80 fold resistance to endosulfan, dimethoate, chlorpyriphos, methyl-o-demeton and fenazaquin.

d) Non-standard pesticides

The sale of spurious pesticides has resulted in loading the orchards with disastrous chemicals, which besides their toxic hazards have changed the balance between pests and natural controlling factors. Such spurious pesticides besides being ineffective impose a sub lethal dose selection pressure on the key pests.

e) Nutritional imbalance

There exists a direct relationship between levels of nitrogen and mite population. Most of the orchards do not receive balanced dose of fertilizers and urea seems to be predominant fertilizer applied to apple trees allowing succulence and predisposition to sucking pests.

f) Adverse effect on natural enemies

The repeated applications of insecticides to temperate fruits with broad-spectrum pesticides have a destructive effect on beneficial insect populations. Fields and orchards under heavy treatment schedules have become veritable "biological deserts". The re-establishment of adequate complexes of natural enemies in such biological deserts may require few years after the use of insecticides/ acaricides is abandoned.

g) Pesticidal poisoning of pollinators

The poisoning of honey bees by the application of insecticides is a major problem of beekeepers especially where applications are made to crops during bloom period. Because of domestication honey bees have not developed the wide variety of detoxification enzymes as found in many other insects, therefore, are highly susceptible to pesticides. Wild bees such as bumble bees, *Megachile* and other bees are also important for pollination and are highly susceptible to insecticides.

h) Pre-mature leaf fall

The premature leaf fall in apple orchards has become an alarming syndrome in recent years. The problem is enhanced and further aggravated by *Alternaria* leaf blotch particularly under high temperature and higher humidity regimes. Higher mite population in the *Alternaria* affected orchards and concurrent presence of the two in the affected orchards lead to higher disease incidence, intensity and premature leaf fall.

i) Socioeconomic constraints

Participation of people/farmers in taking up the management strategies seriously is very poor.

j) Kashmir as agro-export zone

Declaration of the state as Agro-Export Zone for the apple and walnut fruits was one of the major challenge before growers, development officers and the agricultural scientists to figure in export map. The fruits for export have to be qualitative, competitive, standard grade and produced either organically or free of all pesticide residues. Apple concentrate being exported have been found free from objectionable pesticide residues. There is strong need to promote biological and genetic methods of pest control as vital components of IPM to further enhance the potential of apple and walnut export. The potential of walnut export from India is to the tune of 700 crores against Rs 100 crores currently being exported to other countries.

k) High-density plantation

High-density plantation with dwarf cultivars is an important strategy for enhancing productivity manifold but it has been found to be highly susceptible to pests. Therefore, new management strategies have to be developed for combating the pest in high-density plantation in order to enhance productivity. High density plantation have been found to be infested by a different pest spectrum in addition to the pest recorded in traditional orchards, therefore the pest and disease management in this system is more management, labour and cost intensive.

References

Anonymous, 1988. Annual Report, Division of Entomology, SKUAST-K Shalimar, *Technical Document No. 3*, pp. 45-56.

Anonymous, 2007. *Economic Survey 2007-2008*. II Edition. Directorate of Economics and Statistics, Planning and Development Department, Jammu and Kashmir Government.

Anonymous, 2008. *Apple Spray Schedule for the Year 2008.* Department of Horticulture, Government of Jammu and Kashmir, Srinagar, J&K. 6p.

Rather, A.Q. and Masoodi, M.A. 1988. Survey of mite pests and their natural enemies. In: *Annual Report*, Division of Entomology, SKUAST-K Shalimar, pp. 71-75.

Sofi, M.A., Zaki, F.A., Dar, M.A. and Khan, Z.H. 2008. New approaches to San Jose scale control by newer oils alone and in combination with insecticides and their effect on grading of apple in Kashmir. *Environment and Ecology*, 26(3A): 1369-1372.

Zaki, F.A. and Chan, S.A. 2003. Management of European red mite (*Panonychus ulmi*) on apple in Kashmir. *SKUAST Journal of Research*, 5: 198-203.

Zaki, F.A. Dhobi, G.M. and Mantoo, M.A. 2004. Interaction of European red mite (*Panonychus ulmi* Koch) infestation with *Alternaria* sp. in premature leaf fall in apple (*Malus domestica* Borkh). In *"Bioresources: Concerns and Conservation"* (Kamili, A. N. and Yousuf, A. R. Eds.) pp. 119-126.

Zaki, F.A., Lone, A. H. and Mantoo, M.A. 2005. Resistance in European red mite (*Panonychus ulmi* Koch) to some pesticides on apple in Kashmir. In *"Proceedings of the 1st Congress on Insect Science: Contributed papers"* (Dilawari, V.K., Doel, G.S., Joia, B.S. and Chhuneja, P.K. Eds.), December 15-17, INSAIS, PAU, Ludhiana, pp. 161-163.

Zaki, F.A., Mantoo, M.A. and Parrey, M.A. 2008a. Bioefficacy of some newer acaricides against European red mite, *Panonychus ulmi* Koch (Acari:Tetranychidae) on apple crop, *Malus domestica* Borkh in Kashmir. *Science For Better Tomorrow*: 1-4.

Zaki, F.A., Mantoo, M.A., Parray, M.A., Khan, A.A., Rather, A.Q., Wani, N.A. and Wani, A.R. 2008b. Paraffinic mineral oils - A new eco-friendly agrochemical strategy for agricultural pest management. IUPAC sponsored *First International Conference on Agrochemicals Protecting Crop, Health and Natural Environment*, Jan.,08-11, 2008, New Delhi, India. Abstr. No. P-77.

□□□

Ecologically Based Integrated Pest Management, 2012
© D.P. Abrol and Uma Shankar (eds.), pp. 729-754
New India Publishing Agency, New Delhi (India)
e-mail : info@nipabooks.com; website : www.nipabooks.com

37

Integrated Disease Management of Fruit Crops under Hill Agroecosystem

DEEPAK KUMAR AND S.K. SINGH

Introduction

Fruits, oldest forms of food known to man, are not only good source of nutrition to the body but also have medicinal property and treat ailments. Fruits like apple, lemon, orange and pomegranate aid in proper functioning of the heart. Nutrients in apple, date and mangoes sharpens memory, prevents exhaustion, hysteria, insomnia and mental tension. All berries are rich in iron, phosphorus and sodium which are essential minerals for blood building and nervous system strengthening. Liver ailments, indigestion and rheumatism can be cured by the use of Lemons. Watermelons are good kidney cleansers. Pineapple, pomegranates help in soothing inflammation of nose, throat, fever and other chronic or bronchial ailments. Grapefruit juice is an effective aid for common cold. Fresh and ripe fruits like grapes, apples, bananas and figs are good for all brain deficiencies. Even the kernel of walnut is a helpful remedy for weakness of the brain.

Crop	Nutritional Value	Action
Fruits	Potassium, Magnesium, Sodium and Vitamins	Hydrating, Diuretic, Alkalinizing, Mineralizing, Laxative effect, Tonic action and Diuretic
Dry Fruits	Calcium and Iron	Strong bones, Good amount of blood
Custard apple	Calcium	Strong bones
Guavas, Custard apples, Papaya and Citrus fruits	Vitamin C	Maintains skin
Mango, Papaya, etc.	Carotene (Precursor of Vitamin A)	Improves vision

India accounts for nearly 10 per cent of total world production of fruit crops. The annual production of fruits in India is about 44 to 46 million tonnes. The major fruits are mango, banana, grape, citrus, apple, guava, papaya, and pineapple. India ranks first in world production of two fruit crops mango and banana, with the contribution of about 42 per cent and 26 per cent, respectively. During 1998-99, production of mango in India was 98 lakh tonnes of mangoes from an area of about 14 lakh ha. Productivity of grapes in India is the highest in world. There has been large growth and export potential of raisin (dried grapes). This is the sector which is upcoming with good potentials for further growth. Ber and aonla are the major fruit crops, which are becoming popular in dry land areas due to its marketability and medicinal properties. Pomegranate is another important crop for dry land areas. Citrus group (orange, lime, lemon etc.) covers an area of 4.88 lakh ha with 46 lakh tonnes of production during 1998-99 is also an important horticultural crop in the country. Apple, mainly grown in temprate areas like, Jammu & Kashmir, Himachal Pradesh and hills of Uttarakhand, has expanded its domestic market in recent time and has provided livelihood to large number of people in hills. The production of apple was 13.8 lakh tones during 1998-99 from acreage of 2.31 lakh ha.

In the last three decades the area under fruit crops has increased, but still it is nearly one per cent of the total cultivated area. The increase in area alone cannot maximize the yields but productivity should also increase. It is possible only through the introduction and popularization of high yielding varieties and the knowledge about the different factors responsible for the crop losses. Among different factors responsible for low productivity, plant diseases especially fungal diseases, play a major role in reducing the horticultural crops yield in the tropical areas. It is estimated that the production could be increased at least by 28% if the crop could be protected well in time against various diseases.

All-India area and production of fruits

(Area: 000' ha., Production: 000' t)

	Area				Production			
Crops	1996-97	1997-98	1998-99	1999-00	1996-97	1997-98	1998-99	1999-00
Apple	222.7	227.7	131.4	238.3	1308.4	1320.6	1380.4	1047.4
Banana	424.7	441.7	464.3	490.7	1244.0	14141.4	15072.7	16813.5
Citrus	474.7	482.7	488.1	526.9	4456.2	4258.5	4575.0	4650.6
Grapes	42.9	40.8	42.6	44.3	1134.7	969.3	1082.7	1137.8
Guava	150.9	151.5	151.3	150.9	1601.2	1631.4	1800.5	1710.5
Litchi	51.2	57.8	56.2	56.4	377.6	454.8	428.9	433.2
Mango	1344.9	1381.2	1401.6	1486.9	9981.3	10157.0	9781.8	10503.5
Papaya	63.0	69.2	67.7	60.5	1299.3	1582.1	1582.4	1666.2
Pineapple	68.7	69.1	74.2	75.5	924.6	946.7	1006.4	1025.4
Sapota	45.7	48.2	50.4	64.4	588.5	629.3	667.6	800.3
Others	690.1	711.8	799.2	602.0	6346.8	7776.8	6663.6	5707.6
Total	**3579.5**	**3681.7**	**3727.0**	**3796.8**	**40458.2**	**42867.9**	**44042.0**	**45496.0**

Source : Indian Horticulture Database, 2001, National Horticulture Board, Ministry of Agriculture, Govt. of India.

Disease

According to American Phytopathological Society and British Mycological Society the most acceptable definition of disease is "Disease is a malfunctioning process that is caused by continuous irritation which results in some suffering produce symptoms".

This continuous Irritation in the host caused by the pathogen may be due to:

i) Utilizing the host cell content by the pathogen

ii) Causing cell death

iii) Interference with the metabolic activities by the pathogens through the production of enzymes, toxins and growth regulators

iv) Tissue weakening by continuous nutrients loss and

v) Interference with the translocation of food, minerals and water of host

Causes of Disease

A) Non-parasitic Causes of Disease (Abiotic, Inanimate or Non-Infectious)

1. Adverse Climatic Condition
2. Adverse Soil Condition
3. Chemical Injuries
4. Atmospheric Impurities

B) Parasitic Causes of Disease (Biotic, Animate or Infectious)

1. Fungi
2. Bacteria
3. Viruses
4. MLOs
5. Algae and Protozoa
6. Flowering Parasitic Plants

Management Practices of the Diseases

Different types of control measures are being used to manage or mitigate the amount of pathogen present in an area, or on the plants or plant parts.

i) *Cultural Management* : This method of disease management depends on certain actions of farmers. Under the cultural management the alteration in the following actions can be made to minimize the disease intensity / incidence.

1. Host eradication
2. Crop rotation
3. Sanitation
4. Proper Fertilization
5. Mulching
6. Drip Irrigation
7. Tillage Practice
8. Soil Amendments

ii) *Physical Management* : This method of disease management depends on the physical energy such as heat and cold. Under the physical management the following actions can be taken to minimize the disease intensity/incidence.

1. Soil Solarization
2. Heat Treatment
3. Refrigeration
4. Radiations

iii) *Chemical Management* : This method of Control measures includes the destruction of pathogens or the source of primary inoculums in the field by chemical. Under the head chemical management chemicals are used as the following methods of application to minimize the disease intensity/incidence.

1. Soil Treatment
2. Soil Fumigation
3. Seed Treatments
4. Treatment of diseased plant

iv) *Biological Method* : This method of disease management includes the use of living organisms to reduce the inoculums. The use of following micro organisms to minimize the disease intensity/incidence comes under the biological methods.

1. Fungi
2. Bacteria
3. Viruses (Cross Protection)

Integrated Disease Management

As the name indicates this method of the disease control is the integration of two or more control methods to manage the disease intensity/incidence percentage under economic injury level.

Major Fruit Crops

The Major Fruits Grown in the Tropical and Sub-tropical areas are mango, litichi, citrus, guava, aonla and ber. Apple, pear and walnut are the main fruit crop of the temperate region. All the fruits commonly grown in this area are highly susceptible to a number of fungal, bacterial and viral diseases. These diseases are the main constraints in the cultivation of these high value crops.

Mango

Mango (*Mangifera indica* L.) is an important fruit due to their taste and export potential. Mango is prune to a number of diseases at all its stages of development. Different pathogens attack all the plant parts, such as trunk, branch, twig, leaf, petiole, flower and fruit starting from nursery to the consumption of fruits. Some of the diseases of mango are so severe on plant and fruits and act as limiting factor in the mango cultivation.

Powdery Mildew (*Oidium mangifera* Bert.)

The disease was first reported from Brazil in 1914 by Berthet (1914), and it has been reported from different states of India like, Uttar Pradesh, Punjab, Maharashtra, Gujarat, Madhya Pradesh, Jammu & Kashmir, Haryana, Andhra Pradesh, Karnataka and Tamil Nadu. The losses caused by this pathogens varies from 5-20 per cent depending upon the weather conditions.

Symptoms

Pathogen attacks the leaves, flower scales, buds of tender flower heads, axils, stalks and fruits. The pathogen manifests itself as wefts of white mycelia mat on the affected plant parts. The whole plant surface is later on covered with a white powdery substance which is blown away by even a slight wind disturbance. The fungus is of ectophytic nature, establishes mostly on the surface and draws its nutrition from the plants through the haustoria which enters the plant cells by penetration of the epidermal layer. Mealy white growth on leaves and inflorescence can be seen. The affected fruits do not attain proper size and pre mature fruit drop may occur before attaining pea size. Heavy morning dew with Warm temperature and cloudy weather favours the disease development (Kulkarni 1924). The maximum, minimum and optimum temperature to be 30.5, 9, and 22°C respectively for conidia germination is stated by Uppal *et al.*, 1941. Verma and Kaur (1997) reported that none of the cultivar was resistant or moderately resistant. According to Sinha *et al.* (2001) Powdery mildew (*Oidium mangiferae*) of mango attacks the young floral branches and severely impairs fruit set. Disease development was highly influenced by weather conditions such as high relative humidity (>80%) and minimum temperature (>10°C). Consecutive occurrence of high relative humidity and minimum temperature above 10°C for about 70-80 h during the flowering season caused the powdery mildew infection.

Management

Timely pruning of diseased leaves and twigs and proper burning of disease infested parts reduce further spread of disease. Dusting with fine sulphur (250-300 mesh) at the rate of 1-3 lb. of sulphur/tree is the most effective method of disease control. The incidence of powdery mildew [*Oidium mangiferae*] decreases with increasing NPK application rates (Cheema and Malhi, 1986). The first application may be soon after flowering, second 15 days later followed by a third one. Several fungicides, such as benlate (0.2%), karathane or morestan, Dinocap or Tridemorph (Joshi and Chauhan, 1985; Sharma *et al.*, 1993) carbendizim (0.1%) (Prakash and Roouf,1982), Baycor, Calixin, (0.05%), Systhane (0.05%), Saprol (0.2%) were found to be effective in controlling the disease (Sohi and Sridhar 1973; Lingraj 1969; Prakash and Singh 1982; Rawal and Ullasa 1985; Rawal and Saxena 1989). Sinha and Verma (2002) reported that two sprays with wettable sulfur (0.2%), dinocap (0.05%), carbendazim (0.1%), tridemorph (0.1%) or thiophanate-methyl (0.1%) at 15 days interval, after the occurrence of high humidity and low temperature for 4-5 days in disease prone areas effectively controls the disease. Nofal and Haggag (2006) found the BCAs (*Verticillium lecanii, Bacillus subtilis* and *Tilletiopsis minor*), phosphate solutions,(KH_2PO_4); antitranspirant (kaolin) and antioxidant (ascorbic acid) in combinations as an integrated management system to control powdery mildew in mango trees.

Anthracnose

Mango anthracnose caused by *Collectotrichum gloeosporioides* is a serious disease for mango cultivation which usually occurs at fruit ripening stage, damaging the fruit production by 10-20 per cent (Li *et al.*, 2006). Anthracnose, caused by *Glomerella cingulata*, is the most prevalent postharvest disease of mango in humid growing areas (Arauz, 2000). Rainfall, high relative humidity and high temp. favours the disease development (Ann *et al.*, 1994). In India, it is widely distributed in Bihar, Punjab, Maharashtra, Gujarat, Kerala, Uttar Pradesh, Rajasthan and Karnataka.

Symptoms

The disease produces many type of symptoms like, leaf spot, blossom blight, wither tip, twig blight and fruit rot symptoms. Disease develops on all the tender plant parts and is serious especially on tender twigs, flowers, flower stems and fruits in storage. Young leaves wither and dry up when gets attacked. Sometimes only the leaf edges are attacked and their margins darken, dry up, may fall out, giving rugged appearance of the leaf. It produces a wither tip symptom on young tender twigs and results in die-back of the twig with a blackish spot on the growing tips. The affected branch ultimately dries up and infection keeps on penetrating down. In case of fruits, most of the infection takes place from the start of the blossoming period until the fruits are near to mature. The spots appear near the stem end as small brown areas that enlarge rapidly and become black. The affected areas usually crack and sink slightly. The decay is limited to the fruit skin but in late stages it penetrates the shallow areas of flesh. The latent infection during ripening is carried from the field and develops during storage which cause fruit rot. Healthy fruits get infected after coming in contact with diseased fruits. The inoculum remains on dried leaves, defoliated branches and mummified flowers and flower brackets. Excessive rains, heavy dews or high humidity during critical infection period influence the severity and prevalence of the disease (Doidge 1932).

Management

Anthracnose disease can be effectively controlled by spraying with Carbendazim (1%) or Topsin-M (0.1%) or Chlorothalonil (0.2%) at 14 days intervals until harvest. Benlate (0.2%) and Dithane Z-78 (0.2%) are extremely toxic to fungus in culture. Spraying of Micop, Blitox and Dithiocarbomate are also effective chemicals for the management of the disease. Bordeaux mixture (1.0%) has been recommended by Gadre (1979) and Li *et al.* (2006). Application of Captan (3 g/l) or Phaltan (2g/l) are also recommended by Oricux 1961. Four sprays of Copper oxychloride gave significantly better control of mango anthracnose, caused by *Colletotrichum gloeosporioides* (Javaid 1998). According to the Venkataravanappa *et al.* (2007) Tricyclazole was highly effective in the management of the disease and among non-systemic and combination products, SAAF was highly effective followed by copper oxychloride. Anthracnose on the harvested fruits would be controlled by dipping the fruits in various fungicides or exposing them to Ammonia, Sulphur dioxide and Carbon dioxide gases and also by hot air treatment (Tandon and Singh 1968). The fruit dipping in Benomyl solution (9500 ppm a.i) or Thiobendazole (1000 ppm) immediately before storage reduces the disease to 5 per cent from 29 per cent (Sohi and Sridhar, 1973). Pre-harvest application of benomyl (250 mg/litre a.i.) significantly reduced the number of conidia on the harvested fruits (Dodd *et al.*,1991). Pre-harvest spraying with Bavistin (0.1%) or Topsin-

M (0.1%) or Prochloraz (0.1%) followed by dipping in the same after harvesting have also been found to be effective. Hot water treatment (52-54°C for 8-10 min) or treatment with Iprodione or Thiabendazole (1000 ppm.) effectively controlled the mango anthracnose (Feng *et al.,* 1991). Yang and Lin (1997) recommended the dipping of Mango fruits into 53° hot water for 5 minutes, and immediately dripped into tap water to reduce the temperature reduce the disease intensity. Prakash and Pandey (2000) observed that hot water treatment at 52°C alone for 30 min was very effective in controlling the disease. Hot water (52° for 15 min) together with Carbendazim (each @ 0.1%) treated fruits could be stored for 26 days at 12 degrees without any anthracnose infection. A hot Benomyl dip (850 mg/litre a.i. at 52-55°C for 10 min) completely eradicated the pathogen on fruits treated on the day of harvest (Dodd *et al.,* 1991). Treatment of hot water + Bavistin at 1000 ppm, followed by Carbendazim, Captaf, hot water and Prochloraz give good control (Ullasa, 1989).

Mango malformation (*Fusarium moniliforme* var. *subglutinans* Sheldon)

Mango malformation disease occurs in many countries of Asia, Africa, and the Americas and was first reported in India in the year 1891 (Marasas *et al.,* 2006) and is the most important disease of mango. The losses caused by this disease vary from 50-80 per cent. Prakash and Raoof (1987) found none of the cultivars as resistant against the disease. Witches broom or bunchy top or vegetative malformation with a proliferation of infected tissue and floral malformation are two types of symptoms of the disease.

Symptoms

The flowering panicles turn into compact and a very hard mass and not soft like normal panicle. Individual flower is greatly enlarged and has a large disc and the no of bisexual flower reduced in penicle. Floral malformation, which is common in northern India, particularly in Punjab, gives malformed panicles with mostly male flowers, resulting in reduced crops (Majumder and Diware, 1989). All flowers and fruitlets from diseased panicles were found infected (Freeman *et al,* 2004). In bunchy top, small and compact leaves are formed in a bunch at the apex of shoot or in the leaf axil. A similar bunch consisting of small rudimentary leaves crowded together of short shootlets is seen in vegetative malformation. Vegetative malformation is more common in young seedling and seedling trees. The malformed heads dry up and converts into black masses and persist on the trees for a long time. Vegetative malformation was more pronounced in young plants than in older plants. Floral malformation was directly responsible for reductions in the yield of mango plants. Malformed panicles showed higher C:N ratio than normal panicles (Singh *et al.,*1999). Singh *et al.* (1992) reported that Malformed plants were shorter and had shallower root systems than healthy plants and also younger plants were more affected by the disease than older trees. The incidence of vegetative malformation (bunchy top) increased with rising N rates but it was reduced by high P and K rates. The lowest incidence of floral malformation was observed with the medium (200g) N rate (Cheema and Malhi, 1986).

Management

Reduced incidence of malformation has been reported by spraying 100-200 ppm NAA during October by Majumdar 1977; Mehta *et al.*, 1986 and Siddiqui *et al.*, 1987. Application of NAA (200 ppm) in Oct., and deblossoming at bud-burst (Dec.-Jan.) to delay flowering, reduced malformation and gave a 5- to 20-fold increase in crop yield (Majumder and Diware, 1989). Use of disease free planting material and prophylactic spray of insecticides and fungicides can keep the orchards healthy. In areas with less than 5-10 per cent infection, pruning of diseased plants should be made compulsory; pruning of diseased parts along the basal 15-20 cm apparently healthy portions followed by the spraying of Bavistin (0.1%) or Captaf (0.2%). Benlate and Folicur proved highly effective in minimizing disease intencity (Iqbal *et al.*, 1998). Mahrous (2004) showed that the incidence of floral **malformation** disease was best reduced in the next flower season by spraying sprayed with NAA at 200 ppm, IBA at 500 ppm, phosphoric acid at 500 ppm, GA3 at 50 ppm, zinc sulphate at 5% and Cuprous oxide (Cu2O)at 0.4%. Sirohi *et al.* (2005) reported the reduction in mango malformation in trees sprayed with 200 ppm NAA in combination with 0.1% carbendazim.

Stem End Rot (*Botryodiplodia Theobromae* Pat.)

It is an important disease of mango which occurs in the ripening stage. Die-back symptoms in orchards are also causes by this fungus. In India, it is reported from all the mango growing areas.

Symptoms

Disease is characterized as the dark epicarp around the base of the pedicel in the initial stage. In the next few hours, the affected area enlarges to form a circular, black patch which in humid atmosphere extends rapidly and turns the whole fruit completely black within two or three days. The disease is carried from the dead twigs and bark of the trees harbouring the pathogen and with the onset of rains the orchards get contaminated.

Management

Thomas and Dalai (1968) reported dipping of mangoes in 6% borax solution at 43°C for 3 minutes reduces the incidence. Dharamvir *et al.* (1967)

found fruit dipping in Aureofungin and wrapping as effective control measure. Hot water treatments after harvesting the fruits at 53°C for 10 minutes (Quimio and Quimio 1974) and at 51 to 55°C for 15 minutes (Liu 1985) have been recommended. Injury to the fruits during harvesting should be avoided and at all stages of handling after harvest. Rawal and Ullasa (1988) recomended that Bavistin (0.1%) or Topsin-M (0.1%) or Chlorothalonil (0.2%) spraying in the field before harvesting give very effective control.

Die-Back (*Botryodiplodia thiobromae* Pat)

It is a most destructive disease of mango and is known to occur in India. Die-back of mango stone grafts caused by *Colletotrichum gloeosporioides* [*Glomerella cingulata*] and *Botryodiplodia theobromae* Pat either alone or in combination is a major disease syndrome widely prevalent in the Konkan region of Maharashtra, India (Savant and Raut, 2000).

Symptoms

Discolouration and darkening of the bark some distance from the tip of the twigs is the main visible symptom of the disease. The darkening of bark advances and with the result young green leaves starts withering first at the base and then extending outwards along the vein. The leaves affected with the pathogen turn brown and the margins roll in upward direction. At this stage, the affected twig or branch dies, leaves shrivel and fall down prematurely. This symptom may be accompanied by exudation of gum. Internal discolouration or brown streaks of vascular tissues can be seen in the infected twigs on splitting the twigs lengthwise. The die-back disease is also caused due to *Colletotrichum gloeosporioides.* Both the pathogens attack fruits also and can cause spoilage during storage or ripening. *Xyleborus affinis,* a beetle, is considered as the vector of the disease (Batista 1947).

Management

Pruning and destruction of infected twigs is the foremost practice for effective control of the disease. Periodical spraying of the trees with copper oxychloride sulphate is recommended by Alvarez and Lopez (1971) and Prakash and Raoof (1989). Sprays of carbendazim (0.1%) or methyl thiophenate (0.1%) or chlorothalonil (0.2%) at fortnightly interval during rainy season is important. Healthy twigs should be selected for grafting of seedling during propagation.

Black Spot (*Xanthomonas compestris* pv *mangiferae indicae*)

It is one of the serious diseases of mango. In India, the disease is reported from many mango growing states like, Maharashtra, Delhi, Uttar Pradesh, Tamil Nadu and Karnataka.

Symptoms

Mango bacterial black spot, caused by *Xanthomonas campestris* pv. *mangiferaeindicae*, is a potentially severe disease in several tropical and subtropical areas (Pruvost *et al.*, 2005) The pathogen affects all tender parts of plant viz; leaves, petioles, fruits and tender stems. Numerous small angular and water soaked lesions appear in groups on the tip of the leaf blade. The lesion are firstly appear as light yellowish spot but later turn dark brown to black and get surrounded by a distinct halo. In case of young fruits, water-soaked lesions develop on the fruits skin which later on turn dark brown to black. Infected fruits may show cracks on skin and badly affected fruits dry prematurely. Infection produces diseased spots on tender stems, petioles and on stalk bearing fruits. These spots turn into deep longitudinal scars along with oozing out of appreciable amount of bacterial gum.

Management

Sprays with fungicides such as copper containing materials and agrimycin (Viljoen and Kotze 1972) have also been advocated. Orchard sanitation by removal of infected materials and treatment of seedling as preventive measures. Agrimycin-100 proved best in arresting the disease development. Plantomycin (200 ppm) followed by agrimycin + bavistin (1000 ppm) was reported to be the best against this disease (Prakash and Raoof, 1985).

Black Tip

This disease is peculiar to India and does not occur in any other country. It has been reported to occur in regions of West Bengal, Uttar Pradesh, Bihar and Punjab in orchards located in close proximity of a brick kiln.

Symptoms

The symptoms manifest as small etiolated area at the distal end of the fruit, which gradually spreads, turns nearly black and covers the tip completely. The brown discolouration spreads to the neighbouring parenchyma and the deposits also appear in the ducts. The browning and deposits gradually spread throughout the mesocarp and the affected cells disintegrate and coalesce into a dead tissue. In severe cases, the necrosis extends to the endocarp.

Management

The practical control measure lies in keeping the brick kilns away from the mango orchards. The use of telescoping brick kiln chimney of 12-15 m height has been recommended by Sen (1943). Spraying of plants with borax (0.6%) at 10-14 day intervals at the time of fruit maturity has been recommended (Reddy and Kapoor 1965).

Phanerogamic Parasites (*Dendrophthoe viscum*) (*Loranthus*)

The mango trees are severely affected by this parasite particularly in neglected orchards. They are commonly present on the trunk or branches of the trees. The foliage of the infected tree is sparse, reduced in size and its bearing capacity and quality of fruits is considerably reduced. The fruits of this parasite are baccate with viscous mesocarp and attract the birds by its attractive colour. There are many other species of parasites which also attack mango trees.

Citrus (*Citrus* spp.)

Citrus groups rank the third position among the fruits industry of India and assumes considerable importance. In tropical and subtropical conditions, the crop grows much better. There is considerable increase in area under citrus cultivation from 90.7 thousand hectares in 1961 to 279.4 thousand hectares in 1988. Some of the fungal diseases like root rot, foot rot, leaf fall and fruit rot, twig blight, gummosis, powdery mildew and scab, etc., are the important diseases of citrus cultivation. Die-back or citrus decline caused by Phytophthora spp. (Faser and Singh, 1966) is the most serious problem at present our county is facing in its cultivation.

Gummosis, Leaf Fall, Root Rot, Foot Rot or Fruit Rot (*Phytophthora* spp.)

This disease is one of the serious disease in all the citrus growing areas especially in the areas of high rainfall. This disease is caused by more than one species of *Phytophthora,* i.e., *P. nicotianae* var. *parasitica* and *P. palmivora. P. nicotianae* var. *parasitica* is widespread in Assam and is often associated with *Fusarium lateritium. P. nicotianae* var. *parasitica* is also predominant cause of the disease in Karnataka, whereas *P. palmivora* (Butl) is prevalent throughout the country. *Phytophthora* produces symptoms of decline in two ways: (1) by rotting of the rootlets, (2) by girdling of the trunk and a dropping of blighted leaves. The discarded fruits lying on the ground are supposed to invasions by the pathogens and development of brown rot. *P. citrophthora* has also been reported to cause the foot and root rot in citrus.

Symptoms

Bark of lower part of trunk near to ground level is more susceptible to infection than the bark below the ground level. The secretion gum like substances is sometime noticed on higher branches also. Initially, the bark is firm but with ageing of the lesion bark of the affected areas shrinks and cracks, leading to shredding in lengthwise strips as it dries. The live bark at the advancing margins lesions frequently develops a callus roll which helps in checking of further spread particularly in an upward direction. Symptoms on leaves of citrus are similar to those produced on the foliage of potato by *Phytophthora infestans*. Blighted leaves drop prematurely, and in case of heavy defoliation the impression is like that the tree is in decline. Penetration of trunk tissue by the pathogen may be through the wounds or through the growth on cracks that develop with the emergence of adventitious shoots near the bud union. The infected plants usually show heavy blossoming and die before maturity of fruits.

Management

Adoption of tolerant rootstocks like sour orange, Cleopatra mandarin, trifoliate orange, Rangpur lime, etc., had been recommended for the management of the disease in the field by Chadha (1970). Use of nursery stock grown under *Phytophthora* free conditions and maintenance of good drainage for prevention of water stagnation is also recommended for ninimization of the disease in field. The disease can further be prevented to some extent by dusting of walls and bottom of planting pits with zinc-copper lime in ratio of 5:1:4 (Ramakrishnan 1954). The exposed wood can also be covered with zinc copper lime (3:1:3.5) wash or bordeaux paste which should be followed by application of white lead or any wound dressing agents. Desai and Pathak (1970) achieved complete control of spraying Khagzi lime trees infected with *P. citrophthora* with aureofungin 3 g/30 ml of liquid soap/30 gallons of water applied 2-3 times. Good control of gummosis incited by *P. nicotianae* var. *nicotianae* by aureofungin as soil drench and spraying of foliage twice at an interval of one month. Drenching of Ridomil MZ (0.2%) has also been found effective against gummosis. Captafol (3 g/l), copper ammonium carbonate and cupric hydroxide are effective for foot rot (Timmer 1977). Dipping of seedlings in captan solution before planting is recommended. Fumigation of nursery bed with methyl bromide has proved most effective to avoid nursery infection. Leaf fall and fruit rot could be controlled by spraying bordeaux mixture (1%) or bordeaux mixture plus tin sulphate or difolatan (0.3%). Spraying of Ridomil MZ and Aliette (0.2%) have also been reported effective (Rawal 1990). Diseased leaves and fruits may be collected from orchard periodically and burnt or burried deeply. Jadeja *et al.* (2000) observed that for effective control of canker and gummosis, citrus trees may be treated with Bordeaux paste on the main trunk or soil drenching around the basal trunk with metalaxyl + mancozeb or fosetyl Al in combination with foliar application of streptomycin sulphate + copper oxychloride three times a year i.e. before monsoon, in August and December.

Anthracnose

The disease is a serious problem and a limiting factor of the citrus industry of our country. In India, the disease has been reported from many states namely Punjab, Madhya Pradesh, Uttar Pradesh, Assam, Rajasthan and South India. The disease is also known as withertip. In recent years this disease was responsible for decline in production of lime in North India. *Colletotrichum gloeosporioides*. Produses symptoms like, Withertip, leaf-spot and fruit rot of citrus.

Symptoms

The pathogen of anthracnose attacks on the the young leaves, shoots, blossoms and fruits of small acid limes. Blightening of Young foliage and blossoms are produced with distinct lesions on leaves and fruits. Frequent premature fruit drop of affected fruit is common. Withertip is characterized by die-back of twigs and shedding of leaves. Black dot like acervuli appears in concentric rings on dead twigs. Branches show die-back and the tree dies in severe cases in a few years. Twigs show a sharp line of separation between healthy and dead tissues with slight gumming.

Management

Proper management of orchards by proper irrigation, manuring, cultural practices and plant protection measures can minimize the disease. Dead twigs should be pruned and destroyed properly and cut ends should be protected by bordeaux paste. The trees after pruning should be sprayed with bavistin (0.1%) or captafol (0.2%) 3 times after pruning. Spraying of bavistin (0.1%) or Topsin-M (0.1%) at fortnightly intervals is also found effective during fruiting period.

Citrus Canker

The first report of the disease is from Japan and now it is known to occur in every citrus growing area of the world. The pathogen is known to infect all citrus species/cultivars of commercial importance and causing heavy loss in yield (Rangaswami, 1975). The disease severity on different species/cultivars of citrus varies from region to region depending on the distribution of various bacterial strains and environmental conditions (Das and Dubey,1989).

Symptoms

The pathogen may attack any part of the tree above the ground level. The leaves, twigs, young branches and fruits are considered most susceptible parts. The lesion first appears as small watery, transluscent spots, usually dark green in colour than the surrounding tissue and with a raised convex surface. The surface of the spots becomes white or greyish and finally ruptures with the advancement of disease, exposing a light brown, spongy, central mass developed in crator like formation. The spots usually surrounded by a yellow halo. Yellowish halo are not found in case of fruit lesions but the crater like appearance is more pronounced. Fruits may crack and drop prematurely in severe cases. Grape fruits, sweet oranges lime and lemon are very much susceptible to canker. Casual organism is of the disease is *Xanthomonas campestris* pv *citri*. Braithwaite *et al.* (2002) reported that canker lesions were yellow/brown, raised and corky, which darkened and developed central depressions with age. The edges of the lesion remained raised and were frequently surrounded by a chlorotic halo. In Severe infections a variety of effects including defoliation, dieback, severely blemished fruit, reduced fruit quality and pre mature fruit drop were reported by Das (2003). Ye Tian Min *et al.* (2007) found small yellow spots appeared first on the leaf blade, then corky and grey with stripes at the centre. Symptoms on infected fruits were similar to these on damaged leaves and shoots.

Management

The appropriate management of the disease can be mainly obtained through the proper following of plant quarantine or planting of the disease free planting material. Eradication and proper destruction of infected trees, shoots or leaves are the important cultural practice. Spraying of neem cake at the rate of 7 kg per acre or any systemic insecticides have been reported to be highly effective in checking citrus canker as well as leaf minor. Spraying of streptomycin sulphate (0.05%) at new flush emergence. Phytomycine (0.02%) has also been reported to be effective in checking the disease, and Spraying with copper oxychloride or bordeaux mixture (1%) along with pruning of infected plant parts were found effective. Pan (2004) observed that spraying 500 times dilution of solution of 77 per cent kocide (copper hydroxide) wettable powder to summer shoots could get 100 per cent of the shoot leaves with out infection. Maher *et al.* (2005) found that Streptomycin sulphate, vitavax, dithane M-45 and agrimycin-100 in order proved effective in reducing the disease intensity. Koller *et al.* (2006) reported that citrus canker can be controlled by spraying Bordeaux mixture or copper oxychloride under the organic or conventional system of cultivation.

Litchi

Litchi chinensis Sonn. does not have much problem of disease as such (Galan Sanco 1989). The tar leaf spots caused by *Pestalotia pauciseta* Sacc., *Botryodiplodia theobromae* Pat. and *Colletotrichum gloeosporioides* occur in litchi orchards of some areas. The leaves occasionally showed mixed infections of *C. gloeosporioides* and *B. theobromae,* and some spots may be due to the *P. pauciseta.* The lesions coalesced at the later stages. The younger leaves show low percentage of infections than older leaves. The diseased leaves serve as the major source of primary inoculum. These leaf spots spread rapidly during humid and rainy months. The fungicides like Carbendazim, Topsin M, Baycor and Dithiocarbmates are used for the management of the disease.

Microdiplodia litchi is another serious leaf spot of litchi. Yellow brown to brick red areas mostly towards margins can be seen on the diseased leaves. The areas gradually become light brown and show black dot like pycnidia. The other disease problems on litchi due to *Clitocybe tabescence.* sudden wilt caused by *Fusarium* sp., die-back due to *Phomopsis* and foliar necrosis by *Gloeosporium* spp. are also reported. Some of the post-harvest diseases are also reported which are carried along with the fruits during transit and storage. The pathogens causing post harvest diseases like *B. theobromae, C. gloeosporioides, Cylindrocorpon forkinense,* and *Phomopsis* spp. and *Pestalotia* species are the most important and need attention.

Apple and Pear

Verma and Thapa hairy root (*Agrobacterium rhizogenes*), collar rot (*Phytophthora cactorum*), smoky blight (*Botryosphaeria quercum*) and canker (*Botryobasidium salmonicolor*) were the major diseases observed at different locations limiting apple production.

Apple Scab *Venturia inaequalis (Spilocea pomi)*

Scab is one of the most serious disease of apple throughout the world. Kellerhals *et al.* (2009) reported that the Apple scab caused by the fungus *Venturia inaequalis* is still the most important disease in apple production. In India it is considered as main constraint of apple cultivation.

Symptoms

Typical symptoms of the disease appear on the leaf and fruit as light brown or olive green, round spots with a velvety and distinct outline appear on either or both sides of the young leaves in spring. The lesion later turns brownish (musty) black and margins become distinct but may get obscured when several lesions gets coalesce. In severe cases the premature yellowing and defoliation of leaves is observed. Young infected fruits develop dull, brownish black round spots with a serrate outline. Severe early infection results in the formation of deformed, cracked, knotty fruits which provides the path for the entry of other organisms causing fruit rot. Apparently healthy fruits which get infected in late summer (monsoon) develop small rough, black circular lesions on skin in the storage. It affects shape, size, shelf-life and quality of fruits, reducing yield up to 70–80% in endemic years.

Management

Spray 5% urea shortly before general leaf fall for hastening the decomposition of apple leaf and to reduce the primary inoculums (Gupta, 1985). Collection and burning of fallen leaves during winter. The following protective spray schedule may be followed to protect trees. At silver tip to green tip stage spray Mancozeb (0.3%) or Dodine (0.1%) or Captan (0.3%). At pink bud stage spray Fenarimol (0.04%) or Bitertanol (0.05%) or Mancozeb (0.3%) + sulphur (0.2%). At petal fall stage spray Carbendazim or Thiophanate methyl (0.05%). At (pea-sized) fruits spray Dodine (0.075%) or Fenarimol (0.04%) or Mancozeb (0.3%) or Bitertanol (0.05%). At walnut-sized fruits spray Carbendazim or Thiophanate methyl (0.025%) + Mancozeb (0.25%) or Captan (0.3%). At fruit development spray Captan (0.3%) or Bitertanol (0.05%) or Mancozeb (0.3%) or Fenarimol (0.04%). Spray Mancozeb (0.3%) or Captan (0.3%) 20–25 days before harvesting. Grow resistant cultivars—Prima, Prisilla, Sir Prize, Macfree, Red free, Nova, Essygro, Liberty and Freedom.

Cankers and Die-Back Diseases (Pink Canker)
(Botryobasidium salmonicolor)

Symptoms

Vanneste *et al.* (2002) reported that in early summer the leaves appear chlorotic and they later **die**. The necrosis moves down the stem during the growing period. Most of these symptoms were found in close proximity to apple and pear trees affected by fire blight. Different types of symptoms of canker disease are produced on trunk and branches. Cankers usually start from the open wounds and produce deep sunken lesions on the bark. The bark portions above or below the wound is killed. The disease weakens the trees and reduces life of plants and fruit yield also by drawing nutrition from the plant.

Management

Cut and burn the badly cankered portions of the tree. During dormancy, scarify the cankered portions up to the healthy portions and cover such wounds with Chaubatia paste or Copper oxychloride paint or cowdung paste (1 part fresh cowdung + one part clay soil and water to make thick paste) which provide good healing to healthy wounds. Spray Copper oxychloride (0.3%) or Carbendazim (0.05%) or Captan 75WP (0.2%) after harvesting and at bud swell stage. Spray schedule adopted for apple scab also helps in reducing its incidence. Disease can be controlled effectively by application of 2% tridemefon (Bayleton) paste on affected area and spraying of 0.5% tridemefon (Bayleton)soon after harvesting of crop.

Powdery Mildew
(Podosphaera Leucotricha)

Gupta and Sharma (2005) reported that the Powdery mildew of apple caused by *Podosphaera leucotricha* is a serious nursery disease in Himachal Pradesh and is fast spreading in the orchards under warm climatic conditions.

Dunemann and Schuster (2009) reprted that Powdery mildew of apple, caused by the obligate biotrophic ascomycete fungus *Podosphaera leucotricha*, is one of the main fungal diseases in commercial apple production.

Symptoms

Infaction on leaves first appear as whitish, felt like patches of fungal mycelium and spore commonly on lower surface of leaves. The lesions may appear on upper surface as chlorotic spots which may sprayed to cover entire leaf with a white powdery mass of spores and mycelium. The disease affects the growth of buds, new shoots and leaves, by producing white powdery growth on these plant parts. Rossetting of fruits of certain varieties like Jonathan also takes place.

Management

Spray during late dormancy at bud swell and petal fall, and weeks later with wettable sulphur (0.2–0.3%) or Carbendazim/Thiophanate methyl (0.05%) is recommended by Verma and Gupta (1986). In nurseries, spraying of fungicides should be done at 7 days interval. Stickers like Selwet E, Selwet 99, Teepol, Sandovit or Triton, must be added in fungicidal suspension. Prune off terminals covered by powdery mildew and destroy them. Apple scab spray schedule also reduces the disease intensity. Gupta and Sharma (2005) Myclobutanil (0.05%), penconazole (0.05%) and hexaconazole (0.05%) were the most effective fungicides in reducing the powdery mildew disease of apple significantly over check.

White Root Rot *(Dematophora necatrix)*

White root rot (*Dematophora necatrix*) is a pernicious soil borne disease of apple in India and is highly damaging in nurseries, as well as to mature trees (Gupta and Sharma, 2004).

Symptoms

Mainly in nursery plants globular, elongated or irregular (variously shaped) galls/tumors are formed at or near the graft-union. Plant growth is affected. Affected trees show yellowing and bronzing of leaves, sparse foliage and retarded growth rate. Roots turn brown and remain covered with white cottony mycelial mat of the fungus in rainy season. The trees ultimately die.

Management

Maintain proper drainage in orchard. Unearth and remove infected roots of trees during dormancy. Apply Chaubatia paste (red lead, Copper carbonate and linseed oil 1:1:250 on cut ends and healthy portions of roots). For curing the ailing trees, give at least 3 drenchings of Carbendazim (0.1%) or

Aureofungin (0.05 + copper sulphate 0.05%) 15–20 days intervals during monsoon. Application of Carbendazim (0.1%) or Aureofungin (0.05 + copper sulphate 0.05%) should be made in 15–25 cm deep holes, 30 cm apart throughout the drip area of tree. Sharma and Bhardwaj (2005) found that P. fluorescens, B. subtilis, and Bacillus sp. were effective in checking the root rot disease of apple. Gupta and Sharma (2004) found that application of *E. aerogenes* + Bavistin (0.1 or 0.05%) showed more than 92 % disease control when applied as pre-inoculation to the pathogen. In simultaneous inoculation, 0.1% Bavistin in combination, completely prevented the appearance of the disease, however, 0.05 % in combination with *E. aerogenes, T. viride* and *G. virens* gave 90.4, 86.9 and 81.0 % disease control, respectively.

Collar-Rot
(Phytophthora cactorum)
Symptoms

Collar region of affected trees near the ground level turns brown, soft and spongy. Complete girdling of stem leads to killing of trees. Spread of disease is maximum in poorly drained soil. Affects trees yield fruits adversely in the beginning and trees ultimately die.

Management

During dormant period remove the soil around the collar region and expose the affected portion. Remove the affected bark and apply Chaubatia paint or Copper oxychloride paint. Drench 30 cm radius area around tree trunk with Mancozeb (0.3–04%) or copper oxychloride (0.5%) or Metalaxyl (0.3%). Keep graft union 20–25 cm above the soil level while planting. Use resistant clonal rootstocks M2, M4, M9, M26, M113 and M114. Sharma and Sharma (2005) reported that symptom appearance of collar rot of apple was delayed by oil cakes ammendment. Neem cake was highly effective with minimum disease incidence followed by mustard cake giving 31.65 and 21.33 per cent disease control under pot culture conditions.

Leaf spots : Mycosphaerella* spp., *Alternaria mali, A. alternata, Coniothyrium pyrinum, Phyllosticta spp.* and *Botryosphaeria quercum.

Symptoms

Spots of various types, colour, size and shape are of common occurrence on the leaves during late summer and rainy season. Heavily spotted leaves turn yellow and fall prematurely.

Management

Spray schedule adopted for apple scab also controls leaf spots. Spray 2-3 times, a mixture of Carbendazim (0.03%) + Mancozeb (0.25%) at 15 days interval during rainy season.

Disease and Causative Agents	Symptoms	Management
AONLA		
Rust *Ravenelia emblica*	• Conspicuous black coloured rust pustules on leaflets and fruits. • Affected fruits drop off prematurely.	Two sprays of wettable sulphur or Chlorothalonil at the rate of 2g / litre of water during December.
BER		
Powdery mildew *Oidium zizyphi var. indicae*	• White mealy growth on leaves and fruits. • Infected fruits appear scating.	Spray of wettable sulphur (2g / litre) Karathane / Bayleton / Systhane (1g / litre) and Topas / Rubigon (0.5ml / litre) during flowering to fruiting.
Leaf-spot *Alternaria sp.*	• Dark brown patches on leaves which later turn yellow and fall off. • On fruits brown depressed elongated spots are prominent	Spray Kavach / Rovral / Mancozole (2g / litre) during December-January.
Rust *Phakospora zizyphi*	• Brown rusty patches on lower surface of leaves. • The upper surface of infected part of leaf remains green, whereas rest of the leaf turns yellow and defoliates.	Apply wettable sulphur/Kavach (2g/litre) Baycor (1g/litre) immediately after disease appearance.
Black-spot *Isariopsis zizyphi*	• Dark black sooty spots on lower leaf surface. • Such leaves loose turgidity and lustre and defoliate.	Baycor / Topsin-M (1g/litre) and Kavach / Rovral / Mancozit (2g / litre) during rainy season.

Sclerotius Blight *(Sclerotium rolfsii)*

Symptoms

Infected seedlings are killed rapidly and small mustard coloured sclerotia appear at the collar region during monsoon. Seedlings die.

Management

Uproot affected seedlings and destroy. Drench the nursery with Thiram (300 g/100 litres) or Aureofungin (40 g/100 litres). Nursery site should be changed regularly and fumigated.

Crown Gall *(Agrobacterium tumefaciens)*

Symptoms

Crown gall disease is an economically significant problem in fruit and nut orchards, vineyards, and nurseries worldwide. Tumors on stems and leaves result from excessive production of the phytohormones auxin and cytokinin in plant cells genetically transformed by *Agrobacterium tumefaciens* (Walter *et al.*, 2003 and Viss *et al.*, 2003). Globular, elongated or irregular (variously shaped) galls/tumors are formed at or near the graft-union mainly in nursery plants. Plant growth is affected. In contrast to other tissues, roots do not respond to high cytokinin levels, and auxin overproduction is sufficient to cause tumor growth on roots (Walter *et al.*, 2003).

Management

Biocontrol with a related strain of bacterium is being practiced.

References

Agrios, G.N. 2005. Plant Pathology. Academic Press, California.

Alveraze, G.L.A. and J.G. Lopez, 1971. Gummosis die-back and fruit rot of mango (*Mangifera indica* L.) caused by *Physalospora rhodina* (B&C) Cke. in Puerto Rico. *J. Agric Univ Res.* 55 (4):435-450.

Ann, P.J.; R.C. Huang, and M.F. Chen, 1994. Effects of environmental factors on disease incidence of mango anthracnose. *Plant Pathology Bulletin.* 3 (1): 34-44.

Arauz, L.F. 2000. Mango anthracnose: economic impact and current options for integrated management. *Plant Disease.* 84 (6): 600-611.

Batista, A.C. 1947. A serious disease of mango. Thesis, *Pemambuco College of Agric. Pemambuco* 19:212-215.

Berthet, I.A. 1914. Molestia de Mangucira. *Boletin de Agric. Sao Paulo XV*, 8-10 : 818-819.

Braithwaite, M., R.P. Leite ; J.J. Smith; E. Boa and G.S. Saddler, 2002. First report of citrus canker caused by *Xanthomonas compestrisa* pv. *citri* on *citrus sinensis* in Bolivia. *Plant Pathology.* 51: 383.

Chadha, K.L. 1970. Rootstocks and Citrus Decline in India (K.L. Chadha, M.S. Randhawa, J.S. Chauhan and L.C. Knorr, eds.).

Chadha, K.L. 2001. Handbook of Horticulture. ICAR Publication, New Delhi.

Cheema, S.S. and C.S. Malhi, 1986. Effect of N, P and K fertilizers on the incidence of mango malformation, anthracnose and powdery mildew. *Punjab Horticultural Journal.* 26 (1/4): 6-7.

Das, A.K. 2003. Citrus canker a review. *J. of Applied Horti.* 5: 52-60.

Das, B.C. and L.N. Dubey, 1989. The incidence of citrus canker *Xanthomonas citri* (Dawson) in Assam. *J. Res. Assam Agril. Uni.* 10: 80-82.

Dharamvir, S.P. Ray Chaudhury and M.J. Thirumalachar, 1967. Aureofungin as fruit dip and wrap treatment for the control of *Diplodia* rot of mango and *Atlernaria* rot of tomato fruits during transit and storage. *Indian Phytopath.* 20 : 301-303.

Dodd, J.C.; R. Bugante, I. Koomen, P. Jeffries, and M.J. Jeger, 1991. Pre- and post-harvest control of mango anthracnose in the Philippines. *Plant Pathology.* 40 (4): 576-583.

Doidge, E.M. 1932. Black spot of mangoes. *Farming South Africa* 7 : 89-91.

Dunemann F. and M. Schuster, 2009. Genetic Characterization and Mapping of the Major Powdery Mildew Resistance Gene Plbj from Malus Baccata Jackii. *Acta Horticulturae.* (814).

Feng, S.Q.; Y.X. Chen; H.Z. Wu and S.T. Zhou, 1991. The methods for delaying ripening and controlling postharvest diseases of mango. *Acta Agriculturae Universitatis Pekinensis.* 17 (4): 61-65.

Faser, L.R. and D. Singh, 1966. Root rot of citrus in India. *Indian Hort. Oct.-Dec.,* 3.

Freeman, S.; D. Klein Gueta; N. Korolev and A. Sztejnberg, 2004. Epidemiology and survival of Fusarium mangiferae, the causal agent of mango malformation disease. *Acta Horticulturae.* 645: 487-491.

Gadre, U.A. 1979. Control of mango Anthracnose (*Colletotrichum gloeosporioides*). *Mango Workers Meeting. Panaji, Goa (AICFIP), 2-5 May,* 197-198.

Galan Sanco, V. 1989. Lychee Cultivation. *FAO Plant Protection and Production Paper 83, FAO, Rome.*

Gupta, D. and J.N.L Sharma, 2005. Chemical Control of Powdery Mildew of Apple in Warmer Climates of Himachal Pradesh, India. *Acta Horticulturae.* (696).

Gupta, G.K. 1985. Apple scab (*Ventura inaequalis*) *E. Merk (India) Ltd.,* Bombay, 36pp.

Gupta, V.K. and K. Sharma, 2004. Integration of Chemicals and Biocontrol Agents for Managing White Root Rot of Apple. *Acta Horticulturae.* (635).

Iqbal, M.Z.; M.A. Halimie; M. Ismail and M. Ayub, 1998. Chemical control of mango malformation. *Pakistan Journal of Phytopathology.* 10 (2): 56-58.

Jadeja, K.B.; N.G. Mayani; V.A. Patel and M.T. Ghodasara, 2000. Chemical control of canker and gummosis of citrus in Gujarat. *Journal of Mycology and Plant Pathology.* 30 (1): 87-88.

Javaid, I. 1998. Evaluation of copper oxychloride sprays for the protection of mango fruits against anthracnose. *Tests of Agrochemicals and Cultivars.* 19: 16-17.

Joshi, H.U. and H.L. Chauhan, 1985. Effective control of powdery mildew of mango. *Second Int. Sym. Mango, Bangalore* : 63 pp.

Kellerhals, M.; T. Szekely; C. Sauer; J.E. Frey and A. Patocchi, 2009. Pyramiding Scab Resistances in Apple Breeding. *Erwerbs Obstbau.* 51 (1) : 21-28.

Koller, O.C., R.B.L. de Oliveira; D.S. Nunes; F.D. Soglio; N.V. Panzenhagen; I.A. Sartori and F. Manteze, 2006. Chemical control of citrus canker in young plants managed in conventional organic systems. *Ciencia Rural*. 36: 1043-1048.

Kulkarni, G.S. 1924. Report of the work done in Plant Pathology section during the year 1922 to 1923. Pp. 167-171. *In* Ann. *Rept. Agri Bombay Presidency for year* 1922-23.

Li, GuiZhen; DaKui Qin and GuiLi Li. 2006. The integrated methods for control of mango anthracnose disease. *South China Fruits*. 4: 42-43.

Lingaraj, D.S. 1969. Mango diseases and pests (*Mangifera indica* L.). *J. Mysore Hort. Soc.* 14:15-18.

Liu, X.J. 1985. Control of infection of mango by latent pathogenic fungi during storage. *Chinese J. Trop. Crops* 5 (2) : 107-112.

Maher, S.M., Sahi, S.T., Ghazanfar, M.U., Inam-ul-Haq., Imran-ul-Haq., Yasir, Iftikhar., Sarwar, M.S. and Tauqir, Ahmad, 2005. Evaluation of different toxicants against *Xanthomonas axonopodis* pv. *citri* (Hasse) Dows. *Inter. J. of Agri. and Bio* 7: 121-124.

Mahrous, H.A.H. 2004. Effect of Spraying some Chemical Substances and a Fungicide on Floral Malformation Disease in Mango. *Acta Horticulturae*. 645.

Marasas, W.F.O.; R.C. Ploetz; M.J. Wingfield; B.D. Wingfield and E.T. Steenkamp, 2006. Mango malformation disease and the associated Fusarium species. *Phytopathology*. 96 (6): 667-672.

Majumdar, P.J. 1977. The Statesman, 24th April. New Delhi.

Majumder, P.K. and D.V. Diware, 1989. Studies on horticultural aspects of mango malformation. *Acta Horticulturae*. 231: 840-845.

Mehta, N., J.K. Sandooja, R.L. Madaan and R. Yamdagni, 1986. Role of different chemicals in mango malformation and related physiological factors. *Pesticides*. 20 (4) : 17-18.

Nofal, M.A. and W.M. Haggag, 2006. Integrated management of powdery mildew of mango in Egypt. *Crop Protection*. 25 (5): 480-486.

Orieux, L. 1961. The control of plant diseases. *Rev. Agric. Suer. Maurice* 39 (3) : 128-130.

Pan, Xiao Qiang, 2004. Experiment of control of citrus canker disease by kocide fungicide. *China Fruits. 5:* 57.

Prakash, O. and B.K. Pandey, 2000. Control of mango anthracnose by hot water and fungicide treatment. *Indian Phytopathology*. 53 (1): 92-94.

Prakash, O. and M.A. Raoof, 1982. Economic spray schedule for the control of powdery mildew disease (*Oidium mangifarae*). *Proc. AICFIP workshop, February, 9-14, Nagpur:* 251-252 pp.

Prakash, O. and M.A. Raoof, 1985. Bacterial canker in mango (abst.). *2nd Inter. Symp. Mango, Bangalore*. 59 p.

Prakash, O. and M.A. Raoof, 1987. Incidence of malformation in mango cultivars. *Indian Journal of Plant Pathology*. 5 (1): 14-18.

Prakash, O. and M.A. Raoof, 1989. Die-back disease of mango (*Mangifera indica*), its distribution, incidance, cause and management. *Fitopathology. Bras.* 14 (2):207-215.

Prakash, O. and U.N. Singh, 1982. Evaluation of various fungicides for the control of powdery mildew of mango caused by *Oidium mangiferae. Pesticides* XVI :171-178.

Pruvost, O.; P. Roumagnac; C. Gaube; F. Chiroleu and L Gagnevin, 2005. New media for the semiselective isolation and enumeration of *Xanthomonas campestris* pv. *mangiferaeindicae*, the causal agent of mango bacterial black spot. *Journal of Applied Microbiology*. 99 (4) : 803-815.

Quimio, A.J. and T.H. Quimio, 1974. Control of *Diplodia* rot of mango by hot water treatment. *Philippines Phytopath*. 10:16-18.

Ramakrishnan, T.S. 1954. Common Diseases of Citrus in Madras State. *Publication of the Govt. of Madras,* Madras.

Rangaswami, G. 1975. Effect of some antibiotics on *Xanthomonas malvacearum* (E.F.Smith) Dowson and on the micro flora of seed cotton. *Indian Phytopathology*. 10: 8-10.

Rawal, R.D. 1990. Fungal and bacterial diseases of fruits. A decade of research on diseases of horticultural crops under AICRP (1980-1989). *Group discussion of plant pahtologists working in the co-ordinated project of horticultural crops held at IIHR, Bangalore during 14-15 June 1990.*

Rawal, R.D. and B.A. Ullasa, 1985. Control of mango powdery mildew (*Oidium mangiferae* Berth) by fungicides. *Second Inter. Symp. Mango (abst.), Bangalore*. 62 p.

Rawal, R.D. and B.A. Ullasa, 1988. Efficacy of pre-harvest field sprays of fungicides against stem end rot of mango. *Acta Hort.* 231 : 537-539.

Rawal, R.D. and A.K. Saxena, 1989. Control of powdery mildew of mango through fungicidal sprays. *Proc. 3rd Inter Mango Symp. held at Darwin, Australia during Sept. 25-29, 1989.*

Reddy, D.B. and S.P. Kapoor, 1965. Black tip in mango. *Intensive Agric.,* Delhi. 9-10 p.

Savant, N.V and S.P. Raut. 2000. Studies on symptomatology of die-back of mango stone grafts. *Acta Horticulturae*. (509): 823-832.

Sen, P.M.K. 1943. Further studies on black tip of mango. *Indian J. Agric. Sci.* 13 : 300-333.

Sharma, I.M. and S.S. Bhardwaj, 2005. Bacterial Antagonists in Management of Collar Rot and Root Rot Diseases in Apple. *Acta Horticulturae*. (696).

Sharma, I.M., H. Raj and N.K. Sharma, 1993. Field evaluation of some new systemic fungicides against powdery mildew of mango in Himachal Pradesh. *Pestology* XVII : 25-27.

Siddiqui, S., J.K. Sandooja, N. Mehta, and R. Yamdagni, 1987. Biochemical changes during malformation in mango cultivars as influenced by various chemicals. *Pesticides* 21 (30) : 17-19.

Singh, J.N.; C.B.S. Rajput and A.N. Maurya, 1992. Survey on the intensity of mango malformation in Varanasi area. *Progressive Horticulture*. 24 (1-2): 9-12.

Singh, N.D.; A.R. Singh; B. Singh and R.K. Singh, 1999. Effect of age and cultivars on the susceptibility of mango malformation. *Advances in Horticulture and Forestry.* 6: 9-21.

Singh, R.S. 2000. Plant Diseases. Oxford & IBH Publishing Co. Pvt. Ltd., New Delhi.

Singh, R.S. 2001. Principles of Plant Pathology. Oxford & IBH Publishing Co. Pvt. Ltd., New Delhi.

Sinha, P. and A. Varma, 2002. Chemical management of mango powdery mildew. *Annals of Plant Protection Sciences.* 10 (2): 291-294.

Sinha, P.; U. Chakrabarti and A. Varma, 2001. Critical factors for mango powdery mildew development. *Annals of Plant Protection Sciences.* 9 (2): 264-267.

Sirohi, S.C.; S. Prakash; P. Rana and R. Singh, 2005. Response of mango malformation to foliar application of NAA and pesticides. *Indian Journal of Horticulture.* 62 (1): 79-80.

Sohi, H.S. and T.S. Sridhar, 1973. Powdery mildew of mango and its control. *Indian Phytopath* 26 : 361-362.

Tandon, I.N. and B.B. Singh, 1968. Control of mango anthracnose by fungicides. *Indian Phytopath.* 21 : 212-216.

Thomas, P. and V.B. Dalai, 1968. A note on the effect of some chemicals on control of stem end rot in mangoes after harvest. *Sci. and Cult.* 34 (7) : 311.

Timmer, L.W. 1977. Preventive and curative trunk tratments for control of *Phytophthora* foot rot of citrus. *Phytopath.* 67 (9) : 1149-1154.

Ullasa, B.A. 1989. Influence of time of application of some fungicides on their efficacy against mango anthracnose caused by Colletotrichum gloeosporioides. *Plant Disease Research.* 4 (2): 196-197.

Uppal, B.M., M.K. Patel and M.N. Kamat, 1941. Powdery mildew of mango. *J. Univ. Bombay* 9 :12-16.

Vanneste, J.L.; S. Lex; M. Vermeulen and F. Berger, 2002. Isolation of Erwinia amylovora from Blighted Plums (*Prunus domestica*) and Potato Roses (*Rosa rugosa*). *Acta Horticulturae.* 590.

Venkataravanappa, V.; V.B. Nargund; V.I. Benagi; Somasekar; C.N. Lakshminarayana-reddy; D.S. Aswathnarayana and M.K.P. Kumar, 2007. In-vitro and in-vivo evaluation of different fungicides against Colletotrichum gloeosporioides - causal agent of mango anthracnose. *Environment and Ecology.* 25S (Special 2) : 318-321.

Verma, K.D. and G.K. Gupta, 1986. Field evaluation of fungicides for the control of powdery mildew (*Podosphaera leucotricha*) of Apple. *In Advances in Research in Temprate Fruits. (T. R. Chadha, V. P. Bhutani and J. L. Koul, eds) UHF. Solan* : 407-409.

Verma, K.S. and D.R. Kaur, 1997. Observations on severity of mango powdery mildew in relation to location and germplasm. *Plant Disease Research.* 12 (1): 28-29.

Viljoen, N.M. and J.M. Kotze, 1972. Bacterial black spot of mango. *Phytoparasitica* 4 (3) : 93-94.

Viss, W.J.; J. Pitrak; J. Humann; M. Cook; J. Driver; and W. Ream, 2003. Crown-gall-resistant transgenic apple trees that Silence *Agrobacterium tumefaciens* oncogenes. *Molecular Breeding: New Strategies in Plant Improvement*. 12 (4) : 283-295.

Yang, H.R. and Y.T. Lin, 1997. Effect of hot water treatment for controlling anthracnose of mango fruits of Tainung No.1 variety. *Plant Protection Bulletin Taipei*. 39 (3): 241-249.

Ye, Tian Min., Lin, Meilin., Zeng, Rui Qin and Hut, Dong Sheng, 2007. The occurrence of citrus canker on pummelo and its control. *South China Fruits*. 30: 24.

□□□

Ecologically Based Integrated Pest Management, 2012
© D.P. Abrol and Uma Shankar (eds.), pp. 755-763
New India Publishing Agency, New Delhi (India)
e-mail : info@nipabooks.com; website : www.nipabooks.com

38

Nematode Pests of Fruit Crops and their Management

VIRENDRA KUMAR SINGH AND UMA SHANKAR

Introduction

Plant parasitic nematodes occur wherever plants grow. These nematodes are referred to as plant pathogenic when they cause crop losses by direct injury to roots, stems, leaves and seeds. A large number of plant parasitic nematodes have been found associated with fruit crops in all over India but pathogenisity has not been proved for all. Most of commonly occurring nematodes belong to genera *Meloidogyne* spp., *Pratylenchus* spp., *Rotylenchulus* spp., *Hoplolaimus* spp., *Tylenchulus* spp., *Xiphinema* spp., and *Radopholus* spp. etc. The nematodes damage the crops not only by feeding on plants but also by interacting with various other organisms like fungi, bacteria and viruses etc. Estimated overall average annual yield loss of the world's major crops due to damage by plant parasitic nematodes was 12.3%. Estimated annual crop losses in India have been accounted to Rs. 242.1 billion. The crop losses due to nematode pests are not only in the form of reduced plant growth and yield but are also in the marketable quality of the produce.

Root-Knot Nematodes (*Meloidogyne* spp.)

Root knot nematode infested papaya plant

Root-knot nematodes are more widely distributed throughout the world than any other group of plant parasitic nematodes. Nematodes are most active in warm weather in moist, but well aerated, sandy soils in the presence of host plants. Root-knot nematodes are important pests of several fruit crops. India about 12 species have been reported to occur. But, of all these species, predominating populations encountered are of *M. incognita, M. javanica, M. arenaria* and *M. hapla*. Bio-chemical differences in protein and enzyme content among species have also been used for taxonomic identification. These four species account for more than 95% of the root-knot nematode population world over (Sasser & Carter, 1985). The above ground symptoms: Root knot nematode damage results in poor growth, a decline in quality and yield of the crops. Nematode damaged roots do not utilize water and fertilizers as effectively leading to additional losses for the grower. Below ground symptoms: The most distinctive nematode symptoms on roots are galls caused by *Meloidogyne* spp. They are small, individual, bead like or fusiform swellings in some roots. In other plants, galls may be massive lumps of fleshy tissue more than 1 inch in diameter containing dozens of nematodes.. The male and female root-knot nematodes are easily distinguishable morphologically. The males are wormlike and about 1.2 to 1.5 mm long by 30 to 36μm in diameter. The females are pear shaped and about 0.40 to 1.30 mm long by 0.27 to 0.75 mm wide. The life cycle includes egg, juvenile and adult stages.

Lesion Nematodes (Pratylenchus spp.)

The root lesion nematodes occur in all parts of the world. Assessment of exact loss by lesion nematodes has not been made possible under field conditions due to presence of mixed population of the nematode in the field. Some species are economically very important on fruit crops in Europe, Canada, USA, and India. Root lesion nematodes feed upon cells in the root cortex. Cells are killed and in many instances small roots are killed. The plants show chlorosis, stunting and general lack of vigour resulting into wilt. The roots show necrosis and lesions which become ideal for infection of other microorganisms. The presence of small brown to black lesions on the root surface is the most important symptoms or damage produced by the lesion nematode. Both male and female of these nematodes are wormlike, 0.4 to 0.7 mm long and 20 to 25 m in diameter. They are migratory endo parasitic nematodes. The life cycle of the various species of Pratylenchus is completed within 45 to 65 days. Temperature, soil type, moisture and tillage operations are important environment factors which greatly affect the development and reproduction of nematode species as well as disease development.

Reniform Nematodes (Rotylenchulus Reniformis)

Reniform nematode is pest of a large number of crops in India, it is considered as pest of great significance after root-knot nematode. The nematode is a semi-endoparasite and remains attached to roots. Infected plants show stunting in growth with reduced and discolored root system. Damage during pre and post emergence of seedlings leads to reduction in germination and crop stand. The total duration of life cycle is about 4-5 weeks under optimum conditions. Soil PH is also an important factor affecting reproduction of the nematode. Soil moisture and temperature have profound influence on infectivity, penetration and the biology of the nematode.

Lance Nematodes (Hoplolaimus spp.)

Lance nematodes are ectoparasites. Some times they feed at a particular site for a long time with nearly half of the body inside the rot system (sedentary ectoparasite). In many cases, juveniles of the lance nematode completely enter the cortical tissue (endoparasite). Hoplolaimus spp. multiplies slowly in comparison to endoparasitic nematodes but they inflict significant crop damage at a lower level of infection. Several workers have reported significant reduction by the Hoplolaimus spp. on various crops in the different parts of India (Haider *et al.*, 1978, Gaur and Mishra, 1981; Das, 1982; Mulk and Jairajpuri, 1975; Sharma, 1985). The availability of feeder roots and temperature are important factors for population build up of this nematode (Haider and Nath, 1992).

Burrowing Nematode (Radopholus Similis)

The burrowing nematode occurs in most of the tropical and subtropical areas of the world. Though, there have been some reports about the occurrence of the burrowing nematode in South India (Goodey, 1933, Mayne & Subramanyam, 1933) the first authentic report was from banana roots in Kerala (Nair *et al.*, 1966). Several cultivars of banana, citrus black pepper, coconut, cardamom, Tea and other spices are known to have originated in the several parts of India. The borrowing nematode is a migratary endoparasitic capable of spending its entire life in the roots. All the juvniles stages and females infest healthy, succulent feeder root tips. Sexual diamorphism is apparent with degenerate, non parasitic adult males. The burrowing nematode causes typical lesions and rotting of roots. The life cycle takes 18-20 days at 24-26°C in citrus race and 20-25 days at 24-32°C in banana race.

Citrus Nematode (Tylenchulus Semipenetranas)

The citrus nematode has been reported from almost every citrus growing country in the world (Thorne, 1961). In India, the nematode was first recorded by Siddiqi (1961) and is now widely distributed in most of the citrus growing states. The trees showing initial die-back symptoms are observed to support a very high nematode population and the population decreases gradually with an increase in the intensity or advancement of die back indicating that trees in early stages of decline support relatively high nematode population (Swarup *et al*; 1964). General decline of infested trees, defoliation of shoots, smalling and yellowing of leaves and fruits, sign of starvation in leaves, necrotic stripes on roots, sloughing off of roots ultimately leading to root decay are some of broad effects recorded under pathogenisity tests. The juveniles can survive for fairly long periods of time even in the absence of host. According to some report, these can survive without host for 9 years and can withstand as high a temperature at 45°C for several hours. The juveniles can survive also in water at 100°C for 24 months.

Cultural Management

One of the most effective eco-friendly management practices is the use of proper combination of cultural practices. The affectivity of cultural method depends upon the proper knowledge of life history, population dynamics host range of the plant parasitic nematodes infesting crops. The cultural practices include crop rotation, fallowing, flooding, sanitation, ploughing during summer season, mulching, organic manure, spacing of plants in the field, time of sowing, resistant varieties etc. Nematode management through inter cropping and crop rotation is based on the fact that some species of nematodes are able to feed and multiply only on host crop. Rotation is a very old practices for reducing soil borne problems. Rotation to non-host crops may cause many of those pests to cease reproduction and allow natural mortality factors to reduce their numbers. Flooding or following may be used to help reduce numbers of nematode pests. The period of flooding appears to vary with several factors such as kind of soil, season etc. Warm conditions are said to reduce the period of time required for control. Juveniles (larvae) are more easily killed by flooding than eggs. The period of flooding needs to be worked out for each condition. The addition of inorganic fertilizers alone without organic manure usually increases the nematode population and disease intensity. NH_4-N reduces the disease incidence while NO_3-N may increase the same. Particular forms as well as dose and proportion of NPK may also reduce or increase the incidence of disease. Use of tolerant/resistant varieties is most practical approach for the management of nematode diseases. Crop cultivars resistant to phytonematodes can be the most useful and cheapest means of nematode control for the small-scale farmers. Nematicidal plants with roots containing nematicidal substances have been investigated. These toxic substances reduce the population level of some nematode species. Application of certain indigenous plant products, non edible oil cakes and other organic waste for the control of nematode has received attention in recent years. Use of organic manures is of great value since the decomposition products and production of natural enemies decrease nematode populations. (Mankau, 1963; Singh and Sitaramaiah 1973, Mishra and Prasad 1974, Haseeb *et al.*, 1984, Alam 1990, Singh and Singh, 1991; Singh and Singh, 1992; Singh and Singh, 2001). Soil amendments with green manure, compost, oil cakes of neem, mahua, mustard, groundnut, cotton, linseed, karanj and saw dust etc. have been found to reduce nematode populations. The amount of oil cakes or any organic matter to be incorporated depends on various factors like, soil type and texture, crop to be planted, the predominant nematode fauna of the soil and the amount of soil moisture present in the soil. Applications of botanicals are easy, environmentally safe, and having no phytotoxic effect on crops. The inhibition of root-knot development may be due to the accumulation of toxic by products of decomposition (Alam *et al.*, 1979) to increased phenolic contents resulting in host resistance (Alam *et al.*, 1979, 1980), or to changed physical and chemical properties of soil inimical to the nematodes (Ahmad *et al.*, 1972). One of the most economical and effective ways to control plant parasitic nematodes is the growing of nematode resistant plant cultivars. The identification of a dominant Mi gene importing resistance to the root-knot nematodes and its linkage to an acid phosphates gene is a major break through in this area of research. The use of resistant root stocks can play a major role in combating nematode problems. Few workers determined the reaction of some citrus varieties to species of Meloidogyne.

Biological Management

Application of natural enemies of plant parasitic nematodes for controlling nematode population is an essential component of eco-friendly management. Nematologists in all over the world are working very hard to identify and lean to manipulate natural enemies of nematodes so they can be used as biological control agents. Nematodes have many natural enemies including fungi, bacteria and predacious nematodes. Certain fungi capture and kill nematodes in the soil. *Arthrobotrys* spp., *Dactylaria* spp., *Dactylella* spp., *Catenaria* spp., and *Trichothecium* spp., are the genera most commonly represented. Some fungi capture nematode by adhesion, but many employ specialized devices that include networks of adhesive branches, stalked adhesive knobs, non constricting rings and constricting rings. The surface of the nematode is penetrated and the fungus hyphae grow throughout the nematode body, digesting and absorbing its contents. Under favourable conditions, large numbers of nematodes may be captured, and killed especially by those fungi that form adhesive net works or hyphal loops. *Trichoderma harzianum, Trichoderma virens, Aspergillus niger, Paecilomytes lilacinus, Verticillium chlamydosporium* are found promising biocontrol agents. The biological control of root knot nematode on various crops under green house conditions by using several fungi have been reported by Pandey and Trivedi (1992); Zuckerman *et al.,* (1993) and Bandyopadhayay *et al.,* (2001). Now mycorrhiza is not restricted to its use only as biofertilizers, its potential role in the biological control of plant parasitic nematodes is reported by many workers. Sikora (1979) found that prior presence of VAM fungi Glomus mosseae has resulted into an increase in plant resistance against *Meloidogyne* spp. A bacterial parasite of nematodes Pasteuria penetrans, has received much attention and research effort in recent years, *P. penetrans* is probably the most specific obligate parasite of nematodes, with a life cycle remarkably well adapted to parasitism of certain phytonematodes. It directly parasitizes juvenile nematode, thus affects penetration and reproduction. Seed bacterization, soil drenching and bare root dip application with Pseudomonas fluroscenes, *Bacillus subtilis, B. polymyxa* effectively controls plant parasitic nematodes was also reported by many workers. Among the predatory nematodes, monarchs may be proved efficient predators because of stronger predatory potential, high rate of predacity and high strike rate (Bilgrami, 1998).

Chemical Management

Nematicides are chemicals used to control plant parasitic nematodes. Two major groups of nematicides are distinguished by the manner in which they spread through the soil. Soil fumigants are gases in cylinders or liquids which spread as gases from the point where they are infected into the soil. Non fumigant nematicides include a variety of water soluble compounds which are applied to the soil as liquid or granular formulations. Most belong to the carbamate or organophosphate families of pesticides. These distributions in the soil depend on physical mixing during application and moving in solution in soil water. There are no perfect nematicides for all purposes. Nematicides vary in their effectiveness against different kinds of nematodes, ease to handling, cost effect on other classes of pests (weeds, disease organisms, insects) behavior in different soils, toxicity to different plants and availability. The performance of the nematicides will depend on soil conditions, temperatures and rainfall. A yield benefit is not guaranteed and nematicides are expensive.The older nematicides are mostly fumigants which are applied in soil. These are seldom recommended because of their hazardous nature and high toxicity to non target organisms. Recently, large number of non fumigant and systemic nematicides are available which are safe on plant. DBCP (nemagon) has been found very effective for maximum crops against root-knot nematodes but its use has been suspended due to adverse effect on human beings. Carbofuran 3G (furadon), phorate 10G' (thimet), fenamiphos (nemacur), fensulphothion (dasanit) have been recommended for a variety of crops. The dose and method of application would vary with crop. Chemicals and phytoalexins that have implications in the nematode control (Prasad and Swarup 1986), synthetic pheromones to disrupt nematode reproduction (Bone, 1987), and chemicals that directly interfere with nematode chemoreceptors (Janson, 1987) should be exploited for nematode population management. Application of carbofuran and phorate at 6 kg a.i /ha to sweet orange trees reduced plant parasitic nematodes including *T. semipenetrans* by 76.0 and 66.0% respectively.

Regulatory Management

Numerous attempts have been made to prevent the introduction of nematodes into countries or provinces by means of quarantine. Quarantines are established by legislative action in parliament, etc., and usually give quarantine authorities power to make and enforce regulations to accomplish the purpose. Such regulations usually prohibit bringing infected seeds, plant and plant materials into protected areas where similar crops might become infected.

Physical Management

Physical means of nematode control includes heat treatment of soil, solar drying, steam sterilization, hot water treatment and soil solarization. Hot water treatment at 45.5°C for 15-20 minute kills the nematode in apple roots, 48°C for 10 minute in strawberry roots. Soil solarization with transparent polyethylene sheet has been attended as a means of raising the soil temperature to lethal levels to control soil pathogens (Sharma and Nene, 1990) Soil solarization, a method of pasteurization can effectively suppress most species of nematode along with other microbes and weeds under field conditions (Rao and Krishappa, 1995, Kumar *et al.* 1993). Generally, sheets of 50-100 μm thickness are most suitable for raising the soil temperature. The use of transparent polyethylene had yielded better results than black sheets, since transparent sheet transmit most of the incident radiation to soil (Mehrer, 1979, Mehrer and Katan, 1980). Additions of salt, sugar, charcoal etc. create osmotic stress on nematodes and can be used for controlling nematodes.

Integrated Management

An integrated management strategy prevents the excessive build up any single nematode, insect, or disease population and minimizes the development of pest resistance to any single tactic. Integrated pest management systems require flexibility and depend upon the specific pest problem and locally available management options. A fixed set of recommendations may keep a pest complex in check for a limited period of time, but as the pest population shifts, recommendations will have to change also. Therefore, system development takes into account many factors including the species and race of pests present, the availability of resistant host plants, the longevity of the pest, and the crops, cropping systems, and climate of the geographical region. The end result is a management strategy tailored to fit the unique circumstances of each pest situation.

Application of Molecular Biology and Biotechnology

Advances in molecular biology and biotechnology provide means to develop novel strategies for nematode management. Nematode diagnostics is essential for success of any nematode management programme even more when the control method is highly specific to the control species/pathotypes/race. A number of new techniques for analyzing nucleic acids, proteins, carbohydrates and lipids can be helpful in the identification of pests of those allozyme; monoclonal antibody and DNA based systems are most well developed for nematodes. Polyacrylamide gel electrophoresis (PAGE) has mainly focused on cyst and root-knot nematodes (Davies, 1977). PCR techniques or RFLP analysis used as supplementary tools wherever necessary. Nematode resistant transgenic plants can be designed by various approaches. The simple method is to introduce resistant genes effective against plant parasitic nematodes from wild species to commercial cultivars.

References

Ahmad, R., Khan, A.M and Saxena, S.K. 1972. Changes resulting from amending the soil with oil cakes and analysis of oil cakes (Abstract) *Proc. 59th Ses. Indian Sci. Con.Calcutta*. Part III, pp. 164.

Alam, M.M., Kahn, A.M., and Saxena, S.K. (1979). Mechanism of control of plant parasitic nematodes as a result of the application of organic amendments to the soil. V. role of phenolic compounds. *Indian J. Nematol*. 9: 136-142.

Alam, M.M. 1990. Control of plant parasitic nematodes with organic amendments and nematicides in nurseries of annual plants. *J.Bangladesh Acad. Sci*. 14:107-113.

Bandyopadhyay, P. Kumar, D. Singh, V.K. and Singh, K.P. 2001. Eco-friendly management of root-knot nematode of tomato by Arthrobotrys oligospora and Dactylaria brochopaga. *Indian J. Nematol*. 31 (2): 153-156.

Bilgrami, A.L. (1981). Predatory nematodes and protozoans as biopesticides of plant parasitic nematodes. In plant nematode management a biological approach, (Ed.) by P.C. Trivedi. pp 4-23.

Bone, I.W. 1987. Pheromone communications in nematodes. Pages 147-152 In *vistas on Nematology*. A commemoration of the twenty fifth anniversary of the Society of

Nematologists. (Veech, J.A. and Dickson, D.W., eds.), Hyattsville, Maryland, U.S.A. Society of Nematologists.

Haider, M.G. and Nath, R.P. 1992. *Curr. Nematol.* 3: 173-177.

Haseeb, A., Alam, M.M. and Khan, A.M. 1984a. Control of plant parasitic nematodes with chopped plant leaves. *Indian J. Plant Pathol.* 2: 180-181.

Jansson, H.B. 1987. Receptors and recognition in nematodes. Pages 153-158 in vistas on Nematology. A commemoration of the twenty fifth anniversary of the Society of Nematologists. (Veech, J.A. and Dickson, D.W., eds.), Hyattsville, Maryland, U.S.A. Society of Nematologists.

Kumar, B.N., Yaduraju, T., Ahuja, K.N. and Prasad, D.P. 1993. Effect of soil solarization on weeds and nematodes under tropical Indian conditions. *Weed Research*, 33:423-429.

Katan, J. 1981. Solar heating (solarization) of soil for control of soil borne pests. *Ann.Rev. Phytopathol.* 19:211-236.

Mehrer, Y. and Katan, J. 1980. Prediction of soil temperature under polythene mulch. *HASS.* 60: 1384-1387.

Mankau, R. 1963. Effect of organic soil amendments on nematode population. *Phytopathology.* 53: 881-882.

Mishra, S.D. and Prasad, S.K. 1974. Effect of soil amendments on nematodes and crop yields. *Indian J. Nematol.* 4:1-19.

Mulk, M.M. and Jairajpuri, M.S. 1975. Nematodes of leguminous crops in India III three new species of Hoplolaimus. *Indian J. Nematol.* 5 (1): 1-8.

Sasser, J.N. and Carter, C.C. 1985. An advanced treatise on Meloidogyne. Vol I: *Biology and Control.* North Carolina State University, Graphics pp.1-422.

Sikora, R.A. 1979. Predisposition to Meloidogyne infection by the endo tropic mycorrhizal fungus Glomusmosseae. In root-knot nematode Meloidogyne spp. systematic biology and control (eds.F.Lamberti and C .Taylor Academic Press London) pp. 399-404.

Singh, R.S. and K. Sitaramaiah, 1973. Res. Bull. No. 6, G.B. Pant University of Agril. Sci. & Tech., pp 1-289.

Singh, K.P. and Singh, V.K. 1991. Nematicidal natures of arjun bark Terminelia arjuna on cyst nematode, Heterodera cajani of pigeon pea. *New Agric.* 2(1) 77-78.

Singh, K.P. and Singh, V.K. 1992. Terminelia arjuna leaf powder reduces population density of Heterodera cajani. *International pigeonpea newsletter* 16:17-18.

Singh, V.K. and Singh, K.P. 2001. Effect of some medicinal plant leaves on the biology of Heterodera cajani. *Indian J. Nematol.* 31(2):143-147.

Swarup, G.; Sethi, S.L. and Gill, J.S. 1964. Some records of plant parasitic nematodes in India. *Current Science.* 33:593.

❑❑❑

Ecologically Based Integrated Pest Management, 2012
© D.P. Abrol and Uma Shankar (eds.), pp. 765-785
New India Publishing Agency, New Delhi (India)
e-mail : info@nipabooks.com; website : www.nipabooks.com

39

Pest Management in Walnut : An Overview

R.M. SHARMA, M.K. PANDEY AND UMA SHANKAR

Introduction

Walnut is the oldest cultivated fruit in the world (Ozkan and Koyuncu, 2005). *Juglans regia* L., the Persian walnut, is probably native to a wide region extending from the Carpathian Mountains across Turkey, Iraq, Iran, Afghanisan and southern Russia to northern India. Species of walnuts, genus *Juglans,* are found in the forests of the eastern and southern parts of the USA, in Mexico and Central America, the Andean parts of Southern America from Columbia to Argentina, the West Indies, Japan, China, southern Asia from India to Turkey and southern Europe to the Carpathian Mountains of Poland (McGranahan and Leslie,1990). The trees are large and tall with round headed canopy. The rather smooth, light coloured, large nuts are free of the hull at maturity. Shell thickness varies from paper–thin to very thick and hard. The Greeks probably obtained *J. regia* from Persia from where, the seeds were transported to Rome, where they were known as *Jovis Glans*, or Jupiter's acorn, from which comes *Juglans*. From Italy, it spreaded to what is now France, Spain, Portugal and southern Germany. Trees of the species were in England by 1562, and were bought to America by the earlier settelers. The American colonists are said to have called the species English walnut to distinguish it from the eastern black walnut, *J. nigra*. It is generally prefer to call it Persian walnut as it is associated with the place of origin bit commercially, it is called as English walnut. The term Carpathian walnut is sometimes used to refer to cold hardy Persian walnuts (Forde and McGranahan,1996). The other important species are *J. ailantifolia* Carr. syn. *J. sieboldiana* Maxim. (Japanese walnut), *J. nigra* L. (American black walnut), *J. cinerea* L. (butternut), *J. microcarpa* Berlander (Texus Black walnut), *J. major* Heller (Arizona black walnut), *J. californica* S. Wats (Southern California black walnut) and *J. hindsii* Jeps. (Northern California black walnut).

The walnut plant has a high nutritional value and a high quality wood. In addition, walnuts have significant economic value and medicinal importance for human health because of their biochemical composition of polyunsaturated fatty acids, especially 18:2 and 18:3 and protein value (Savage *et al.*, 2001). People consume it in large quantities; therefore, it has a very important place in public nutritional habits. Walnut has a high calorie level, rich nutrient composition and a special value in traditional foods (Sen, 1986). Wild walnuts have been utilized as food for humans and animals since prehistoric times. Some of its species have been used for timber probably for as long as wood has been used as building materials. Walnut is a rich source of alpha linolenic acid which has substantial cardio protective effects in human beings (Hu *et al.*, 1999; Djousse *et al.*, 2001). Besides, being a source of powerful antioxidant named melatonin, which is produced by pineal gland and inducting and regulating sleep (Reiter *et al.*, 2005), walnut is a perfect evening food for the sound sleep during night. The benefical action of walnut oil on skin is known for centuries and it is widely used in cosmetic manufacturing industry. The walnut oil is a component of dry skin creams, antiwrinkle and anti aging products, because it has moisturizing properties as well as free radical scavenging capacity (Espin *et al.*, 2000). Apparently the use of walnut oil for cosmetic purposes may be due to its hig content in essential fatty acids and in particular linoleic and linolenic acids (Aarleskind, 1996). These acids are referred as crucial agents of the most important function of the skin, which is the regulation of the transepidermal water loss (Qiang *et al.*, 1993). Liposomes are hollow spheres of lipid bilayers, which are formed mainly of phospholipids molecules. Liposome formulation is widely used as media for active ingredients and lipid transfer to the skin (Hatziantoniou *et al.*, 2000).

Jammu and Kashmir state of India has the monopoly on the production of export quality walnuts. This state alone produces more than 98 percent of the nation's crop with an average productivity of 1.14 MT/ha. Presently walnut is being cultivated on 81393 ha in J&K with an annual production of

114926 MT. Walnut industry has the great role in states' economy. Approximately 7000 MT kernel and 100 MT inshelled walnuts are being exported to about 23 countries including France, Germany, Spain, Portugal, Austria, United Kingdom, Kuwait, Behrain, Dubai and Soudi Arabia, worth of rupees 150 crores annually. Besides, the domestic demand of approximately 1000 MT and 1500 MT of inshelled walnut is being catered by the state per annum.

Pest management is really a big constraint in the walnut because of the difficulty faced to sparying the giant trees. In fact, walnut (*Juglans regia*) competes with many insect and disease pests to produce a marketable and acceptable crop. Traditional control strategies work reasonably well for most pests. Walnuts are susceptible to several diseases and pests, and even though most of them threaten the life of plants and they may ruin a sizable portion of the edible nuts and the visual appearance of the fruits. However, environmental concerns, loss of certain pesticides and new or impending regulations threaten the use of many traditional techniques for control of many of the pests.

Major Diseases of Walnut and their Management

1. Walnut Blight

In USA, more than 50 per cent of the crop is damaged when disease reach in epiphytotic form (Miller, 1953). In India, it was first reported from UP hills infecting 16 to 47 per cent nuts of different exotic varieties (Adhikari *et al.*, 1988). Walnut blight is also known as bacterial blight.

The disease attacks leaves, nuts, catkins, bud and shoots of current growth. The disease first appears in the spring on the young leaves as small reddish brown spots at the margins or in the tissues between the veins. The spots gradually increase in size, coalesce and form circular to irregular patches. Severe infection leads to defoliation. On the stem and young shoots the disease caused slightly depressed black spots which often girdle the shoots and cause dieback symptoms. Individual florets die in the distorted infected catkins. Nut infection occurs right from the early spring to summer. Early infection causes nut drop while late infection results in nut shriveling and discolouration.

The disease is caused by gram negative bacterium *Xanthomonas arboricola* pv. *Juglandis* (Pierce). Walnut Blight is the most damaging when it occurs during cool and wet periods, around flowering time (so it is most dangerous to species which leaf early). The bacteria will overwinter in seemingly healthy buds, launching a new attack when new growth is at hand and disseminated by rain drops. The bacterium enters the tissues of the current growth through stomata and injuries. Moisture is the essential for the infection to take place, so high rainfall, fog, dew and sprinkler irrigation favors faster multiplication and spread of the pathogen and thus aggravates the blight problem.

Management

In India, local cultivars are reported to be free while exotic collections recorded 16 to 47 per cent fruit infection. Cultivation of resistant cultivars such as Sorrento, Marbot, Meylannaise, Plovdisvskii and Parisienne is recommended (Tamponi, 1988; Tamponi and Donati, 1990).

Blight can be checked by applying timely spray of copper based fungicides. The first spray should be applied at bloom stage at the opening of first pistillate flowers followed by subsequent sprays at 7 to 14 days interval. Treatment consists of cutting out damaged parts (cutting too much is better than cutting too little) and burning them. Prevention may be attained by using a soil with a pH above 6, avoiding excessive nitrogen, excessive wetness and allowing for good aeration by pruning.

With the emergence of copper tolerant strains of bacterium from France (Gardan *et al.*, 1993), use of iron amended (50 µµg/ml) copper containing bactericides (Kocide 101 or Champion) is advised for copper tolerant and sensitive strains (Lee *et al.*, 1993). Application of a mixture of streptocycline (1.0g) and copper oxychloride (20 g) per 10 liters of water, immediately after the appearance of symptoms has also been recommended (Jindal *et al.*, 1994). In nursery, 1.4 meter inter-row distance with two sprays of Bordeaux mixture at foliage flush stage checks the disease effectively (Belisario and Zoina, 1995).

2. Leaf Blotch and Anthracnose

The disease is wide spread in occurrence and destructive at young nut stage causing premature nut drop and extensive yield loss. It was observed in almost all the walnut growing localities of Kashmir (Kaul, 1962).

The pathogen attacks leaves, young shoots and nuts. On leaves initially it produces round reddish brown spots that have grayish brown centers and are up to three-fourth inch in diameter which later developed into large blotches involving major part of lamina (Pathak, 1980). On young shoots of the current year growth, the lesions are oval, sunken, grayish brown surrounded by reddish brown margins. Black acervuli develop on the lower surface of the affected leaf. Brown coloured perithecia appear mainly on the affected portions of dead, fallen and over wintered leaves. Blotch causes depressed circular to irregularly circular dead spots on the husk of the nuts. These spots are generally smaller than those which appear on the leaves. Early infection on nuts, result in premature fruit drop.

The disease is caused by *Marssonina juglandis* (Lib) Magn. [Tel.: *Gnomonia leptostyla*, (Fr.) Ces and De Not]. The fungus overwinters on infected fallen leaves, nuts as well as on infected twigs. Infection starts with the onset of spring rains with the release of ascospores discharged forcefully from the perithecia and are carried by wind current to new shoots where they cause primary infection. The subsequent generation of summer spores (conidia) produced on primary and secondary lesions further spread the disease. The process of disease spread continues until late autumn, however, it is most conspicuous during July and August (Kaul, 1962; Veghelyi and Penzes, 1990).

Management

The fallen leaves should be collected and burnt. Spray zineb or captan (0.25%) or Bordeaux mixture (0.8%) at (i) leafing stage i.e. when leaf starts unfolding, (ii) at full leaf stage and (iii) two weeks after the second spray. Successful disease control has also been achieved by the application of copper oxychloride, mancozeb and dithianon (Cimanowski *et al.*, 1991). None of the walnut cultivars exhibit complete resistance, however, Corne, Franquette and Paradox have shown lower incidence of the disease.

3. Root and Crown Rot

The disease is widespread in occurrence and quite disastrous than other diseases because it kills the tree out rightly at any age. Infected trees may produce low yields of poor quality nuts.

Yellowish sparse foliage with poor shoot growth serve as the disease indicator at the first instance. The tree appears drought stricken, leaves turn yellow and drop prematurely. Twigs and small branches exhibit dieback symptoms. Affected trees often set an unusual heavey crop of nuts and trees generally succumb to the disease pressure in the following years.

Bark at the crown becomes brown to black and irregular cankers or decayed areas develop. In advanced stages, the disease extends to lateral roots and kills the main roots. A black fluid often accumulates in cavities in the cambial region and eventually finds its way to the surface. The disease symptoms appear gradually in a number of years depending upon the species of fungus, temperature, soil moisture and relative resistance of rootstock.

Many species of Phyrophthora are involved but *P. cactorum* (Lebert & Cohn) Schrot is the most common fungus associated with the disease.

The fungus survives in the soil. Presence of free water is necessary for the release of zoospores from sporangia. Free water carries the zoospores on to the soil surface where they lose their flagella, encyst, germinate and produce hypha which infects the root or collar region. Injury on these parts hastens the infection process. Poorly drained soils with prolonged water stagnation, low level of budding and injuries to root as well as collar region are some of the predisposing factors.

Management

Selection of well drained sites for establishing new orchards ie recommended. If soil drainage is good, it can be improved by installing drainage tiles or by planting trees on ridges. Graft union should be kept at least 30 cm above ground. Diseased tissues near the ground level should be scarified during dormant stage and broad spectrum fungicidal pastes like Bordeaux paint or paste or Chaubatia paste should be applied. Use of Paradox rootstock to minimise the infection is also recommended.

4. Shoestring Root and Crown Rot

The most reliable diagnosis can be done by its mycelium strands resembling flattened shoestring. Rhizomorphs adhere to the outer bark of roots which are brown to black from outside and white from inside.

The leaves of infected trees turn yellowish and sometimes fall prematurely. Dead bark exposing the heartwood appears near or at the crown and root inner bark. With late rains in the season or during fall, abundant fruiting bodies grow in groups, These mushroom like fruit bodies are yellow to orange in colour, the caps of which measures 5 to 12 cm in diameter.

The disease is caused by fungus *Armillaria mella* (Varhl. Ex Fr.) Kumm., the fruiting bodies of which are honey coloured or light brown toadstools or mushrooms which appear at the base of infected trees in October and November. The causal fungus may be brought into the orchard on infected nursery stocks or it may be present in the soil. The fungus can live for many years on old, rotten pieces of roots and spreads from one spot or tree to another through Rhizomorphs which grow through the soil by contaminated irrigation water or cultivating tools. The fungus produces fruiting structures called a toadstool which produces spores. On germination, the spores produce mycelium which lead saprophytic life cycle by feeding on rotten roots and dead trees stumps. Later it produces Rhizomorphs which travel through soils and gain access to the plant or root either through wounds or directly through penetration of epidermis.

Management

Plantation of new orchards in freshly cleared forest soils should be avoided. Care should also be taken not to injure the basal part of the trunk or the main roots of walnut trees during cultivation; as such wounds increase the chance of infection. Application of fumigants like carbon disulfide by making holes 20 to 25 cm deep at a distance of 45 cm apart is recommended (Cimanowski *et al.*, 1991). Use of resistant rootstocks such as Eastern Black Walnut and Hinds Black California Walnut are recommended.

5. Crown Gall

Crown gall is a common problem in nurseries of walnut. The disease has been known since 1853 and is worldwide in distribution. It was also reported on walnut from India by Jindal and Dwivedi (1994).

The disease first appears as small overgrowths like tumor or wart like aberrations on the roots and at the base of the trunk near the soil line. In early stages these tumors are more or less spherical, which or flesh coloured and soft.

Late as these aberrations become larger their surfaces become dark brown of black acquiring rough and convoluted surface due to death and decay of the peripheral cells. They may be on one side of the tree only or many extend around the circumference of the crown or collar. The galls are hard and woody and they increase in size from year to year and persists for many years. The affected plants may become stunted, produce small chlorotic leaves and generally become more susceptible to adverse environmental conditions like winter injury.

Crown gall is caused by a bacterium *Agrobacterium tumefaciens* (Smith & Townsend) Conn. The bacterium overwinters in infested soils, where it can live as a saprophyte for several years. When plants of susceptible hosts are grown in such infested soils, the bacterium enters the roots or stems near the ground through fresh wounds caused by cultural practices, grafting and insects. Once inside the tissue, the bacteria stimulate the surrounding cells to divide. Once or more groups or whorls of hyperplastic cells appear in the cortex or in the cambial layer depending on the depth of the wound. These cells may contain one to several nuclei. They divide at a very fast rate, producing cells that show no differentiation. The cells continue to divide and elongate so fast that 10 to 14 days after inoculation a small tumor can be seen with the naked eye.

When very young and expanding tissues are infected, in addition to the primary tumor which develops at the point of infection, secondary tumors appear. These usually are formed below but are often seen above the primary tumor. In several cases secondary tumors seem to be connected with the primary tumors by tumor strands, causing hyperplastic condition.

Management

Susceptible nursery stock known to be infested with the pathogen should not be planted in fields. All nursery stock with symptomatic roots and visible galls should not be destroyed. Nursery planting sites should be rotated frequently i.e. at least in 4 to 5 years. Infested fields should be planted with corn or other grain crops for several years before they are planted with walnut nursery stock. Nursery stock should be budded rather than grafted because of the much greater incidence of galls on graft than on bud unions. Biological control has been achieved with a non-pathogenic strain K-1026 isolate of A. radiobacter (Kerr, 1989). Root dips of nursery stock into a suspension containing strain of K-1026 have proved very successful in preventing this disease on nut fruits.

6. Stem Cankers and Dieback Diseases

In India, 63 per cent population of 6 to 9 years old budded plants were found to be infected with Cytospora canker in Kashmir valley (Puttoo and Chaudhary, 1984). It has also been reported from hills of Uttar Pradesh. More than 84 per cent black walnut plantations in North America were infected with stem canker caused by *Fusarium solani.*

The bark of almost entire length of the trunk initially exhibit depressions which turns dark brown to black in cilour. Bark dries up and splits, cracks also develop in the underneath wood which leads to the drying of foliage and twigs, resulting in defoliation. As the disease progresses fruiting bodies of causal fungi develop.

There are several canker causing:

a. Cytospora spp.

Different species of Cytospora have been found to cause canker in different nut crops. In India, C. juglandis has been reported to infect 6 to 9 years seedlings of walnut (Bose and Sindhan, 1976; Putto and Chaudhary, 1984). In addition *C. juglandicola, C. leucostoma* and *C. cincta* have also been reported. These fungi produce pycnidia bearing hyaline conidia. Perithecia and pycnidia are found in the same stroma, the former is less frequently found than the later. Perithecia are generally produced only in old stroma. These contain numerous asci which contain ascospores.

b. Fusarium solani

The fungus has been reported to cause canker on black walnut (*J. nigra*) in North America and more than 84 per cent plantation exhibited stem cankers (Carlson *et al.*, 1993). The incidence of this canker was higher in bottom land sites (12.2%) than on upland sites (7.2%). Unprotected pruning wounds have been found to aggravate the canker incidence.

c. Botryosphaeria spp.

Two species of the fungus namely [*B. dothidea* (Moug. Ex Fr.) Ces. & de Not. And *B. Ribis* Gross & Duggar] have been reported to cause stem canker and dieback of walnut (Lee *et al.*, 1992; Frisullo *et al.*, 1994).

In addition to above fungi *Melanconis jugandis* (Ell. & Everh) Graves on Persian and Japanese on Juglans sinensis in Korea (Lee *et al.*, 1992), *Sphaeropsis camarosporium*, *Marssonina juglandis* (Lib.) *Magnus* and *Phomosis sp.* On *J. regia* and *J. nigra*, *Melanconium juglandium* and *Eutypa lata* from Greece (Rumbos, 1993) have been found associated with canker and dieback diseases.

These canker fungi are generally wound parasites and require wounded or injured host tissues to penetrate. Unprotected pruning cuts, insect punctures and injuries caused during various horticultural operations predispose the host to infection.

These fungi survive in infected tissues and mycelium starts growing when host growth starts. Asexual or sexual spores produced on host surface and disseminated and initiate the infection in susceptible hosts.

Management

Vigorous, healthy trees are better able to cope with cankers than are trees in poor health or under stress. Inadequate soil moisture is the most common stress factor in urban and suburban sites, but it can be corrected with regular watering. Trees should not be planted where soil conditions are poor or where structures such as sidewalks and stone walls limit proper root development. Annual fertilization in late fall or early spring will also help improve vigor.

Proper pruning will help to reduce canker problems. Pruning should be done in early spring, prior to bud break, or in mid-June after leaf expansion. Remove dead and weak branches and those which are rubbing against others creating wounds. Cuts should be made such that no stubs are left and the resulting wound is as small as possible.

For cankers on trunks, tracing may help to reduce damage if done soon after the canker is incited. Cut outside the cankered area and into healthy bark. Shape the cut to an ellipse. Extreme care must be taken to avoid infection of this fresh wound with canker pathogens. Sterilize pruning tools between cuts by dipping them in 10% household bleach or 70% alcohol for several minutes. If the canker is more than one-half way around the circumference of the main trunk, the tree development should be carefully watched for internal decay in subsequent years. If structural integrity is threatened, the tree may have to be removed.

After pruning and training practices cut ends should be treated with fungicide pastes like Bordeaux paste/ paint and Chaubatia paste. In Kashmir valley, management of Cytospora canker has been achieved by application of copper oxychloride paste on diseased parts after scrapping the affected tissues followed by foliage spray (0.2%) of the same fungicide (Puttoo and Chaudhary, 1984).

7. *Mistletoe*

Mistletoe (*Visum album* L.) is a parasitic flowering plant with characteristic formation of clumps (swelling of host tissues) that are visible on the host plant walnut (Fig. 1). It forms witches broom on the tree and feeds on walnut plant reducing its yield in Kashmir Valley (Motial and Gilkar, 1970). Its seeds are disseminated by birds that eat or transport the berries (fruits) and deposit the seeds on walnut trees. Mistletoes absorb both water and mineral nutrients from its host trees. The effects of mistletoes on their hosts include reductions in growth, vigour and fruiting. Severe infection by mistletoes is often associated with premature mortality of host trees (Srivastava *et al.*, 2010).

Management

Prevention of mistletoe infection is very difficult since birds disseminate the seeds over large areas. The most effective method for controlling mistletoe and also preventing its spread is to prune out infected branches, if possible as soon as the parasite appears. Severe heading (topping) is often used to remove heavy tree infestation. Mistletoe shoots may be controlled by the application of ethephon (ethylene) @ 2% a.i. through spraying on mistletoe foliage at the time of dormancy of walnut tree. Mistletoes are best controlled by boring of two rows of holes along the trunk with a carpenter's auger, at an interval of 15cm at an angle of 45° to reach the sapwood. Filling with the mixture of 8g $CuSO_4$ and 1g fernoxone (83%. 2,4-D Sodium salt) into each hole during dormancy period is the most effective treatment to control this parasite (Srivastava *et al.*, 2010).

Major Insect Pests of Walnut and their Management

Occasionally, the walnut trees are found infected with insect pests, including walnut weevil (*Alcides porrectirostris* Marsha), walnut chaffer beetle, san jose scale (*Diaspidiotus* (=*Quadraspidiotus*) *perniciosus*), tree trunk and shoot borers and walnut green aphid (*Chromaphis juglandicola* Kalt). Of the various insect pests, walnut weevil and chaffer beetles are considered serious. The pest of national significance comprises Stem borer, *Acolesthes sarta*; pin hole borer, *Scolytis nitidus*; hairy caterpillar, *Lymantria obfuscata*; leaf roller, *Archips argyrospilus*; mealy bug, *Drosicha dalbergia*; grey weevil, *Mylocerus sp.*; defoliator, *Chaetoprota sp.* and walnut weevil, *Alcides porrectirostris* while walnut aphid, *Chromaphis juglandicola* and *Panaphis juglandis*; codling moth, *Cydia pomonella* and san jose scale, *Quadraspidiotus perniosus* falls under the pest of regional significance (DPPQS, 2003).

IPM technology manages the pest population in such a manner that economic loss is avoided and adverse side effects of chemical pesticides are minimized. The IPM package encompasses various management strategies for containing the pest and disease problems. Pest monitoring is one of the important components of IPM to take proper decision to manage the pest problem. It can be done through agro-ecosystem analysis (AESA), field scouting, light, sticky and pheromone traps. The determination of economic threshold level (ETL) against major pests helps in taking appropriate control measures when pest population crosses ETL (DPPQS, 2003). Pest management is really a big constraint in the walnut because of the difficulty faced to sparaying the giant trees. English walnut (*Juglans regia*) producers in California compete with many insect and disease pests to produce an acceptable crop. Traditional control strategies work reasonably well for most pests. However, Environmental concerns, loss of certain pesticides and new or impending regulations threaten the use of many traditional techniques for control of many of the pests. Control strategies that use integrated pest management programs, beneficial insects, mating disruption, insect growth regulators, improved monitoring techniques and precise treatment timing based on the insect's life cycle is leading edge techniques currently available for insect control in walnuts (Olson and Buchner, 2002). Integrated control of walnut pests has been practised in China against *Cnidocampa flavescens, Atrijuglans hetaohei, Botocera horsjieldii* [Batocera horsfieldi] and *Drosicha corpulenta* (Jia *et al.*, 2001).

1. Walnut weevil, *Alcidodes porrectirostris* Marshall (Curculinoidae: Coleoptera)

Walnut weevil (Fig. 2) has been recorded in all the walnut growing regions throughout the world. Female weevil lay eggs in the pits excavated on the fruits. On hatching the grubs bore deeper, feed on the kernels and reducing the kernels into black mass (Fig. 3). The grubs also cause extensive premature dropping of the fruits. The affected fruits also show dark brown spots which are dried resinous excretions. The adult weevil comes out by biting circular holes. The adult weevils feed on petioles, female floral buds, tender shoots and even on young fruits. Leaves, fruits are damaged by weevils during June-July. Fall of nuts to the extent of 50-60 per cent are observed. Jet black weevils appearing in early May are noticed, which deposit eggs after mating on the green nutlets which after hatching bore the nuts and cause the damage.

Management

Proper sanitation should be maintained. Collection and destruction of fallen nuts during the month of May and June prevent buildup the pest populations. Smaller and medium plants can be sprayed with 0.02% chlorpyriphos 20EC or 0.05% dichlorvos during the month of April-May. It can also be controlled by spraying carbaryl (0.2 %) twice at an interval of 7 days commencing from the time of first attack (Rathore, 1991).

2. Indian gypsy moth, *Lymantria obfuscata* Walker (Lymentriidae: Lepidoptera)

Indian Gypsy moth, *Lymantria obfuscata* is a common pest of apple, poplar, willow, walnut, peach, plum, cherry, etc. in Jammu and Kashmir. It has been found to cause severe damage to many tree and fruit crops in temperate regions. The caterpillars are gregarious and nocturnal in habit and defoliate the trees completely which results in failure of fruit formation. The larvae cause damage on leaves, green fruits. The leaves are eaten away from the margin. Eggs hatch in April, small grey to blackish colour larvae having hairs on their body (Fig. 4) feeds voraciously on the leaves up to ending June, then undergo pupation and adults emerge by end of July-August and after mating deposit eggs on the crevices, bark of the tree for next year infestation.

Management

Collection and destruction of egg masses, pupae, larvae and even female adults. Use of Burlap bands (30 cm wide) stapled around the tree trunk during May-June may be useful for the destruction of larvae/pupae (Masoodi *et al.*, 1987). Egg parasitoid *Anastatus kashmirensis*; larval parasitoids *Glyptapanteles indiensis, G. flavicoxis, G. liparidis, Cotesia melanoscela, Hyposoter lymantriae, Rogas indiscretus, Exorista rossica, Palexorista conspicua*; pupal parasitoids *Brachymeria intermedia* and *B. lasus* are the dominant parasitoids of Indian gypsy moth. In addition to these pupal parasitoids, *Pimpla laothoe, Theronia atalantae* and *Calosoma himalayanum* have been recorded as population regulatory factors of the Indian gypsy moth (Rao *et al.*, 1971; Rishi and Shah, 1985). Spraying of 0.02% chlorpyriphos 20EC or 0.07% endosulfan 35EC at petal fall stage provide the good control.

3.Codling moth, *Cydia pomonella* (Tortricidae: Lepidoptera)

It is a serious pest of apple which also attacks early blooming walnuts but it is confined only to Ladakh regions in India (Pawar *et al.,* 1984). Three generations of moths are recorded damaging the fruits. Codling moth (*Cydia pomonella*), walnut husk fly (*Rhagoletis completa*), and walnut aphid (*Chromaphis juglandicola*) are also recorded as the major insects that affecting California walnut production (Olson and Buchner, 2002).

Management

Tebufenozide and diflubenzuron can be used effectively where codling moth populations are moderate, and where preservation of beneficials is very important (Hendricks, 2001). Preferential heating of insects in dry nuts and fruits at radio frequencies can be used in developing thermal treatments to control codling moth larvae without adversely affecting product quality (Wang *et al.,* 2003).

4. Chaffer beetle, *Anomala* sp. and *Holotrichia* sp. (Melolonthidae: Coleoptera)

Both the grub and adult cause damage to roots and leaves, respectively. In case of severe infestation the entire tree is defoliated. The grub attack the plant throughout the year, the adult beetles is active during June to July (Fig. 5, 5a, 5b).

Management

Collection and destruction of adult beetles. Light trap for trapping beetles of white grub has been developed and evaluated in low, mid and high altitudes of Uttarakhand and microbial pathogens *Bacilus cereus* WGPSB-2 may be promising in controlling white grub adults (VPKAS - Perspective Plan Vision 2025, 2007).

5. San jose scale, *Quadraspidiotus perniciosus* (Comstock) (Diaspididae: Hemiptera)

Nature of damage : These tiny insects suck the sap of leaves, tender shoots, branch, trunk. Grayish encrustation exists on the infested tree parts. The bark showing reddish spots wherever inscets under the scales exists. The market value of such fruits is depreciated. Heavily infested tree lack vigour, its foliage turn thin and yellow speckled. In the beginning only growth is checked but as the infestation increases, the attacked tree dies due to devitalization. More often, certain limbs become infested and die before the whole tree is affected. Even after successful control of scale on heavily infested trees/limbs the branches and the buds may be so weakened that no fruit is borne for one or two seasons.

Management

Heavily infested plants may be uprooted and burnt. Use 6.30% diesel oil emulsion during dormancy in the ratio of 1:7. In case of heavy infestation, Spray 0.02% chlorpyriphos 20EC or 0.07% endosulfan 35EC to control scales population.

6. Walnut green aphid, *Chromaphis juglandicola* Kaltenbach) (Aphididae: hemiptera)

This is a serious pest of nursery plantation and medium trees. Walnut aphid (*Chromaphis juglandicola*) are the major insects that affect California walnut production (Olson and Buchner, 2002). The colonies of nymphs and adults suck the sap confining themselves to ventral surface of the leaves and tender shoots (Fig. 6). The affected part become dry, pale, curl and disfigured. In case of severe infestation, even the growth of the tree is affected and plant is stunted. The aphids also secrete honeydew, on which sooty mould develops causing hindrance in photosynthetic activity of the plant and further retarding its growth. Numerous green dusky veined soft bodied aphids (*Callaphis julandis*) in characteristics colonies on the upper surface of leaves along the midribs suck the sap and devitalize them. Plant growth becomes stunted. Nymphs and adults of Leaf curling aphid, *Aphis* spp. suck the sap from the leaves, petioles, blossoms and fruits causing leaf curl and distortion. Blossom withers and fruits do not develop and drop if formed. Green peach aphid, *Myzus persicae* Sulz. is a major pest of almond. It feed on the leaves, blossoms and fruitlets resulting in fruit drop and sometimes leaf curling. The young shoots are infested by colonies of aphids as a result of which the terminal shoots become bunchy, malformed and sticky. Sooty mould develops on the leaves owing to the accumulation of honey dew on them.

Management

The introduction of a climatically adapted strain of *Trioxys pallidus* (Hal.) from France for the control of *C. juglandicola* and of another strain of *T. pallidus* from Iran have virtually eliminated the need for chemical control of aphid in USA (Riedl *et al.*, 1979). Spray 0.02% chlorpyriphos 20EC or 0.07% endosulfan 35EC or 0.05% dimethoate during late May or early June. If infetstation is heavy, repeat the spray after 3 weeks.

7. Walnut tree trunk borer, *Aeolesthes sarta* Solsky (Cerambycidae: Coleoptera)

Tree trunk borer is highly destructive pest of temperate fruits viz., apple, walnut, almond, cherry etc. (Sheikh, 1985). The pest is active throughout the year and is serious problem in old trees (Fig. 7). The eggs are laid in the cuts or wounds on the bark of the trees. The grubs on hatching bore in the bark and sapwood and tunnel downwards. The pest makes zig-zag tunnels into trunk and main branches therby reducing the life of trees and fruit yield (Bhat *et al.*, 2010). Big holes measuring 1.50 to 2.00 inches in length in the main stem or old branches of trees from which saw dust comes out causing drying of trees or branches.

Management

Tree trunk should be plastered with 10% malathion dust and soil in the ratio of 1:6. Plugging of the holes with petrol and napthlene (crushed) in the ratio of 1 litre and 200g. Push cotton soaked in the above mixture in the holes and plug with mud. Placement of odonil iinside the tunnels and closing the holes with mud proved most effective (80.45-81.55% control within 15 days after treatment on walnut (Sheikh and Bhat, 1991). Bhat *et al.* (2010) recorded that aluminium phosphide (78.78% reduction) and dichlorovos are quite effective in controlling trunk borer for more than 45 days after the treatment.

8. Shoot hole borer, *Scolytoplatypus* spp. (Scolytidae: Coleoptera)

This pest is serious in semi dried trees where attack of san jose scale has been noticed for 2-3 years. Adult and grubs tunnel into sapwood and hardwood of branches making pin holes. Round pin holes can be seen on main trunk, branches, tender shoots of the trees containing small brown insects. Surface of infested branches get perforated followed by yellowing and wilting of leaves. In severe cases, the tree may also die. On breaking the infested branched emits foul fermenting odour.

Management

Severly infested branches be cut and burnt. Plaster the main trunk and branches with 10% malathion dust and soil in the ratio of 1:6. Prune branches in the month of June to trap the egg laying females.

9. Flat headed tree borer, *Chrysobothris* spp. (Cerambycidae: Coleoptera)

The infestation of the pest is mild in stem and branches. Presence of big galleries under bark, deep in to cambium, die back of branches and drying of bark and stem. Grubs are medium sized, elongate, bore on infested parts.

Management

White wash the trunk and plaster the trunk with malathion 10% dust and soil in the ratio of 1:6 during June –July. Spray 0.02% chlorpyriphos 20EC or 0.07% endosulfan 35EC or 0.05% dimethoate.

10. European red mite, *Panonychus ulmi* and two spotted mite, *Tetranychus urticae* (Tetranychidae: Acarina)

Species that have reached pest status because of a disturbance of the complex of their natural enemies include *Chromaphis juglandicola* (Kalt.), *Eulecanium pruinosum* (Coq.), several species of scales and various species of mites (Riedl *et al.*, 1979). Mite feed on the cell contents of leaf tissue, Intially feeding cause light stippling of the leaves prolonged feeding by the heavy population work gradually give the leaves a bronzed appearance which can reduce the nut yield significantly. Mite feeding causes browning of leaves cluster of brown leaves are often the first sign in mite population. Heavy population produce copious webbing and their feeding causes leaves to desiccate and rot. Defoliation in early season will greatly reduce the nut yield in respect of quality and quantity.

Management

Spray the 0.05% dicofol or 0.02% chlorpyriphos 20EC or 0.05% dimethoate at bud bursting stage. Repeat the spray if noticed (>5mites/leaf) with alternate miticides during middle of June. *Amblyseius finlandicus* was the most common predator on mite *Tetranychus urticae* attacking apple in Kashmir. The life cycle of the host synchronised with that of the predator (Rao *et al.*, 1971).

References

Adhikari, R.S., Bora, S.S. and Singh, S.B. 1988. *Xanthomonas campestris* pv. *Juglandis* – A new report from India. *Curr. Sci.*, 57: 728.

Belisario, A. and Zoina, A. 1995. Occurrence of Persian (English) walnut blight and its control in the nursery. *Eur. J. For. Pathol.*, 25: 224-231.

Bhat, M.A., Mantoo, M.A. and Zaki, F.A. 2010. Management of tree trunk borer, *Aeolesthes sarta* (Solsky) infesting apple in Kashmir. *Journal of Insect Science*, 23(2):124-128.

Bose, S.K. and Sindhan, G.S. 1976. AICAR Fruit Res. Workshop. *Tech. Doc.* 9: 491-499.

Carlson, J.C., Mielke, M.E., Appleby, J.E., Hatcher, R., Hayer, F.M., Luley, C.J., Brien, J.G. and Rugg, D.J. 1993. Survey of black walnut canker in plantation in five central states. *N.J. Appl. For.*, 10:10-13.

Cimanowski, J., Dlugowolski, B. and Olszak, M. 1991. Fungicide evaluation in the control of walnut tree diseases. *Sadowinctwa*, 30: 95-98.

Djousse, L., Pankow, J.S., Eckfeldt, J.H., Folsom, A.R., Hopkins, P.N., Province, M.A., Hong, Y. and Ellison, 2001. Relation between dietary linoleic acid and coronary artery diseases in the National Heart, Lung and Blood Institute Family Heart Study. *Amer. J. Clin. Nutr.*, 74:612-619.

DPPQS, 2003. Directorate of Plant Protection, Quarantine and Storage. Integrated pest management package for walnut, Government of India. (*IPM package No. 57*).

Espin, J.C., Soler Rivas, C. and Wichers, H.J. 2000. Characterization of the total free radical scavenger capacity of vegetable oils and oil fractions using 2, 2- diphenyl-1- picrylhydrazyl radical. *J. Agric. Food Chem.*, 48:648-656.

Forde, H.I. and McGranahan, G.H. 1996. Walnut pp:241-273. In:Fruit Breeding, Vol III. Nuts (Janic, J. and James M. M. eds). John Wiley & Sons, Inc.

Frisullo, S., Lops, F., Sisto, D. and Trombetta, N.M. 1994. Pathogenic fungi in southern Italy X. *Botryosphareia* ribis Grossenb et Dugger on walnut. *Inf. Fitopatol.* 44: 61-64.

Gaffar, S.A. and Bhat, A.A. 1991. Management of stem borer, *Aeolesthes sarta* (Solsky), infesting walnut trees in Kashmir. *Indian Journal of Forestry*, 14(2): 138-141.

Gardan, L., Brault, T. and Germain, E. 1993. Copper resistance of *Xanthomonas campestris* pv. *Juglandis* in French walnut. *Acta Hortic.*, 311: 259-265.

Hatziantoniou, S., Rallis, M., Demetzos, C. and Papaioannou, G. 2000. Pharmacological activity of natural lipids on a skin barrier disruption Model. *Pharmacol. Res.*, 42: 55-59.

Hendricks, L.C. 2001. A comparison of tebufenozide (Confirm), diflubenzuron (Dimilin), esfenvalerate (Asana) and chlorpyrifos (Lorsban) for control of codling moth, *Cydia pomonella* in walnut, *Juglans regia*. *Acta Horticulturae*, 544:395-397.

Hu, F.B., Stamfer, M.J., Manson, J.E., Rimm, E.B., Wolk, A., Colditz, G.A., Hennekens, C.H. and Willet, W.C. 1999. Dietary intake of alpha-linolenic acid and risk of fatal ischemic heart disease among women. *Amer. J. Clin. Nutr.*, 69: 890-897.

Jia, X.Y., Ma, G.Y., Wang L.G., Liang, W. and Wen H. 2001. Integrated control of walnut pests. *China Fruits*, 1:39-40.

Jindal, K.K. and Dwivedi, M.P. 1994. Occurrence of crown gall on walnut plants in India. *Plant Dis. Res.*, 9:67-68.

Jindal, K.K., Gupta, V.K. and Sharma, S.K. 1994. Bacterial diseases of stone and nut fruits, their management. *Indian Hortic.*, 39: 15-16.

Karleskind, A. 1996. Oils and Fats Manual: A Comprehensive Treatise, Vol.-1. *Lavoisier Publ. Paris*: 165-168.

Kaul, T.N. 1962. Occurrence of *Gnomonia letostyla* (Fr.) Ces. et de Not. On walnut in India. *Curr. Sci.*, 31:349.

Kerr, A. 1989. Commercial release of a genetically engineered bacterium for the control of crown gall. *Agric. Sci.*, 2:41-44.

Lee, M.O., Lee, Y.H., Cho, W.D. and Lee, K.J. 1992. Identification of three fungi associated with stem and twig diseases of *Juglans sinenensis* in Korea and characterization of factors affecting their growth. *J. Korean For. Soc.*, 81:190-199.

Lee, Y.A., Schroth, M.N., Hendoson, M., Lindow, S.E., Wang, X.L., Olson, B., Buchner, R.P. and Teviotdale, B. 1993. Increased toxicity of iron amended copper-containing bactericides to the walnut blight pathogen *Xanthomonas campestris* pv. *Juglandis*. *Phytopathology*, 83:1460-1465.

Masoodi, A.M., Koul, V.K., Bhagat, K.C. and Singh, T. 1987. Control of *Lymantria obfuscata* Walker on walnut in Kashmir. *Indian Journal of Plant Protection*, 15(1):15-17.

McGranahan, G. and Leslie, C. 1990. Walnut (*Juglans regia* L.) pp: 907-951., In: Genetic Resources of Temperate Fruits and Nut Crops Vol.2 (Moore, J. M. and Ballington, J.R. eds.). *Int. Soc. Hort. Sci.*, Wageningen.

Miller, P.W. 1953. Filberts and Persian walnuts. pp.800-8008. In: Plant Diseases : The Year Book of Agriculture, Oxford & IBH Publ. Co., New Deldi.

Motial, V.S. and Gilkar, A.M. 1970. *Indian Farming*, 20: 31-33.

Olson, W.H. and Buchner, R.P. 2002. Leading edge of plant protection for walnuts. *Hort. Technol.*, 12 (4): 615–618.

Ozkan, G. and Koyuncu, M.A. 2005. Physical and chemical composition of some walnut (*Juglans regia* L.) genotypes grown in Turkey. *Grasas Aceites*, 56: 141-146.

Pathak, V.N. 1980. Diseases of Fruit Crops, Oxford & IBH Publ. Co., New Delhi. 309 pp.

Pawar, A.D., Tuhan, N.C. and Parry, M. 1984. Management of codling moth *Cydia pomonella* (L.) (Lepidoptera: Olethreutidae) at Khalsi (Ladakh), Jammu and Kashmir (India). *Pesticides*, 18: 39-42.

Putto, B.L. and Chaudhary, K.C.B. 1984. Canker of walnut trees and its management in Kashmir. *Pesticides*, 10: 21-22.

Qiang, M.; Elias, P.M. and Feingold, K.R. 1993. Fatty acids are required for epidermal permeability barrier function. *J. Clinical Invest.*, 92: 791-798.

Rao, V.P., Ghani, M.A., Sankaran, T. and Mathur, K.C. 1971. A review of the biological control of insects and other pests in South East Asia and the pacific region. *CIBC Tech. Commun.* 6:149.

Reiter, R.J., Manchester, L.C. and Tan, D.C. 2005. Melatonin in walnuts: influences on levels of Melatonin and total antioxidant capacity of blood. *Nutrition*, 21:920-924.

Riedl, H., Barnes, M.M. and Davis, C.S. 1979. Walnut pest management; historical perspective and present status. *Pest management programs for deciduous tree fruits and nuts*, 15-80, 10 pp.

Rishi, N.D. and Shah, K.A. 1985. Survey and bioecological studies on the natural enemy complex of Indian gypsy moth, *Lymantria obfuscata* Walker (Lepidoptera: Lymantriidae). *Journal of Entomological Research*, 9(1):82-93.

Rumbos, I.C. 1993. Dieback symptom of olive trees caused by the fungus. *Eutypa lata. EPPO Bull.* 23: 441-445.

Savage, G.P., McNeil, D.L. and Dutta, P.C. 2001. Some nutritional advantages of walnuts. *Acta Hort.*, 544: 557-563.

Sen , S.M. 1986. Ceviz Yetistiriciligi. Eser Matbaasi, *Samsun*, 109-229.

Sheikh, A.G. 1985. Insect pests of temperate fruits and their management. Proceedings of National Workshop-cum-Seminar on temperate fruits, SKUAST, Malangpora, Kashmir, pp.95-98.

Sheikh, A.G. and Bhat, A.A. 1991. Management of tree trunk borer, *Aeolesthes sarta* (Solsky) infesting walnut in Kashmir. *Indian Journal of Forestry*, 14 (2):138-141.

Srivastava, J.N., Sharma, R.M., Singh, B. and Ahmad, N. 2010. Mistletoe (*Viscum album* L.): A serious parasitic threat to walnut trees. *Technical Bulletin*: 1/2010, S. K. University of Agricultural Sciences & Technology, Jammu.

Tamponi, G. 1988. Fruit sensitivity to *Xanthomonas juglandis* of walnut cultivars. *Annali dell Insituto superimentale per la Frutticoltura* 19: 19-22.

Tamponi, G. and Donati, G.P. 1990. Walnut cultivars susceptibility to *Xanthomonas juglandis. Acta Hortic.* 284: 301.

Veghelyi, K. and Penzes, T.T. 1990. Life cycle, forecast and control of *Gnomonia leptostyla* (Fr.) Ces. et de Not. *Acta Hortic.* 284: 284-303.

VPKAS - Perspective Plan Vision 2025. 2007. Vivekananda Parvatiya Krishi Anusandhan Sansthan, Almora, Uttarakhand.

Fig. 1 : Mistletoe

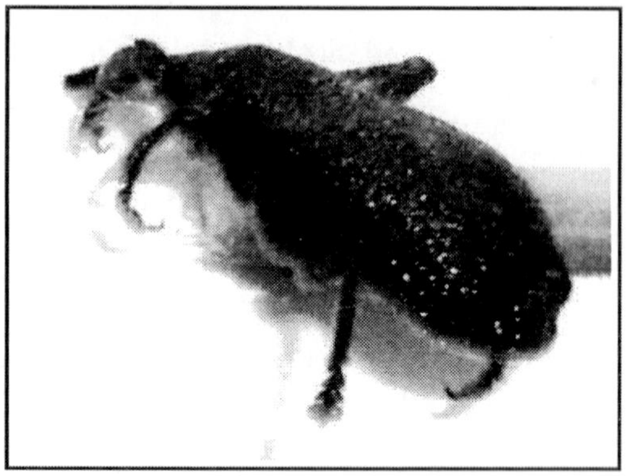

Fig. 2 : Walnut weevil

Fig. 3 : Walnut grub

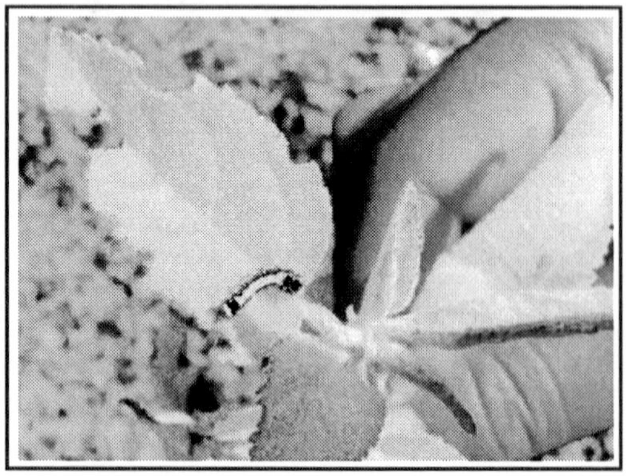

Fig. 4 : Indian gypsy moth larva

Fig. 5 : Chaffer beetle damage on walnut

Fig. 5a : *Anomala* sp.

Fig. 5b : *Holotrichia* sp.

Fig. 6 : Walnut aphid

Fig. 7 : Walnut tree trunk borer

Ecologically Based Integrated Pest Management, 2012
© D.P. Abrol and Uma Shankar (eds.), pp. 787-820
New India Publishing Agency, New Delhi (India)
e-mail : info@nipabooks.com; website : www.nipabooks.com

40

Sustainable Strawberry Production in Sub-Tropical Plains

M.K. PANDEY, UMA SHANKAR AND R.M. SHARMA

Introduction

The cultivated strawberry (*Fragaria* x *ananassa* Duch.), octoploid in nature is a result of hybridization of two American species, the Chilean *F. chiloensis* and Virginian *F. virginiana*. The name strawberry does not incidentally, seems to have anything to do with the practice of putting straw around the plants and long predates their cultivation. It probably comes from 'strayberry' for the runners cause young plants to stray from the parent. Strawberry (*Fragaria* x *ananassa* Duch.) is one of the most important soft fruits, and its cultivated area has increased significantly during the last few years in sub-tropical areas despite of its temperate nature. Being a high input requiring crop, the profit of margin is reduced in sub-tropical areas due to lack of runnering caused by high temperature and high light intensity. Strawberry is one of the most popular rosaceous vegetable used as preferable fruit has a great value for local consumption as well as for export. Attractive appearance, unique taste, availability of fresh fruits during the period of fresh fruits scarcity in the market, short duration nature, high return per unit area, twice earnings in a year (through fruits during spring and runners during autumn) and wider acceptability of its processed products, the strawberry has become a choicest crop of growers of sub tropical areas, and being grown commercially around the various cities of northern India. Although it is a quick and high return giving fruit crop, but the margin of profit in strawberry cultivation under sub-tropical areas is greatly reduced due to regular annual investment (Rs. 0.60 lakhs/ha) by the growers for the procurement of runner plants in every planting season (October) from temperate areas, as high temperature (>35^{0}C) and intense light following fruit harvesting restrict the runnering as well as survival of mother plants in these areas (Sharma and Yamdagni, 2000).

It has been found that human serum antioxidant capacity increases specially after its consumption. Although, earlier it was considered as temperate fruit but due to availability of fertile lands, irrigation facility, semi protected structures and better marketing opportunities near the place of its cultivation, this crop has been gaining popularity to be grown around the various cities of northern India for last few years. Being a short duration crop, it is a valuable option of crop diversification even to the marginal farmers who show reluctance for perennial fruit plantations due to long gestation period.

Varietals suitability, Huge annual investment for planting materials, Frost injury, Production of mid season crop instead of early and late strawberry, Integrated nutrient management, Weed management, Post harvest management including handling, transportation, marketing and value addition and Integrated pests and diseases management are some of the major problems being faced by the strawberry growers of sub-tropical areas. To address the above constraints, the specific technologies have been refined/ generated, and are being transferred at farmers fields for making the strawberry growers to be self sustainable. These technologies are being summarized in brief as follows:

Principally Chandler strawberry is being grown in plains which is a mid season cultivar. The harvesting at one time creates glut of fruit in the market and reduces profit of growers. Oso Grand and Gorella have been identified as early varieties, while Belrubi ripens later than Chandler which can be commercialized in sub-tropical plains. During 2007, a newly introduced cultivar (Kimberly) has been found quite promising having very typical shape, attractive colour and good yield potential, and deserves to be commercialized as an additional option for the growers.

The most serious problem in strawberry cultivation in the warm areas is the regular investment by the growers for procurement of runner plants every year from temperate areas as mother plants do not survive under high temperature following fruit harvesting, thus growers have to invest at least Rs.0.60 lakhs/ha every year for planting materials which greatly reduces the profit margin and makes its cultivation less remunerative. It has been found that covering with green coloured shade net can save mother and runner plants in areas where temperature goes > 42^{0}C during summers. The use of 50% shading net has been found effective for Oso Grand, Chandler and Gorella, while 75% shading has been found best Belrubi for runner formation in subtropical areas.

In frost prone areas, first crop (early season) face serious losses due to low temperature, and leads to kill the primary blossoms. Besides, growing early and late strawberry is a very good alternative to fetch relatively higher price than the mid season crop. Use of polyethylene row covers and black polyethylene mulch with drip irrigation (Fig. 1) can advance the harvesting season by 15-20 days besides protecting the crop from winter rains and frost injury. On the other hand, paddy straw mulch in combination with sprinkler/ drip irrigation can extend the fruiting season by 20-25 days.

The application of 150 kg N, 80 kg P_2O_5 and 45 kg K_2O/has shown to produce very high yield (22t/ha). These doses of inorganic fertilizers can be replaced by poultry manure @ 18 t/ha or FYM @ 72 t/ha without affecting the fruit yield of strawberry. These replacements are of immense importance for producing strawberry as an organic produce.

Weed control is also one of the most serious problems in strawberry production. Because of its low growth habit, it is not able to compete as successfully as other fruits with weeds that deprive the plants of moisture, sunshine and nutrients. To keep fields relatively clean, growers need both a 'base programme' and a 'cleanup programme'. The base programme refers to the primary method relied on to control most weeds. The cleanup programme consists of some specific weedicides to control specific weeds that escape the base programme.

Non-climacteric nature and high rate of metabolic activities tend to make this fruit highly perishable, and thus requires special attention for its post harvest handling, transportation and value addition. Rapid removal of field heat, maintenance of low temperature during grading and packing and refrigerated transportation not only reduce the spoilage of fruits but provide the opportunity to capture market of some distant cities, too. Improper shape, harvesting stage (poor coloured fruits) and packing technique, and mixing of low grade fruits in punnets have been found as the major problem in strawberry marketing. Unmarketable fruits can be successfully processed in to a variety of products like jam, chutney, squash, pickles etc. It has been found that blending of over-ripe and unripe (small or unmarketable) fruits in the ratio of 70:30 can be done successfully without any deleterious effect on the quality of product. One of the major problems encountered in strawberry processing is their susceptibility to textural damage at almost every stage of handling which needs special care. The hierarchy of sensory evaluation has shown that sweet 'n' sour pickle (along with dry fruits) scored the highest numerical rating followed by jam, chutney, squash and preserve.

Integrated Pest and Diseases Management

Integrated Pest Management (IPM) is a long-standing, science-based, decision-making process that identifies and reduces risks from pests and pest management related strategies. It is the use of all suitable tactics to maintain a pest population below an economically damaging level. Effective insect pest management depends on a combination of cultural (modifying crop production procedures to suppress problems), physical control (exclusion and hand-picking), sanitation, biological (use of predators, parasites and pathogens), and resistant varieties and pesticide application practices. Biointensive Integrated Pest Management (BIPM) is a systems approach to pest management that is based on an understanding of pest ecology which begins with the accurate diagnosis of the nature and source of pest problems and biological controls to keep pest populations within acceptable limits. Reduced risk pesticides are used if other tactics have not been adequately effective, as a last resort and with care to minimize risks. Of the $28 billion in pesticides used globally in agriculture and public health, approximately 1% are biologically-based or biological pesticides. Biological pesticides provide safe, environmentally friendly and effective solutions by incorporated into IPM systems.

Strawberries are a unique and high value small fruit crops of subtropics of Jammu & Kashmir and due to seasonal and perishable nature of crop, nearly all fruits are sold and consumed fresh in the markets. However, insect pests along with diseases are the major constraints in obtaining a profitable crop of strawberries. In subtropical conditions, strawberry insect pests complex comprises of defoliators, sucking pests, strawberry sap beetle, soil arthropods and mammalian pests like birds and squirrels. During the vegetative growth i.e. runner production, crops are completely defoliated by the infestation of *Spodoptera litura.* Aphids, jassids, spidermites, thrips, lygus bug and whitefliy populations are mostly recorded during reproductive growth of strawberry crop. They suck the sap from leaves, decreases vitality of crops and causing leaves to curl, and thereby secreting sticky honey dew which eventually favours the growth of sooty mold. Further, they are capable of spreading virus diseases as their populations may build up rapidly. Soil insects include ants, white grubs, wireworms, cut worms, root weevils, mole crickets, termites and nematodes that feed on the roots and the lower part of the plants. Termites and cut worms are the greatest menace in subtropical conditions where water table is low. The entire insect pests complexes are responsible to cause 15-40% damage in strawberry production (Sharma *et al.,* 2008). Strawberry also suffers huge losses due to a variety of diseases inflicted by diverse pathogens. Leaf blight, fruit rot and root rot are of commercial importance in sub-tropical areas. Diseases caused by fungi, bacteria, nematode, viruses and virus like organisms are affecting the cultivation of strawberry, which reduces the yield and quality of fruits considerably. The fungal pathogens are main problem in all area of cultivation and occurring on all parts of infected plant.

The success of integrated pests and diseases management in strawberry depends upon a combination of sanitation, cultural, physical, biological and chemical control measures. There are many new molecules available which are substantially less toxic, selective and more eco-friendly than the conventional pesticides.

Disease Management in Strawberry

Powdery Mildew

This is the most important foliar disease of strawberry caused by fungus *Sphaerotheca macularis* (Wall ex. Fr.) Jazc., and reported from all strawberry grown areas in world. The disease recently occurred in epidemic form in Tiawan and cause severe loss to crop.

The symptoms occurred at all aerial parts including flower calyx and fruits. Initially fungal infection starts from underside of leaves which become developed into white powdery growth. These leaves upward cup like and show radish blotches. In severe infection leaves become burnt at margins to a greater extent. The white powdery mass also develops on infected flower stalk, flower and fruits. Delayed reddening reported in infected fruits (Wilhelm, 1961). The pathogen *Sphaerothecca macularis* (Wall. ex. Fr.) Jacz fsp. *fragariae* produces conidia from conidiophores and ascospores. The pathogen is air borne and overwinters through mycelium in old infected plants. The optimum temperature and relative humidity ranges from 16-18°C and 80-100 per cent for germination of conidia but maximum germination occurs at 20°C temperature. It was reported that the free water has lethal effect on conidial germination and rain has deleterious effect on spore dispersal (Peris, 1962; Jhooty and Mc Kean, 1965). Amsalem, *et al.* (2006) also reported that the optimal environmental conditions for conidial germination and conidial germ tube length ranged between 15 and 25°C with relative humidity (RH) higher than 75%, but less than 98%. High light intensity reduced germination and hyphal growth. The viability of conidia on infected leaves was examined at temperatures ranging from 15 to 35°C at 80-85% RH. Conidia survival declined over time, but a certain percentage of conidia remained active after 5 months incubation. The rate of conidial germination was significantly higher on young leaves than on older leaves. Cleistothecia as sexual fruiting bodies formed in nature (Haward and Alberts, 1973), while the role of cleistothecia as primary source of inoculum is not important because this stage is rarely formed.

Management

Cultivar Milsei-Tudla showed high resistance while cultivars Camarosa, Aromas, Oso Grande, Sweet Charlie and Campinas exhibited moderate resistance. Dover was slightly resistant, whereas Burkley, Vila Nova and Santa Clara were highly susceptible (Brahm *et al.*, 2005). Khanizadeh *et al.* (2005) reported that Harmonie is a new June-bearing strawberry cultivar derived from Yamaska x Joliette, was resistant to powdery mildew disease.

The disease can manage by the use of many fungicides i.e. flusilazole, mycobutonil, tridemifon, fenarimol and buprimate were found effective. The sulphure containing fungicides and benomyl were found most effective but they cause phtotoxicity, which burn the leaves at above 27°C temperature. Two to three sprays of buprimate or tridemifon at 15 days interval give good result at nurshery stage. The fungicides give effective control when they applied after transplanting and just before flowering. Cultivars like ARKING and REAL are reported to be resistant against disease (Moore, 1981). Hukkanen *et al.* (2007) reported that Benzothiadiazole can enhance the accumulation of phenolics in strawberry plants which may then be involved in the BTH-induced resistance to powdery mildew. Foliar sprays to the cultivars 'Aromas' and 'Selva', Calcium chloride (Stopit Reg., Phosyn Plc) and potassium silicate (Kasil 2040 Reg., PQ Corporation) reduced powdery mildew symptoms when both fungicide and potassium silicate or calcium chloride alone had no significant effect (Palmer *et al.*, 2006). Hua-Xiu and Chen-Xian, (2004) reported that Kresoxim-methyl at a dilution of 3000 times obtained the highest control without phytotoxic symptoms compression to Dry flowable formulation (DF) (50%), thiophanate-methyl, emulsifiable (12.5%) myclobutanil, (50%), sulfur and triadimefon (20%).

Anthracnose

Anthracnose disease of strawberries caused by *Colletrotichum fragariae* Brooks was first reported from Florida (1926-29) by Brooks in 1931. Dai *et al.* (2006) was reported of anthracnose fruit rot of strawberry caused by *Colletotrichum acutatum* in China. In India it was first reported by Singh (1974) from Banglore. The disease is fairly worldwide spread in strawberry growing regions. Strawberry anthracnose disease is a complex of three *Colletotricum* species, *C. acutatum*, *C. fragariae* and *C. gloeosporioides*, causes a variety of symptoms that make anthracnose the most impotant disease of strawberry.

C. acutatum causes an anthracnose fruit rot and black leaf spots (Inada and Yamaguchi, 2006), whereas *C. gleosporiods* Penz and *C. fragariae* Brooks infect primarily the crown of the plants and causes crown rot and wilt.

Symptoms may appear as sunken, dark brown, circular lesions on petioles, stolon (runners), which may then be girdled, resulting in wilting and death of the leaf or of the daughter plant beyond the lesion on the stolon. The fungus often spreads into the crowns of young plants, which it rots, and the plants then die in the nursery or after they are transplanted in the field.

On fruits, rot is initially restricted to only ripe fruits and characterized by one or more light brown soaked lesion (Fig. 2). Two or more lesions collapse and occasionally covered by buff colored spores (Haward, 1972). On the leaf, minute dull, violet-black or black spots appear and surrounded by yellow region. In later stages spots become coaleps and forming bigger spots causing defoliation. Other symptoms include bud rot and flower blight also appears. The conidia of *C. acutatum* were hyaline, straight and fusiform, and produced melanized, ellipsoid, ball- or pear-shaped appressoria at the end of germ tubes (Dai *et al.*, 2006).

The fungus mostly survived on infected or contaminated transplants, stolons in the soil, on weed hosts, and possibly in plant debris in the soil. The fungus was able to survive in buried strawberry tissues for a month (Eastburm and Gubler, 1990). *Cassia obtusifolia*, a weed can serve as source of primary inoculum of *C. fragariae* (Howard and Albregts, 1973). In warm humid weather (>21°C), during flowering and fruiting favoured the disease but extensive losses are being caused after rainy periods (Howard, 1972). The fungus produces fruiting body like acervuli in lesions which contain numerous spores i.e. conidia borne in a mucilioginous matrix, these spores are splashed by irrigation or rain to nearby plants or are carried by insects, animals and humans moving among the plants.

Management

Use of straw mulch (Maiden *et al.*, 1993) and use of planting stock obtained from disease free nurseries may provide effective control of the disease. Growing nursheries plants with little fertilizer as possible and removing all infected fruits at each harvest (Agrios, 2005). Using resistant cultivars like Sequoia, Florida balle, Doner (Delp, 1981) and FL 79-1126 (Chandler *et al.*, 1988). Hot water treatment (for 7 minutes at 35°C, followed by 2 or 3 minutes at 50°C) was successful in reducing the proportion of cuttings infected with *C. acutatum* (Johnson *et al.*, 2006). The infections by *C. acutatum* completely manage by the fumigation of soil and planting material with methyle bromide and chloropicrine. Use of soil disinfectants like chloropicrine and dazomet effectively control the disease in nurshery bed (Matsuo *et al.*, 1994).

The twelve resistant varieties of strawberry against anthracnose were reported by Denoyes and Guerin (1996). 'Honeoye' and 'Pandora' cultivars showed a useful level of resistance, even against the most aggressive isolates (Simpson *et al.,* 2006). 'Sweet Charlie', 'Carmine', and 'Earlibrite' were the most resistant cultivars. 'Strawberry Festival' was intermediate susceptibility; while 'Camarosa' and 'Treasure' were highly susceptible (Chandler *et al.,* 2006).

Prochloraz and difenoconazole were most effective fungicides for control of strawberry black flower (Domingues *et al.,* 2001). In 1984 Cheah and Soterosi reported that the use of benomyl with captan gave good results. Ergosterol biosynthesis inhibitors (EBI) a systemic fungicide reduced infection of *C. fragariae* than other fungicides (Smith and Black, 1993). OctaveReg. (462 g/kg prochloraz as the $MnCl_2$ complex) was highly effective in reducing the incidence of colletotrichum crown and stolon rot in runner production (Hutton, 2006).

Verticillium Wilt

Verticillium wilt or vascular wilt occurs worldwide but most important in temperate zone of the world. In United States and England, this disease has gained much importance due to its heavy losses (Wilhelm and Paulus, 1980; Simpson *et al.,* 1994). Verticillium wilt disease is caused by fungus *Verticillium spp.* In England *V. dahliae* Kleb. and *V. albo-atrum* both are reported as pathogen (Keyworth and Benett, 1951) but in United States only *V. albo-atrum* is reported as pathogen (Wilhelm, 1952).

The symptoms of Verticillium wilts are almost identical to those of Fusarium wilt but Verticillium induces wilt at lower temperatures than Fusarium. Verticillium wilt develops primarily in seedlings, which usually die shortly after infection (Fig. 3). More common are late infections, which cause upper leaves to drop and other leaves to develop irregular chlorotic patches that become necrotic. Older plants infected with Verticillium are usually stunted and their vascular tissues show characteristic brownish discoloration. Verticillium infection may result in defoliation, gradual wilting and death of successive branches, or abrupt collapse and death of the entire plant. From India, Mitra (1993) reported that the symptoms of Verticillium wilt in strawberry characterized by browning of wood at the base of crown wilting decrease in runner production and reddening of oetiole and stolon late in season. This disease causes total collapse of stawberry plants during peak of the first year's growth. Development of emerging leaves in related and plants showed stunted growth.

Both species of Verticillium, *V. dahliae* and *V. albo-atrum* produces short lived conidia. *V. albo-atrum* produces dark, thick-walled mycelium but not microsclerotia. Fungus overwinters in the soil as mycelium or as microsclerotia and can survive upto 15 years. *V. albo-atrum* grows best at 20 to 25°C, whereas *V. dahliae* prefers slightly higher temperatures (25-28°C) and sometime more common in warmer regions (Agrios, 2005).

Management

It is difficult to control but adoption of long crop rotation, use disease free planting material in disease free soil, use of soil disinfactant and resistant cultivars to combat the disease (Ercole, 1976 and Agrios, 2005). The thermal inactivation via soil solerization is providing usefull for the control of Verticillium wilt in regions with high summer temperature and low rainfall. Use black mulch with low dose of nitrogenous fertilizers. Best control of disease was achieved by the use of chloropicrine (8 kg/ha), where 2.5 times more yield than land fumigated with ethyle dibromide was obtained (Wilhelm and Paulus, 1980). In addition to chloropicrine, dozomat was also found effective against *V. dahliae* infection (Harris 1980a). The pre-planting treatment of soil with dozomat and chloropicrine followed by covering the land with polyethylene sheet or nylon laminate (Harris 1980b). The total fumigation by chloropropene followed by strip fumigation with methyle bromide control the disease effectively (Meesters, *et al.,* 1996). Soil drenching with fungicides i.e. benomyl and thiophanate M @ 0.1% in the holes and soil solerization during hot months followed by fumigation with chloropicrine reduce the infection of Verticillium in strawberry (Hartz *et. al.*, 1993). Resistant cultivars like Etna (Faedi *et al.,* 1986), Calypso, Tang and Eros (Simpson and Blanke, 1996), Pegasus (Simpson *et al.*, 1994) are reported as resistant against the disease.

Black Root Rot

Black root rot is a serious and common problem of strawberries. The term 'Black root rot' is the general name for several root disorders that produce similar symptoms. This disorders are generally referred as a root rot complex disease of strawberry and it is reported to be incited by a number of soil inhabiting fungi i.e. *Laptosphaeria coniothyruim* (Fckl). Sacc. *Fusarium ortocereras* App. and Wollenw, *Pezizella lytheri* Shear and Dodge, *Ramullaria* species, *Corticium vagum* Berk and Cut., *Packybasidium cadium* Sacc., *Pythium* species Trow, *P. irrigulare* Buis, *Pyrenochaeta* spp., *Cylindrocarpon destrctans* and *Ceratobasidium* species. (Bhardwas and Sharma, 1999). Besides the fungal pathogens certain nematodes, winter injury, fertilizer burn, soil compaction, drought and excess salt, water or improper soil pH are also envolved in this disease complex.

The affected plants show an uneven patchy appearance due to dwarfing and death of affected plants. The root system is much smaller than normal

(Fig. 4). The main roots develop brown areas and small, fibrous rootlets are killed (Fig. 5). Lesions extend and the entire root is blackened. Leaves remain small with short prtioles and turn yellowish, old leaves wither. Berries, if formed, remain small or wither before ripening. The disease is more prevalent in high clay content and waterlogged soil.

Management

Avoid continuous cultivation of stawberries on a same land, planting material should be free from diseases and avoid water logged or poor drained areas, for plantation. Soil treatment with chemicals can reduce the population of undesirable nematodes and fungi in the soil. The soil fumigation with methyl bromode and chloropicrine reduce the populatin of fungi involved in the black root rot disease (Drozdowski, 1987).

Red Stele

Red stele also called as red core, red root rot, brown stele and black stele. This is the most important disease of strawberry caused by the soil-borne fungus *Phytophthora fragariae* Hickman. The disease first of all noticed in Scotland in 1920 (Alcock *et al.*, 1930) and the fungus isolated and reported by Hickman in 1940 for the first time. Fifteen physiological races of *P. fragarea* have been reported from worldwide (Mass, 1984).

Infected plants are stunted, lose their shiny green luster and produce few runners. Younger leaves often have a metallic bluish-green cast. The very small size of fruits formed on infected plants with poor quality, sometimes fruits are not formed. Older leaves turn prematurely yellow or red and with the first hot, dry weather of early summer, diseased plants wilt rapidly and die. Diseased plants have very few new roots compared to healthy plants that have thick, bushy white roots with many secondary feeder roots. Infected strawberry roots usually appear gray, while the new roots of a healthy plant are yellowish-white. The inside or central portion of the root of wilted plant stele becomes pink to brick red or brownish red and on this symptom basis disease is reefers as red stele disease while stele of normal plants is yellowish-white. This red color may show near the tip of the root or it may be extend the length of the root. The red stele disease is best seen in the spring up to the time of fruiting.

Heavy rain fall, frequent irrigation and water logged conditions helps in libration of the Zoospores. Oospores are survives in soil and cause root infection primarily. Germinated oospores to produce sporangia and sporangia germinate indirectly to produce many zoospores. The zoospore germination on the host roots and oospore development in the host tissues occurred at 15°C to 20°C temperature. Root exudates attract the zoospores which encyst around root tips and produces germtube for penetrate epidermis (Anderson and Colby, 1942). Fungus reaches the phloem and pericycle, within three days of infection (George and Milholland, 1986, 1987).

Management

The disease can be controlled by use of disease free planting materials and taken from disease free areas. Acidic soil should be avoided for the planting. Soil fumigation with ethyle bromide and chloropicrine significantly reduces, inoculum potential (Joffens, 1957). Plant drench with systemic fungicide metalexyle effectively manages the disease (Mcintyre and Walton, 1981).

The disease can control by use of resistant varieties but pressure of physiological races it is not easy to maintain resistance. The cultivars of strawberry loke Delmarvel, Mohawk, Primetime and Erose have been found resistant against *P. fragarea* (Igarashi *et. al.*, 1994; Gallete *et. el.*, 1995a, 1995b; Simpson and Blanke 1996).

Fruit Rot Diseases

Rhizopus Rot

A fungal disease caused by *Rhozopus stolonifer* (Ehrenb. Fr.), Vuill. Infected fruits turn light brown which become soft, watery and give exudates juice even at slight pressure. The whole fluffy mycelial growth with black sporangia formed on severely infected fruit during transportation and storage. The pathogen is a wound parasite and can be prevented by pre - cooling of fruits before shipment.

Grey Mould Rot

Grey mould rot disease is caused by *Botrytis cinereae.* It is appear as fruit rot under humid conditions, fungus produces a noticeable grey mould (Fig. 6) fruiting layer on the affected tissues (Agrios, 2005). In high humidity, white cottony mycelial growth may be formed (Bhardwas and sharma 1999).

Management

Remove of infected and infested debris from the field and storage rooms by providing conditions for proper aeration and quick drying of plants and plants products (Agrios, 2005). Proper drainage and aeration in field also reduces the incidence of grey mould rot (Daubeny and Pepin, 1977; Mass, 1978 and Barritt 1980).

Biological control of Botrytis gray mould can obtain by spraying the flowers and fruits with spore suspensions of certain antagonistic fungi and bacteria.

Trichoderma harzianum inhibit the growth of *B. cinereae in vitro* and reduced it infection on fruits *in vivo* under greenhouse and field conditions (EL - Zayat *et. al.*, 1993).

Management of *Botrytis* through chemical has been partially successful. Regular spray of fungicides like captan or thiram and benomyl or thiophanate methyle was found effective against gray mould in strawberry (Horn, 1961; Borecka *et al.*, 1973; Gourley, 1974). Spray of vncolozolin wqas found most effective (Paulus *et al.*, 1978). The fruits decay due to grey mould can reduce by fumigation with acetaldehyde vapour (Prasad and Stadelbacher, 1974).

Leaf Diseases

Leaf Spot

Mycosphaerella leaf spot

White centered purple margined leaf spot is a fungal disease (Fig. 7) caused by *Mycosphaerella fragariae* Tul. (= *Ramularia tulasnii* Sacc, imperfect stage). In India this disease was first reported from Niglar, Bhowadi hill of Uttar Pradesh during October 1952 (bosh, 1970). In initial stage, circular purple scattered spots appear at upper surface of young leaves with an average size of 2-6 mm in diameter. The spots most frequent on blades of leaflets but also appear on petioles, fruits and fruit stems. The leaf spots enlarge, then turn to white and are surrounded by dark purple margins, rendering birds eye effect. On lower surface of leaves, prominent veins touching the spots become radish purple and entire leaf may die in advanced stages. The lesion on stem, fruit stalks are elongated as well as circular. On berries infection is not common however, the pulp of berry becomes discolored and render the fruits unmarketable under high pressure of disease (Fulton, 1958).

This pathogenic fungus belongs to class-ascomycetes. The ascomata are usually very small, mostly emerged in the host tissues commonly in leaves. The single septate hyaline ascospores are measuring 11-45x2-3 μm dimensions, and usually formed on the upper surface of the leaf, sometimes formed on petioles and calyx also. The sclerotia on dead leaves may also produce conidia.

Management

In cultural practices, includes planting the strawberry plants in well drained soil, keeping out the weeds and maintaining proper spacing, which can reduce disease incidence. Avoid excessive use of nitrogenous fertilizers (Harnendo and Casada, 1976). During sprouting in spring and budding, the spray of micronutrients like manganese, copper and boron can reduce the infection and enhance the yield upto 13.20 Per cent (Dorozhkin and Grisanovich, 1972). Chemically this disease manage by spraying of Bordeaux mixture (0.8%), Zeneb, Ferbam, Captan, Cuprex (0.2%) and by benomyl (Stubbs, 1956). The use of benomyl and thiophanate - M as dips to nursery plants or as spray after planting render best control (Paulus *et al.*, 1974). The varieties like, Combridge (Festic and Delkic, 1977), Takane Kurumal 103, Horida's wonder, Hukuda and Benzuru (Kim *et al.*, 1978), Arking (Moore, 1981), Premier, Covaliar, Dilpasand and Albitron (Sindhan and roy, 1981), Elista and Tioga (Saharan and Badiyala, 1985), Tarda vicoda and Marmalada (Zurawicz and Daminikowski, 1995), Joliette (Khanizadeh *et al.*, 1996) were reported as resistant against disease.

Pestaliopsis leaf spot

This is a common leaf spot disease of strawberry caused by fungus *Pestaliopsis disseminate* (Theum) stey. The disease found all strawberry grown areas in India and first time recorded by Singh *et al.* (1975) from Bangalore on cv. Pusa early dwarf.

Symptoms characterized by circular, dark brown to chocolate coloured spots surrounded by radish brown or yellowish margins. Spots may coalesce and shows patchy appearance. The infection normally starts from the leaf margin and extends towards midrib which covered the whole lea. The infected areas become brittle and get detached from the healthy ones. Severely infected leaves get defoliated. Dot like fruiting bodies (acervuli) of the pathogenic fungus appeared on the necrotic areas of the infected leaf. The acervuli are amphiginous, scattered, distinct sometime gregarious, globose to centricular, conic or pyramidal, puntiform, erumpent rupturing the epidermis or diplacing it completely, exposing the black conidial mass. Short conidiophores emerge from the basal cells, slightly broader at the base with tapering end. Conidia elliptical or clavate, fusiform, tampering at the bae, straight or slightly constricted at septa, 5 celled, measuring 18.5x26x5.5 μm (Singh *et al.*, 1975).

Yellow leaf spot

The leaf is caused by fungus *Dendrophoma obscurans* (Ell. And Ev.) Anderson, and prevalent in every strawberry grown areas. In India it has been reported from hills of Uttar-Pradesh and Himanchal Pradesh (Bosh, 1970). It has been found in moderate to severe from the altitude of 1,000 to 24,00 meters above sea level. The first symptoms appear on leaves with onset of the rains and the maximum damage occurs, during July to September. In initial stage of infection minute, circular, radish to purple spots appear on the upper surface of leaves and calyces. The spots rapidly increase in size with yellow or yellow brown centre surrounded by a purple margin. The spots may coalesce and from irregular or oblong patches, covering more than half of leaflets surface. The large spots can distinctly divided in three ; outer purple, inner light brown and centre dark brown zones, and spots often extend towards the edge of lamina, which become "V" or fan shaped. The affected areas with disease may dries up and become puckered or crumpled. Sometimes, shot holes are formed. Pycnidia found in necrotic area of spot, which are deeply embedded in host tissue with a distinct neck extending through the epidermis and act as a source of primary infection (Anderson, 1956; Bose, 1970).

The disease can be control by the use of Phenyle mercury acetate or captan. Spraying captan @ 0.2% start with the onset of rains. Albitron, Premier, Cavalier and Diplasand varieties are reported as resistant against disease (Sindhan and Roy, 1981).

Angular Leaf spot

This is a most important bacterial disease of strawberry and recorded from Carolina, Euthopia, Argentina and Australlia (Mc Gechan and Fahy, 1976; Alipi *et al.*, 1989; Navarthan *et al.*, 1991 and Ritchie *et al.*, 1993). Angular Leaf spot disease is caused by *Xanthomonas fragariae.*

The water soaked spots appear on upper surface of leaves in initial stage, later on which become dark brown and angular in shape (Fig. 8). In severe condition of disease the affected plants may be die. Disease can be effectively control by spraying of streptomycine or streptocyclene, start with the appearance of symptoms (Alipi *et al.*, 1989).

Leaf blight

This disease was first reported by Lele and Pathak (1965) from India. The disease caused by *Rhizoctonia bataticola* (Taub). Bull. [= *Macrophomina phaseolina* (Tassi.) Goid].

The initial symptoms of this disease found on all plant parts during September to October months. More or less circular to spreading spots with ash grey centre and purplish dark margins were found on leaves. In advanced stage these spots become oval to irregular in shape and lesion started spreading from margin of the leaves towards centre. Infected runners and stalks turned dark brown or black with irregular lesions. Due to infection in plants new growth very much suppressed. The poor development in new root and shredded appearance in older ones at their distal portion was found, but no irregular rot was seen.

Leaf Scorch

The disease leaf scorch was first reported by Wolf (1924) worldwide and in India recorded by Bose (1970) from Kumaon hills of Uttar Pradesh. This disease is caused by fungus *Marssonina fragariae* (Sacc.) Kleb *[(Diplocarpon carliana)* (E & E) Wolf]. The disease mostly occurred in month of April or May and symptoms appear on leaves, petioles, peduncle and runners. Irregular, oval or angular and purple colored symptoms surrounded by reddish or pinkish halo are found on leaves (Fig. 9). Elongated, reddish brown or purple streaks developed on petiole and peduncle. Similar symptoms appear on runners which become discolored and girdled (Bose, 1970). Disease can be effectively managed by the spraying of carbendazim.

Insect Pests of Strawberry and their Management

A. Sucking insect pests

1. Strawberry aphid, *Chaetosiphon fragaefolii*

Aphids are small sucking insects and causes damage by sucking the saps of plant leaves. Populations may continue to increase to damaging levels when spring temperatures are moderate and humidity is high. They are capable of spreading virus diseases and their populations may build up rapidly which eventually leads to significant economic losses in strawberries. The aphids winter over in the crowns of the plants. They develop wings in April and migrate to other plants. There are certain types of aphids have been recorded infesting the strawberry at various crop growth stages viz., strawberry aphid (Fig. 10), *Chaetosiphon fragaefolii;* melon aphid, *Aphis gossypii; and* green peach aphid, *Myzus persicae.* Although aphids are not a major problem on field strawberries (Mossler and Nesheim 2003), in greenhouse strawberries, the cotton aphid, *Aphis gossypii* Glover (Homoptera: Aphididae), can be a serious problem (Leclant and Deguine 1994). High populations of aphids can reduce the vigor of the plant, making it susceptible to other pests. The honeydew that aphids excrete reduces fruit quality because of the development of a black sooty mold on the substrate. Moreover, this sooty mold reduces photosynthesis production and reduces the quality of the plant causing considerable economic injury.

2. Strawberry Mites

Twospotted spider mite, *Tetranychus urticae* Koch (Acari: Tetranychidae) are the most important and potential pest of field and greenhouse strawberries wherever strawberries are produced (Price and Kring 1991; Stonneveld *et al.,* 1996; Walsh *et al.,* 2002; Sato *et al.,* 2004; Cloyd *et al.,* 2006). Spider mites suck sap from the plant and cause loss of plant vigour and their populations have become resistant to most of the acaricides due their short life cycle and high fecundity (Williams 2000; Cross *et al.,* 2001; Stumpf and Nauen 2001; Sato *et al.,* 2004). Its high populations can reduce foliar and floral development thereby decreasing the quality and quantity of mature fruit (Rhodes *et al.,*

2006). They feed by sucking the chlorophyll from leaf cells, causing the characteristic whitish spotting, eventually turning the leaves a coppery-bronze colour.

3. Strawberry Thrips

Frankliniella occidentalis (Pergande) is one of the major strawberry pests in southern Brazil which causes russeting and wither in flowers and fruits reducing commercial value (Nondillo *et al.,* 2008). They occur during bloom of crops and its problems are increasing especially during mild winters when most of the plants are in flowering or full bloom. Thrips have rasping, sucking mouthparts that cut the skin cells of strawberry leaves (Fig. 11) and fruit, causing the fruit to turn bronze colored or to cause a slight russetting. Presence of grasses in the surrounding of main field tends to result in higher populations build up of thrips.

4. Strawberry whitefly, *Trialeurodes* sp. and *Chaetosiphon* sp.

The greenhouse whitefly, *Trialeurodes sp.* is the major insect pests of strawberries (Fig. 12). It is an increasing problem in most of the vegetable crops although not common in nurseries. Tiny sized whiteflies often settle on the underside of leaves and suck the sap. It can be a source for potential vectors of disease which eventually reduce the quantity and quality of marketable strawberries. Its infestations are usually found in patches rather than over the entire field.

5. Tarnished Plant Bug

Tarnished plant bug has one of the broadest host ranges of any major insect pest, feeding on weeds and grasses in addition to strawberry blossoms (Fig. 13). They eat parts of the blossom by piercing and sucking out plant juices with their beak-like mouthparts. Their incidence is recorded from flowering up to the harvest of the fruits. The continuous feeding by adults and nymphs causes 'cat facing' of berries. Weed control in the berry field is important because weeds may attract bugs into the strawberry planting and allow adults to overwinter in the strawberry field.

Integrated Management

As strawberry is often grown as an annual or short term perennial crop, exploiting natural populations of natural enemies is difficult. Protected cultivation of strawberry provides more favourable conditions for exploitation of biological control including introduction of insect predators and parasites (Cross *et al.,* 2001). Thrips can be monitored at the same time as tarnished plant bug nymphs. Collect ten flower buds from a field (use a V-shaped sampling pattern to determine sampling locations). Naturally occurring minute pirate bugs (*Orius* spp.) feed on thrips. Sprays targeted against thrips may disrupt biological control of other pests such as the two spotted spider mite. Consider treating only if populations reach over 10 thrips per blossom cluster. Potential predators like ladybugs, lacewings, and syrphid flies should be conserved and encouraged. Although biological control keeps aphid populations low, treatments may be necessary, if weather is conducive to their development. Thiodan and malathion can provide good control but they are slightly disruptive to predatory mites and upset IPM programs. Treatments are also applied in strawberry nurseries to prevent aphid buildup and virus spread. Aphid populations tend to be especially large in plants that receive an excess of nitrogen fertilizer. Therefore judicious use of nitrogenous fertilizers may be applied to keep ward off the aphid populations. Cultural and biological controls and sprays of insecticidal soap are acceptable for use on organically certified strawberries. Natural enemies are found effective against the cotton aphid include lady beetles *Coccinella septempunctata* L. and *Hippodamia convergens*, the green lacewing, *Chrysoperla carnea*, and parasitoid wasps, *Lysiphlebus testaceipes* (Cress) and *Aphidius colemani* L. (Howard *et al.*, 1985; Van Driesche and Bellows 1996; Kaplan and Eubanks 2002; Easterbrook *et al.*, 2006).

Outbreaks of TSSM have become more frequent over the last few decades due to increased use of pesticides in modern cultural practices. As a result, more growers are utilizing biological control as an alternative to chemical management (Huffaker *et al.*, 1969; Escudero & Ferragut 2005; Rhodes *et al.*, 2006). However, little is known about the non-target effects of biological control releases on beneficial arthropods in the strawberry ecosystem. *Phytoseiulus persimilis*, spider mite predators have become important elements of integrated pest management (IPM) in Florida strawberry production (Decou 1994; van de Vrie *et al.*, 1994). Two of the most commonly used phytoseiids are *Phytoseiulus persimi* and *Neoseiulus californicus* (McMurtry & Croft 1997; Cloyd *et al.*, 2006). Rondon *et al.* (2004) developed management plan include the use of high quality transplants and the release of predatory mites at predetermined thresholds.

Neonicotinoids are a new insecticide class which act on acetylcholine receptors in the insect central nervous system, with systemic properties and long residual activity against sucking insects such as whiteflies (Kagabu, 1999; Yamamoto, 1999; Ware, 2000). Since 1998, Admire (imidacloprid), Platinum (thiamexthoxam) / Actara and Assail (acetamiprid) were registered on several crops to control the greenhouse whitefly. At present, the insecticides of conventional classes (such as chlorinated hydrocarbons, organophosphates, carbamates and pyrethroids) are only recommended for limited use in rotation with neonicotinoids and/or insect growth regulators (IGRs) to control whiteflies (Liu and Meister, 2001; Palumbo *et al.*, 2001). On strawberries, foliar application of these conventional insecticides are recommended to suppress the high whitefly populations in mid or late season planting or early post planting. In a small garden, it is possible to spray the underside of leaves with insecticidal soap or 1% ultrafine oil. Sticky traps /delta traps can be used as a non-chemical method to monitor these sucking pests efficiently. Weed control will be effective tool to keep the tarnished plant bug populations from building up.

B. Lepidopteran pests

1. Greasy cut worms, *Agroits ipsilon*

The greasy cut worm, *Agroits ipsilon* (Hufn.) is the most important pest on strawberry; larvae of all instars of the cutworm are causing severe injury to strawberry plants. They are just below the surface of the ground beside the cut plants and eating of plants just above at or a short distance below the surface of the soil (Fig. 14). These larvae often cut off the petioles of the leaves of strawberry plants or the stems of the young fruits, also attack the green and full grown fruits from below, often hollowing it out leaving only the shell of the berry (Taha, 1984; Sharma *et al.*, 2008).

2. Tobacco caterpillar, *Spodoptera litura*

Tobacco caterpillars are a big a problem in nurseries and greenhouses where plants are being propagated in offseason production. Their occurrence is noticed in late summer and found damaging the strawberry runners at vegetative growth. At the time of vegetative growth, unavailability of other food sources may be the possible reason for *Spodoptera* infestation on strawberry. Adult gravid female lays eggs in batches. After hatching, from the eggs, the newly hatched larvae are responsible to completely defoliate the leaves leaving only the midribs (Fig. 15).

3. Strawberry leafroller, *Ancylis comptana*

The two halves of strawberry leaflets are often folded and webbed together. Greenish or brownish caterpillars live and feed inside (Fig. 16). These leaves soon turn brown and die. In severe infestations, the fruit fails to mature. The adult moths appear in February and the females lay eggs singly on the undersides of the leaves. The eggs hatch in about a week, and the larvae feed for about a month. When fully grown they are about ½ inch long and vary in color from yellowish green to greenish brown with a brown head. They transform to brown pupae within the folded leaves and 10 days later the moths emerge.

The insect overwinters both as a larva and a pupa. The moth is light reddish brown with forewings marked with wavy lines of darker brown and white and with wingspread of slightly more than half an inch. This insect can be controlled by spraying with rotenone, which is among the compounds registered for control of this pest, in spring about a week after the moths first appear.

Integrated Management

Spinosad combines the best features of synthetic chemical insecticides and naturally derived biological insect control products. In fact, the low toxicity of spinosad toward beneficial allows it to be incorporated into most (IPM) programs that heavily rely on predators and parasites (Bret *et al.*, 1997). It has the efficacy, broad-spectrum activity against lepidopterous pests e.g. *Ostrinia nubilalis, Helicoverpa zea, Trichoplusia ni, Plutella xylostella, Spodoptera spp., Heliothis spp., Pieris rapae* and *Agrotis ipsilon*. Spinosad also has a very favorable safety profile towards mammals (Sparks *et al.*, 1995), reduced environmental risk and phytotoxicity (Harris and Maclean, 1999) and safety to beneficial insects (Schoonover and Larson, 1995; Sparks *et al.*, 1995). Proclaim effectively controls the larvae stages of many pests at low rates and increasing the crop's value. It can be applied by ground air or ground only, giving the growers the flexibility needed for effective integrated pest management (IPM) programs. (Jansson *et al.*, 1997). Application of bacterial formulation *Bt* is also promising in controlling the lepidopteran caterpillars during the vegetative as well as at reproductive stages of strawberry. Egg masses can be controlled by *Telenomus* egg parasitoids in the field condition. Non-chemical pest management include predators like big-eyed bug, damsel bug, minute pirate bug help in reducing the populations of leaf roller.

C. Coleopteran insect pests

1. Strawberry bud weevil or clipper, *Anthonomus signatus*

Strawberry bud weevil or clipper weevil reduces the number of blossoms in the field. They are the direct pest of the berries and their incidence is recorded during pre-bloom and bloom periods (Fig. 17). They lay eggs in the flower bud and then eat the pedicel partly off. In severe infestations, this insect may cause a loss of 50 to 60 per cent of the crop. However, strawberry plants usually compensate for clipped buds and loss of yield is not often detectable. The beetle is about 1/10 inch long, and varies in color from black to reddish brown. The larva feeds almost entirely on pollen and the eggs are laid almost wholly in the buds.

2. Strawberry crown borer, *Tyloderma fragariae*

The grubs of this flightless weevil tunnel downward through the crown of the plant and by the time they are fully grown, the plants are severely injured. From one to three grubs may tunnel in a plant. The mature grub is about 1/5 inch long and is white with a darker head. The grubs mature in July and transform to weevils within the burrow. The adults are chestnut brown in color, have three darker spots on each wing cover and are about 1/ 6 inch long. The weevils emerge and feed for a time, but on the approach of colder periods go into the soil to overwinter. They appear in early spring and the females lay eggs singly in cavities eaten in the plant near the surface of the ground. There is one annual generation. Insect pathogenic nematode applications may help to reduce populations of this pest.

3. White grub

White grubs larvae cause severe injury to strawberry fields by eating the roots of the plants and causing the plants to suddenly die. White grubs species include Japanese beetle, oriental beetle, European chafer and Asiatic garden beetle, all of which also feed on foliage as adults. Japanese beetles feed on foliage during the day whereas Asiatic garden beetles at night. There are no effective control measures to protect the roots from larvae, so the best option is to reduce white grub populations in surrounding weed hosts.

Integrated Management

Scouting the pest activities, removal and destruction of wild berries on borders, sanitation of field and crop rotations are the essential tool for managing Strawberry clippers. Non-chemical pest management includes light traps, water as a monitoring tool, field sanitation and weed control. Installation of light traps is beneficial for collecting adult grubs' population. *Beauvaria* sp. and *Bacillus papillae* provides good control aginst whitegrubs populations. Peters and Vlug (2005) advised to apply nematodes (*Heterorhabditis bacteriophora*) from the end of July to the mid of September can successfully suppress *P. horticola* (white grub) populations, even though the grubs are not yet present in the soil. Abanowska and Olszak (2003) studied the efficacy of Imidacloprid, bensultap, etofenprox and thiametoxan which gave promising results in controlling adult of strawberry root weevils and white grubs. Most of the coleopteran insects can be fairly controlled by spraying carbaryl 50 WP @ 2ml per lit of water when leaf damage is apparent.

D. Soil Arthropods

Soil dwelling arthropods are causing damage to the roots of crops and are best controlled before setting the plants in the bed.

1. Termites

In fact, termites are the greatest menace in sub-tropical conditions where water table are remarkably low. Most of the rainfed areas of Jammu province are prone to termite infestation. Its attack is especially serious where paddy straw is being used as the mulch. They completely cover the fruits touching the ground with earthen mass and consumed the entire fruit (Fig. 18).

Integrated Management

Use of black plastic mulch is also recommended to protect the crops and fruits from termite infestation. Use imidacloprid @ 0.3 -0.5 ml per lit of water or drenching of chlorpyriphos 10 EC @ 1 ml per lit of water is suitable to control the soil dwelling insect pests. Entomopathogenic nematodes in the families Steinernematidae and Heterorhabditidae are soil inhabiting insect pathogens that possess potential as biological control agents (Kaya and Gaugler, 1993) and against termites (Shankar and Kaul, 2010).

Keep the field fallow for some time and follow the crop rotation for few years are the suggested control methods for nematodes infested fields. Planting materials should be raised in soil solarized beds. Cultivation of marigold (African marigold) are promising in suppressing the nematode population from field. Soil application of carbofuron 3 G @ 25 kg per ha is beneficial to suppress the nematode population in infested fields.

2. Snail and slugs

Mostly these pests are the commonly associated with the area where moisture or damp is prevalent. They are voracious leaf feeders and can completely defoliate the entire plants. They feed on fruit on or near the soil and rendering them unfit for the human consumption (Fig. 19).

Integrated Management

Manual collection and destruction of snails. Snails and slugs can be controlled by metaldehyde bait placing between rows and plants at evening hours. The most common way to kill snails and slugs is to bait them with a sugar pesticide mix. Control of the above-mentioned pests during the fruit production season requires close scouting to determine their presence. It may be necessary to spray on a 7 to 10 days schedule from the time of new growth until 14 days before harvest. Non-chemical pest management includes manage mulch and water (excess of either increases slug problems), field sanitation and weed control.

E. Mammalian and Birds Pest

Certain mammelian pests such as squirrel, rodents, porcupine and some pest cattles, dogs are quite troublesome in strawberry fruit production. Besides these, many species of bird's especially insectivorous birds eat agricultural insect pests. However, some species of birds may found eating the fruit or berry and can create problems for agricultural crops and need to be managed with the use of bird-control devices. Birds can be especially troublesome in berry, grapes, litchi, guava, peach and other fruit plantings. After a planting is established, weeds can be suppressed by applying thick layers of straw mulch or black polythin sheet mulches. Bird damage in berries may account for as much as 30% of crop loss depending upon the year and the location. Damage is most frequently caused by just a few species, including robins, bulbul, Indian parrots, crow, grackles and starlings, but other local and migratory birds may also be problematic, especially in a dry year when other food resources are scanty or limited. These birds cause damage by eating the entire berry or puncturing it and/or eating a portion of the fruit leaving it unmarketable and prone to disease. Birds also knock berries off bushes as they forage.

Integrated Management

Mammelian pests can be checked by putting some fences and underground fences for rodents and porcupines. The fruits which have been destroyed by birds can be prevented by using bird reflectors, reflector strips and bird scare device to keep ward off the birds damage in the fields. Bird netting is quite promising and suitable for the masses of growers. Auditory frightening devices are also effective in keeping away the birds. Research has shown that birds quickly habituate to these noises and to be successful growers should make sure that sound is at least 130 decibels; sounds should cover a wide frequency range and be introduced at random intervals. Noise devices, bird scarer, reflecting ribbons, bird nets and many other home remedies have been tried, with varied success. Successful scare tactics depend on the bird species, bird population pressure, and the grower's management of the devices. Fruit growers use sonic and visual scare devices only at critical times in the growing season, such as fruit ripening, and remove them promptly as soon as that period is over (Ames and Kuepper, 2004).

References

Abanowska, B.H. and Olszak, R.W. 2003. Soil pests and their chemical and biological control on strawberry plantations in Poland. *Bulletin-OILB/SROP*, 26(2): 93-99.

Agrios, G.N. 2005. Plant Pathology (fifth edition). Elsevier Academic Press, UK, pp. 922.

Alcock, N.L., Howells, D.V. and Foister, E.C. 1930. Strawberry disease in Lankashire. *Scott. J. Agric.*, 13: 242-251.

Alipi, A.M., Ronco, B.L. and Carrainza, M.R. 1989. Angular leaf spot of strawberry, a new disease in Argentina. Comparative control with antibiotics and fungicides. *Advances Hortic. Sci.*, 1: 3-6.

Ames, G.K. and Kuepper, G. 2004. *Tree Fruits: Organic Production Overview* http://attra.ncat.org/attra-pub/fruitover.html http://attra.ncat.org/attra-pub/PDF/fruitover.pdf

Amsalem, L., Freeman, S; David, R.D., Nitzani, Y., Sztejnberg, A., Pertot, I. and Elad, Y. 2006. Effect of climatic factors on powdery mildew caused by *Sphaerotheca macularis* f. sp. *fragariae* on strawberry. *European Journal of Plant Pathology*, 114(3): 283-292.

Anderson, H.W. 1956. Diseases of fruit crops. McGraw Hill Book Company Inc. New York, Toronto, London, 501pp.

Anderson, H.W. and Colby, A.S. 1942. Red stele of strawberry on Path finder and Aberdeen. *Plant Dis. Rep.*, 26: 291-292.

Barritt, B.H. 1980. Resistance of strawberry clones to Botrytis fruit rot. J. Am. Soc. *Hortic. Sci.*, 105: 160-164.

Bhardwaj, L.N. and Sharma, S.K. 1999. Diseases of strawberry, Gooseberry and raspberry. Diseases of horticultural crops-fruits. Indus Pub. Com., New Delhi, 316-336.

Borecka, H., Borecki, Z. and Millikan, D.F. 1973. Recent investigations on the control of grey mould of strawberries in Poland. *Plant Dis. Rep.*, 57: 31-33.

Bose, S.K. 1970 diseases of valley fruits in Kumaon (III) leaf spot disease of strawberry. *Prog. Hortic.*, 2: 33-53.

Brahm, R.U., Ueno, B., Oliveira, R.P. De. 2005. Reaction of strawberry cultivars to powdery mildew under greenhouse conditions. *Revista-Brasileira-de-Fruticultura.*, 27(2): 219-221.

Bret, B.L., Larson, L.L., Schoonover, J.R., Sparks, T.C. and Thompson, G.D. 1997.Biological properties of Spinosad. *Down to Earth*, 52: 6-13.

Brooks, A.N. 1931. Anthracnose of strawberry caused by *Colletotrichum fragariae*, N. Sp. *Phytopathol.*, 21: 739-744.

Chandler, C.K., Albregts, E.E and Howard, C.M. 1988. Evaluation of strawberry cultivars and advanced selections at Dover Florida, 1986-88. *Adv. Strawberry Prod.*, 8: 19-22.

Chandler, C.K., Mertely, J.C. and Peres, N. 2006 Resistance of selected strawberry cultivars to anthracnose fruit rot and *Botrytis* fruit rot. *Acta Horticulturae*, 708: 123-126.

Cheah, L.H. and Soterosi, J.J. 1984. Control of black fruit rot of strawberry. Proc. 37th N.Z. *Weed Pest Control Conf.*, 37: 160-162.

Cloyd, R.A., Galle, C.L. and Keith, S.R. 2006. Compatibility of three miticides with predatory mites *Neoseiulus californicus* Mcgregor and *Phytoseiulus persimilis* Athias Henriot (Acari: Phytoseiidae). *HortScience*, 41: 707-710.

Cross, J.V., Easterbrook, M.A., Crook, A.M., Crook, D., Fitzgerald, J.D., Innocenzi, P.J., Jay, C.N. and Solomon, M.G. 2001. Review: natural enemies and biocontrol of pests of strawberry in Northern and Central Europe. *Biocontrol Science and Technology*, 11(2): 165-216.

Cross, J.V., M.A. Easterbrook, A.M. Crook, D. Crook, J.D. Fitzgerald, P.J. Innocenzi, C.N. Jay, and M.G. Solomon, 2001. Review: Natural enemies and biocontrol of pests of strawberry in North and Central Europe. *Biocontrol Sci. Tech.*, 11: 165-216.

Dai, F.M., Ren, X.J. and Lu, J.P. 2006. First report of anthracnose fruit rot of strawberry caused by *Colletotrichum acutatum* in China. *Plant Disease*, 90(11): 1460.

Daubeny, H.A. and Pepin, H.S. 1977. Evaluation of strawberry clones for fruit rot resistance. *J.* Am. Soc. *Hortic. Sci.*, 102: 431-435.

Decou, G.C. 1994. Biological control of the two-spotted spider mite (Acarina: Tetranychidae) on commercial strawberries in Florida with *Phytoseiulus persimilis* (Acarina: Phytopseiidae). *Flo. Entomol.*, 77:33-41.

Denoyes, R.B. and Guerin, G. 1996. Comparison of six inoculation techniques with *Colletotrichum acutatum* on cold stored strawberry plants and screening for resistance to this fungus in French strawberry collections. *Eur. J. Plant Pathol.*, 102: 615-621.

Domingues, R.J., Tofoli, J.G., Oliveira, S.H.F. and Garcia Junior, O. 2001. Chemical control of strawberry black flower rot (*Colletotrichum acutatum* Simmonds) under field conditions. *Arquivos do Instituto Biologico Sao Paulo*, 68(2): 37-42.

Dorozhkin, N.A. and Grisanovich, A.K. 1972. Effect of microelements on susceptibility of strawberry to white spot and grey rot. *Khimiyav Sel'Shom Khozyaistve*, 10:50-51.

DPPQS, Directorate of Plant Protection, Quarantine and Storage, 2001. *Integrated Pest Management Package for Strawberry*, Government of India. *(IPM Package No: 64)*.

Drozdowski, J. 1987. A survey of fungi associated with strawberry black root rot in commercial fields. *Adv. Strawberry Prod.*, 6: 47-49.

Eastburm, D.M. and Gubler, W.D. 1990. The detection and survival of *Colletotrichum acutatum*, a causal agent of strawberry anthracnose in soil. *Plant Dis.*, 74: 161-163.

Easterbrook, M.A., Fitzgerald, J.D. and Solomon, M.G., 2006. Suppression of aphids on strawberry by augmentative releases of larvae of the lacewing *Chrysoperla carnea* (Stephens). *Biocontrol Science and Technology*, 16: 893-900.

EL Zayat, M.M., Okashakha, K.H.A., El Tobshy, Z.M., El Kohli, M.M.A. and El neshuary, S. M. 1993. Efficiency of *Trichoderma harziamum* as a biocontrol agent against gray mould rot (*Botrytis cinerea*) of strawberry fruits. *Ann. Agric. Sci. Cairo.*, 38: 283-290.

Ercole, N. 1976. Verticillium wilt of strawberry; a difficult disease to control. *Informattore Agrario.*, 32: 925-926.

Escudero, L.A. and Farragut, F. 2005. Life-history of predatory mites *Neoseiulus californicus* and *Phytoseiulus persimilis* (Acari: Phytoseiidae) on four spider mites species as prey, with special reference to *Tetranychus evansi* (Acari: Tetranychidae). *Biol Control*, 32:378-384.

Festic, H. and Delkic, I. 1977. The reactions of some strawberry varieties to *Mycosphaerella fragariae* (Schw.) Lind at Bosanska Krajina. Jugoslovonsko-Vocarstovo., 10: 633-637.

Fulton, R.H. 1958. Studies on strawberry leaf spot in Michigan. Michigan Expt. Sta Quart. *Bull.* 40: 581-588.

Galletta, G.J., Mass, J.L., Enns, J.M., Draper, A.D., Fiola, J.A., Scheerens, J.C., Archbold, D.D. and Ballington, J.R. 1995b. Delmarvel strawberry. *Hort. Sci.*, 30; 1099-1103.

Galletta, G.J., Mass, J.L., Enns, J.M., Draper; A.D., Dale, A. and Swartz, H. J. 1995a. Mohawk strawberry. *Hort. Sci.*, 30; 631-634.

George, S.W. and Milholland, R.D. 1986. Inoculation and evaluation of strawberry plants with *Phytophthora fragariae*. *Plant Dis.*, 70: 371-375.

George, S.W. and Milholland, R.D. 1987. Seasonal occurrence and recovery of *Phytophthora fragariae* in infected strawberry roots. *Hort. Sci.*, 22: 1254-1255.

Gourley, C.O. 1974. A comparison of benomyl, thiophanate methyl and captan for control of strawberry fruit rot. *Can. Plant Dis. Surv.*, 54: 27-30.

Harnendo, V. and Casada, M. 1976. The relationship between *Mycosphaerella fragariae* infection and the nutritional status of strawberry plants. *Anales de Edafologia-y-Agrobiologia*, 35: 1-2.

Harris, B.M. and Maclean, B. 1999. Spinosad; control of lepidopteeraous pests in vegetable brassiscas. *Proceedings of the 52nd Plant Protection Conf.*, 65-69.

Harris, D.C. 1989. Control of Verticillium wilt of strawberry in Britain by chemical soil disinfestations. *J. Hortic. Sci.*, 64: 683-686.

Haward, C.M. 1972. A strawberry fruit rot caused by *Colletotrichum fragariae*. *Phytopathology*. 62: 600-602.

Horn, N.L. 1961. Control of Botrytis rot of strawberries. *Plant Dis. Rep.*, 45:818-822.

Howard, C.M. and Alberts, E.E. 1973. Cassia obtusifolia, a possible reservoir for inoculation of *Colletotrichum fragariae*. *Phytopathology*, 63: 533-534.

Howard, C.M., Overman, A.J., Price, J.F. and Albregts, E.E. 1985. Diseases, Nematodes, Mites, and Insects Affecting Strawberries in Florida. University of Florida Institute of Food and Agricultural Sciences, Agricultural Experimental Station. Bulletin 857: 52.

Hua Xiu Feng and Chen Xian, 2004. Control effect of 50% kresoxim-methyl dry flowable formulation on strawberry powdery mildew. *Fujian-Agricultural-Science-and-Technology*, (4): 26-27.

Huffaker, C.B., M. Van de Vrie, and J.A. McMURTRY. 1969. The ecology of tetranychid mites and their natural control. *Annu. Rev. Entomol.*, 14: 125-174.

Hukkanen, A.T., Kokko, H.I., Buchala, A.J., McDougall, G.J., Stewart, D., Karenlampi, S.O. and Karjalainen, R.O. 2007. Benzothiadiazole induces the accumulation of phenolics and improves resistance to powdery mildew in strawberries. *Journal-of-Agricultural-and-Food-Chemistry*, 55(5): 1862-1870.

Hutton, D. 2006. Successful management of *Colletotrichum* crown and stolon rot in runner production in sub-tropical Australia. *ActaHort.*, 708: 293-298.

Igarashi, I., Monma. S., Fujino, M., Okimura, M., Okitsu, S., Takada, K. and Nii, T. 1994. Breeding of new everbearing strawberry variety 'Ever Berry'. *Bull. Natl. Res. Inst. Veg. Ornamental plant and Ta Series A. Vegetables and Ornamental Plants*, No. 9: 69-84.

Inada, M. and Yamaguchi, J.I. 2006. Occurrence of anthracnose fruit rot caused by *Colletotrichum acutatum* and *Colletotrichum gloeosporioides* on strawberry in forcing culture. *Kyushu-Plant-Protection-Research*, 52: 11-17.

Jansson, R.K., Peterson, R.F., Mooke, P.K., Halliday, W.R. and Dybas, R.A. 1997. Development of novel soluble granule formulation of enamectin benzoate for control of lepidopterous pest. *Florida Entomologists*, 80: 425-450.

Jhooty, J.S. and Mc Kean, W.E. 1965. Studies on powdery mildew of strawberry caused by *Sphaerotheca macularis*. *Phytopathol.*, 55: 281-285.

Joffens, W.F. 1957. Soil treatments for control of the red stele disease of strawberries. *Plant Dis. Rep.*, 41: 415-418.

Johnson, A.W., Simpson, D.W. and Berrie, A. 2006. Hot water treatment to eliminate *Colletotrichum acutatum* from strawberry runner cuttings. *Acta-Horticulturae*, 708: 255-258.

Kagabu, S. 1999. Studies on the synthesis and insecticidal activity of neonicotinoid compounds. *Pestic. Sci.* 21:231-239.

Kaplan, I. and Eubanks, M.D. 2002. Disruption of cotton aphid (Homoptera: Aphididae) natural enemy dynamics by red imported fire ants (Hymenoptera: Formicidae). *Environ. Entomol.*, 31: 1175-1183.

Kaya, H.K. and Gaugler, R. 1993. Entomopathogenic nematodes. *Ann. Rev. Entomol.*, 38: 181-206.

Keyworth, W. and Benett, M. 1951. Verticillium wilt of the strawberry. *J. Hortic. Sci.*, 26: 304-316.

Khanizadeh, S., Buszard, D., Carrise, O. and Thiobdean, P.O. 1996. Joliette strawberry. *HortiSci.*, 31: 1036-1037.

Khanizadeh, S., Deschenes, M., Levasseur, A., Carisse, O., Charles, M.T., Rekika, D., Tsao Rong, Yang, R., DeEll, J., Thibeault, P., Prive, J.P., Davidson, C. and Bors, B. 2005. 'Harmonie' strawberry. *HortScience*, 40(2): 480-481.

Kim, J.G., Cho, C.T., Bia, T.U., Han, H.S. and Mun, B.J. 1978. Varietal resistance to strawberry to two strains of *Mycosphaerella fragariae* and chemical control of the pathogen. *Korean J. Plant Prot.*, 17: 149-154.

Leclant, F., and Deguine, J.P. 1994. Aphids, pp. 285- 324 *In* G. A. Mathews and Tunstall, J.P. (eds.), Insect Pests of Cotton. CAB, Oxon, UK.

Liu, H. and Qin, L. 1989. The population fluctuations of some dominant species of ladybird beetles in Eastern Hebei Province. *Chinese Journal of Biological Control*, 5: 92.

Madden, L.V., Wilson, L.L and Ellis, M.A. 1993. Field spread of anthracnose fruit rot of strawberry in relation to ground cover and ambient weather conditions. *Plant Dis.*, 77: 861-866.

Mass, J.L. 1978. Screening for resistance to fruit rot in strawberries and red raspberries: A review. *Hort. Science,* 13: 423-426.

Mass, J.L.1984. Compendium of strawberry diseases. *Am. Phytopathol.. Soc.* St. Paul, Minn.

Matsuo, K., Suga, Y. and Jan, M. 1994. Chemical control of strawberry anthracnose caused by *Colletotrichum acutatum. Proc. Assoc. Plant Prot. Kyushu,* 40: 17-21.

McGechan, J.K. and Fahy, P.C. 1976. Angular leaf spot of strawberry, *Xanthomonas fragariae*: first record of its occurrence in Australia and attempts to eradicate the disease. *APPS Newsletter,* 5: 57-59.

Mcintyre, J.L. and Walton, G.S. 1981. Control of strawberry anthracnose caused by *Colletotrichum acutatum. Proc. Assoc. Plant Prot. Kyushu.,* 40: 17-21.

McMurtry, J.A. and Croft, B.A. 1997. Life-styles of phytoseiid mites and their roles in biological control. *Annu. Rev. Entomol.,* 42:291-321.

Meesters, P., Meurrens, F. and Brugmans, W. 1996. Strawberry cultivation. Total soil fumigation and fumigation in strips for growing *Elsanta. Fruit Belge.,* 64:65-69.

Mitra,S.K. 1993. Strawberries. *In* : Mitra, S.K. Bose, T.K. and Rathore, D.S. (eds.). Temperate Fruits. Alert and Allied Publishers 27/3 Chakraberia Lane Calcutta, pp. 549-596.

Moore, J.N. 1981. 'Arking', a late season, disease resistant strawberry variety. *Arkansas Farm Res.,* 30:6.

Mossler, M., and Nesheim, O.N. 2003. Florida Crop/Pest Management Profile: Strawberries. University of Florida, IFAS Cooperative Extension Service, EDIS 129.

Navarthan, S.J., Degefu, T. and haite, M. 1991. outbreak and new records. Ethopia: Angular leaf spot of strawberry in Ethopia. *FAO Plant Prot. Bull.,* 39: 116.

Nondillo, A., Redaelli, L.R., Marcos, B., Pinent, S.M.J., Gitz, R. 2008. Thermal Requirements and Estimate of the Annual Number of Generations of Frankliniella occidentalis (pergande) (thysanoptera: Thripidae) on Strawberry Crop. *Neotropical Entomology.,* 37(6):646-650.

Palmer *et al.,* 2006. Effect of climatic factors on powdery mildew caused by Sphaerotheca macularis f. sp. fragariae on strawberry. *European Journal of Plant Pathology,* 114(3): 283-292.

Palmer, S., Scott, E., Stangoulis, J., Able, A.J. 2006. The effect of foliar-applied Ca and Si on the severity of powdery mildew in two strawberry cultivars. *Acta Horticulturae.,* (708): 135-139.

Palumbo, J.C., A.R. Horowitz, and N. Prabhaker. 2001. Insecticidal control and resistance management for *Bemisia tabaci. Crop Prot.* 20: 739-765.

Pathak, H.C. and Payak, M.M. 1965. *Hainesia lythri* on strawberry: A new record from India. *Indian Phytopath.,* 18: 237-239.

Paulus, A.O., Voth, V., Nelson, J. and Bowen, H. 1978. Control of Botrytis fruit rot of strawberry. *Calif. Agric.,* 32:9.

Peris, O.S. 1962. Studies on strawberry mildew, caused by *Sphaerotheca macularis* (Wallr. Ex Fries) Jaczcwski II: Host parasite relationships on foliage on strawberry varieties. *Ann. Appl. Biol.*, 50:225-233.

Peters, A. and Vlug, H. 2005. Timing of nematode application to control white grubs (Scarabaeidae). *Bulletin-OILB/SROP*, 28(2): 51-55.

Prasad, K. and Stadelbacher, G.J. 1974. Control of postharvest decay of fresh raspberries by acctaldehyde vapor. *Phytopathol.*, 64: 948-951.

Price, J.F. and Kring, J.B. 1991. Response of twospotted spider mite, *Tetranychus urticae* Koch, and fruit yield to new miticides and their use patterns in strawberry. *J. Agric. Entomol.*, 8: 83-91.

Rhodes, E.M., Liburd, O.E., Kelts, C., Rondon, S.I., Francis, R.R. 2006. Comparison of single and combination treatments of Phytoseiulus persimilis, Neoseiulus californicus, and Acramite (bifenazate) for control of two-spotted spider mites in strawberries. *Experimental and Applied Acarology*, 39: 213-225.

Ritchie, P.F., Averre, C.W. and Milholland, R.D. 1993. First report of angular leaf spot, caused by *Xanthomonas fragariae*, on strawberry in north Carolina. *Plant Dis.*, 77: 1263.

Rondon, S. I., Cantliffe, D.J. and Price, J.F. 2004. The Feeding Behavior of the Bigeyed Bug, Minute Pirate Bug, and Pink Spotted Lady Beetle Relative to Main Strawberry Pests. *Environmental Entomology*, 33(4): 1014-1019.

Sato, M.E., Miyata, T., Da Silva, M., Raga, A., De Souza, Filho, M.F. 2004. Selections for fenpyroximate resistance and susceptibility, and inheritance, cross-resistance and stability of fenpyroximate resistance in *Tetranychus urticae* Koch (Acari: Tetranychidae). *Appl. Entomol. Zoo.*, 39(2):293-302.

Schoonover, J.R. and Larson, L.L. 1995. Laboratory activity of spinosad on non target beneficial arthropods. *Arthropod Mang.*, 20: 357.

Shankar, U. and Kaul, V. 2010. Entomopathogenic Nematodes: A Tool in IPM for Vegetable Crops. *In:Quality Seed Production of Vegetable Crops*, 2010, Sharma, J. P. (ed.), Kalyani Publishers, Ludhiana, India.pp. 207-225.

Sharma, R.M., Kher, R., Dogra, J., Sood, M., Shankar, U. and Verma, V.S. 2008. Sustainable Strawberry Production in North Indian Plains. *3rd Indian Horticulture Congress*, November 6-9, 2008, Bhubneswar. Abstracts pp. 168-169.

Simpson, D. and Blanke, M. 1996. Calypso and Tango perpetual flowering strawberry varieties from Horticultural Research International East Malling. *Erwerbsobstbau*, 38: 55-57.

Simpson, D., Bell, J. A., Harris, D.C., Schmidt, H. and Kellaerhols, M. 1994. Breeding to fungal diseases in strawberry. Proc. Eucarpia Fruit Breeding to fungal diseases in strawberry. *Proc. Eucarpia Fruit Breeding Section Meeting*, Wadenswil. Einsiedeln. Switzerland 30 August to 3 September, 1993. pp. 63-66.

Simpson, D., Hammond, K., Lesemann, S. and Whitehouse, A. 2006. Pathogenicity of UK isolates of *Colletotrichum acutatum* and relative resistance among a range of strawberry cultivars. *Acta-Horticulturae*, 708: 281-285.

Sindhan, G.S. and Roy, A.J. 1981. Screening of different varieties and selections of strawberry against leaf diseases and their control. *Indian Phytopathol.*, 34: 304-306.

Singh, S.J. 1974. A ripe fruit rot of strawberry caused by *Colletotrichum fragariae*. *Indian Phytopath.*, 27: 433-434.

Singh, S.J.; Sastry, K.S.M. and Sastry, K.S. 1975. Investigations on mosaic disease of cape gooseberry. *Curr. Sci.*, 44: 95-96.

Smith, B.J. and Black, L.L. 1993. In vitro activity of fungicides against *Colletotrichum fragariae*. *Acta Hortic.*, 348: 509-512.

Sparks, T.C., Thompson, G.D., Larson, L.L., Kirst, H.A., Jantz, O.K., Worden, T.V., Hertlien, M.B. and Busacca, J.D. 1995. Biological characteristic of the spinosyns: new naturally derived insect control agents. In *Proceedings, Beltwide Cotton Conf.s*, National Cotton Council, Meinphis. TN. PP. 903-907.

Stonneveld, T., Wainwright, H. and Labuschagne, L. 1996. Development of twospotted spider mite (Acari:Tetranychidae) populations on strawberry and raspberry cultivars. *Ann. Appl. Biol.*, 129:405-413.

Stubbs, L.L. 1956. Diseases of strawberries. *J. Agric., (Victoria)* 54: 232-236.

Stumpf, N. and Nauen, N. 2001. Cross-resistance, inheritance, and biochemistry of mitochondrial electron transport inhibitor-acaricide resistance in *Tetranychus urticae* (Acari: Tetranychidae). *J. Econ. Entomol.* 94: 1577-1583.

Taha, A.M. 1984. Studies on some pests infesting strawberry plants at Qalubia governorate. M. Sc. Fac. Agric. Al-Azhar Univ., 146 P.

Van de Vrie, M. and Price, J.F. 1994. Manual for Biological Control of Twospotted Spider Mites on Strawberries in Florida. *Univ. Fla. Dover Res. Rept.* DOV 1994-1. 10 pp.

Van driesche, R.G.V. and Bellows, T.S. 1996. Biology and arthropod parasitoids and predators, pp. 309-335 *In* Biological Control. Chapman and Hall, NY. *http://vvv.caes.state.ct.us/PlantPestHandbookFiles/pphS/pphstra.htm*

Walsh, D.B., Zalom, F.G., Shaw, D.V., Larson, K.D. 2002. Yield reduction caused by twospotted spider mite feeding in an advanced-cycle strawberry breeding population. *J. Am. Soc. Hort. Sci.*, 127:230-237.

Ware, G.W. 2000. The pesticide Book, 5th Edition. Thompson Publishing, Fresno, CA, 415pp.

Wilhelm, M. 1952. Verticillium wilt and black root rot of strawberry. *Calif. Agric.*, 6: 8-9.

Wilhelm, M. 1961. Diseases of strawberry. A guide for the commercial growers. Univ. Calif. Expt. Sta. Circular, 494pp.

Wilhelm, S. and Paulus, A.O. 1980. How soil fumigation benefits the California strawberry industry. *Plant Dis.,* 64: 264-269.

Williams, D. 2000. Twospotted spider mite on ornamental plants, pp. 1-2, Agricultural Notes. State of Victoria, Department of Primary Industries, Victoria, Australia.

Yamamoto, I. 1999. Nicotine to nicotinoids: 1962 to 1997, p. 3-27. In: I. Yamamoto and J.E. Casida, (eds.). Nicotinoid insecticides and the nicotinic acetylcholine receptor. Springer, Tokyo.

Zurawicz, E. and Daminikowski, J. 1995. Preliminary evaluation of productivity of several new strawberry (Fragaria x ananassa Duch.) clone and cultivars in central Poland. *Zesszyty Naukouse Instytute-Sadowhictwa-i-Kwiaciarstwa-w-Skiniewicach,* 2: 5-11.

Fig. 1 : Black mulch with drip irrigation

Fig. 2 : *Anthracnose*

Fig. 3 : *Verticillium wilt*

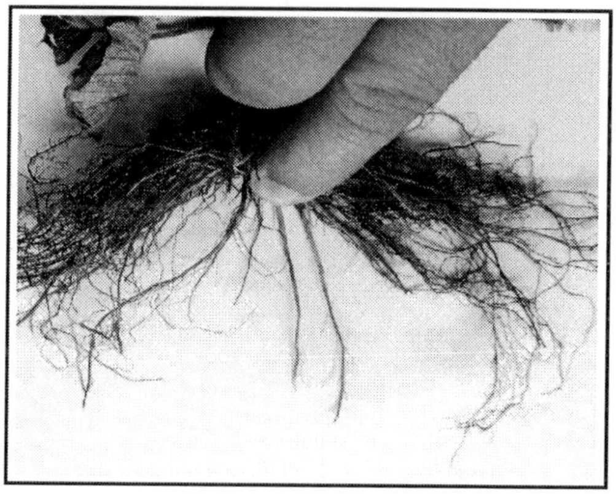

Fig. 4 : Black root rot

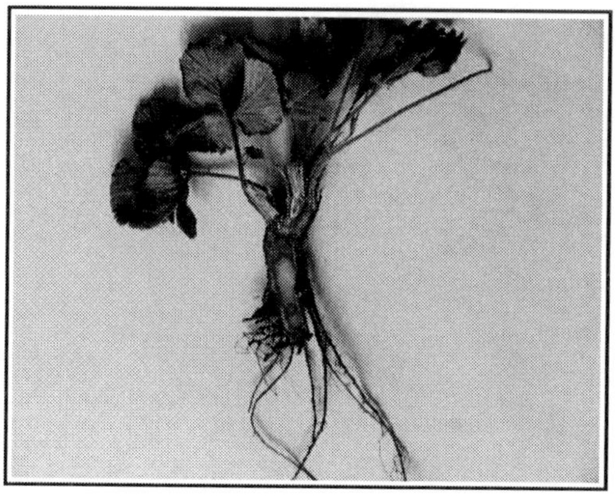

Fig. 5 : Root rot

Fig. 6 : Grey mould rot

Fig. 7 : *Mycosphaerella* leaf spot

Fig. 8 : Angular leaf spot

Fig. 9 : Leaf scorch

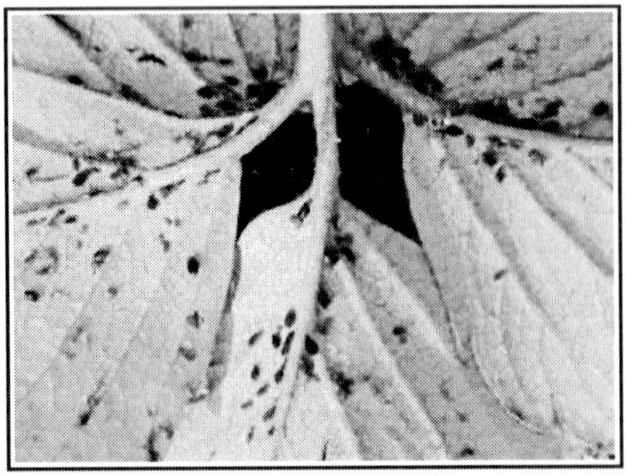

Fig. 10 : Strawberry aphids

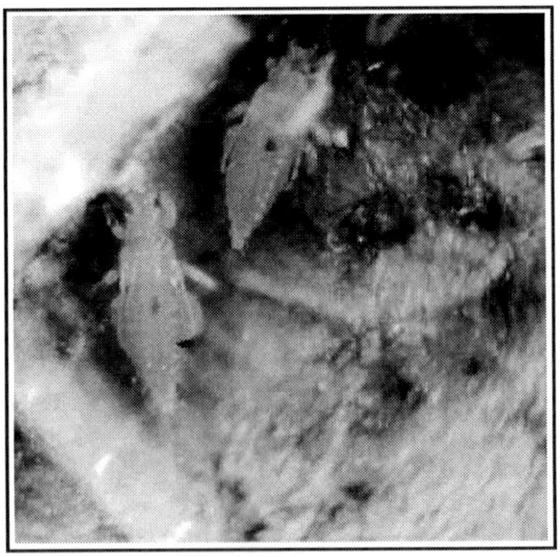

Fig. 11 :

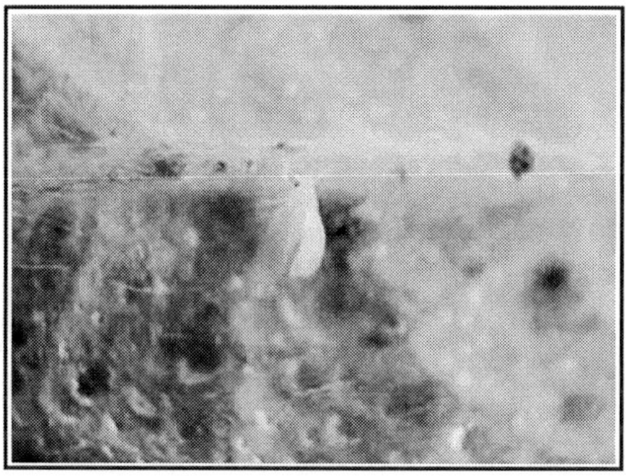

Fig. 12 : Strawberry whitefly

Fig. 13 : Tarnished bug

Fig. 14 : Strawberry cutworm larva

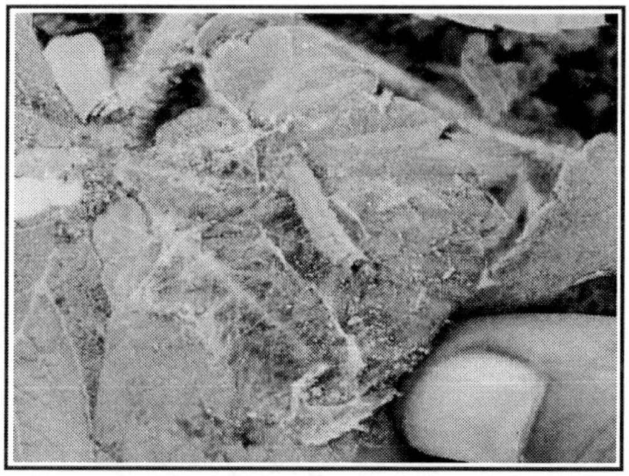

Fig. 15 : Strawberry leaf roller

Fig. 16 :

Fig. 17 : Termite damaged fruit

Fig. 18 : Snail

Fig. 19 : Birds net

Fig. 19a : Birds net

Fig. 20 : Reflecting ribbons

Ecologically Based Integrated Pest Management, 2012
© D.P. Abrol and Uma Shankar (eds.), pp. 821-839
New India Publishing Agency, New Delhi (India)
e-mail : info@nipabooks.com; website : www.nipabooks.com

41

Pest Control and Disinfection in Rural Godowns

A.A. KHAN, Z.H. KHAN AND S.A. GANIE

Introduction

Increased food production is required to ensure food security in the face of the rising world population. The strategy will include cropping of the rising world population. The strategy will include cropping of the new geographical areas and increase in the per hectare crop yields, through better crop management; losses of the food in storage will have to be minimized.

The food grains and their products are susceptible to deterioration by a variety of biotic and abiotic factors. These include the high temperature, moisture, micro-organisms, mites, insects and rodents. Together they account for loss of about 25% of food grains worldwide. The losses are highest in the tropical and sub-tropical areas where conditions are relatively conductive for rapid growth and multiplication of the organisms that cause damage to the food materials, especially insects.Reliable pesticide and storage techniques are required for limiting the damage to food grains in storage in most of the developing countries of the areas.

The Indian sub-continent is bestowed with a variety of agro-climates, produces a range of food grains: cereals, millets, oil seeds and pulses. In India, a large majority of people are vegetarians; the Indian people's diet comprises of cereals, millets and pulses, with minimal processing while rice is used in its polished form, other cereals and millets are used in their powdered form (atta). The pulses are used as whole, split or decoated. The whole grains and their above mentioned directly usable forms are the principle food items that are stored in massive quantities in India in the stores, markets and households. The whole grains are stored by the farmers, and in the wholesale market godowns and the stores of the government; small quantities are also kept by the households. The food materials continue to suffer considerable damage under storage in India, despite the introduction of pesticides and sealable storage bins.

The traditional and modern structures deployed for the food grain storage in India are of both aerated and sealed kinds. The aerated devices include gunny bags and the sealed devices are the earthen mataka's and drums, mataka, pusa bins and Pantnagar kuthalas made of tin etc. In the sealed storage structures, the food materials are now most commonly protected with aluminium phosphate (AIP), ethylene-di-bromide (EDB) and methyl bromide (MB) (Table-1). Depending upon the quantity of materials to be protected, fixed amounts of these chemicals are introduced into the grains and the storage structure is sealed for protection. In a sealeable storage structure of one ton capacity, a 3g tablet of AP or 2 ml ampule of EDB is usually employed. The MB which has been used extensively in other countries has not found much use in India. To protect the grains under the aerated conditions such as in the gunny bags, about 0.1% EC of malathion is sprayed over the gunny bags. Malathion is also being sprayed in the carriers and godowns of the gunny bags to cause disinfestations and prevent fresh infestation. Gunny bags already impregnated with malathion have also been in use as alternative food grains storage containers. Like in other parts of the world, resistance in several insect-pests towards the commonly used chemicals has been noted in India.

Storage Losses

In the post harvest stages, maximum benefits will accrue from the improvement in storage practices. Chickpea seeds in developing countries suffer heavy qualitative and quantitative losses from the attack of pulse beetle, *Callosobruchus chinensis* L. (Alam, 1971). Nearly 8.5% of total annual production is lost during post harvest handling and storage (Agrawal *et al.*, 1988). Various estimates of losses of food grains during post-harvest operations have been made. The Govt. of India had constituted a committee for the assessment of storage losses, according to the interim report of the committee, the post harvest losses have been estimated to the extent of 9.33%(IGSI, 1990) and crop wise given in Table (Lal,1988). The break-up is as under:

Threshing yard losses	1.68%
Transport losses	0.15%
Processing	0.92%
Storage	
i. Rodents	2.50%
ii. Birds	0.85%
iii. Insects	2.55%
iv. Moisture	0.68%
Total	**9.33%**

Table 1 : Estimates of important foodgrain losses at different stages (%)

Stages at which the losses occur	Wheat	Rice	Jowar	Bajra	Maize	Gram	Millets	Pulses	Total
Threshing yard	1.0	2.5	2.0	0.5	0.5	0.6	1.0	0.5	1.68
Transport	0.5	0.5	0.5	0.5	0.5	0.5	0.5	0.5	0.15
Processing	-	2.0	-	-	-	-	-	-	0.92
Storage									
I. Rodents	2.5	2.5	2.5	2.5	2.5	2.5	2.5	2.5	2.50
ii. Birds	0.5	1.0	1.0	1.0	0.5	0.5	2.0	0.5	0.85
iii. Insects	3.0	2.0	2.0	1.0	3.0	5.0	0.5	5.0	2.55
iv. Moisture	0.5	0.5	2.0	0.5	0.5	0.5	0.5	0.5	0.68
Total	**8.00**	**11.0**	**10.0**	**6.0**	**7.5**	**9.5**	**7.0**	**9.5**	**9.33**

[*Source :* Lal, 1988]

Requirement of Storage

Food grains are required to be preserved in storage for following purposes:

1. For human food
2. For animal feed
3. For seed
4. For sale or barter

Factor Responsible for Food Grains Deterioration

During storage the foodgrains deteriorate due to following factors:

1. Physical - i) Temperature
   ii) Relative humidity/moisture
2. Biological - i) Produce (properties)
   ii) Pests
3. Chemical - i) Breakdown of produce
   ii) Pesticides
4. Mechanical - i) Mode of storage & handling
   ii) Mode of preservation

Foodgrains consist of mainly carbohydrates, proteins, vitamins, minerals, fats and fibrous cellulose. In addition they contain some water in the form of moisture content. The degree of deterioration can be checked by controlling physiological, chemical and mechanical factors to a large extent. However, maximum efforts are required to control the biological factors which are as follows:

1. Microflora
2. Insects and Mites
3. Birds
4. Rodents

Microflora

Fungi

i) Field fungi
   Altenaria spp.
   Cladosporium spp.
   Helminthosporium spp.
   Fusarium spp.

ii) Storage fungi
   Aspergillus spp.
   Penicillium spp.
   Rhizopus spp

Bacteria

i) Pseudomonas spp.

ii) Salmonella spp.

Yeasts

i) Monilia spp.
ii) Candida spp.
iii) Torula spp.
iv) Fusidium spp.

Actinomycetes

i) Actinomyces spp.
ii) Streptomyces spp

Birds

Pigeon	*Columba livia* (Gmelin)
House Sparrow	*Passer domesticus* (Linnaeus)
House crow	*Corvus splendens* (Vieillot)
Parakeet (Parrot)	*Psitticula krameri* (Linnaeus)
Common mynah	*Acridotheres tristis* (Linnaeus)

Rodents

House rat	*Rattus rattus* (Linn.)
House mouse	Mus musculus (Linn.)
Brown rat	*Rattus norvegicus* (Birken)
Indian Mole rat	*Bandicota bengalensis* (grey)
Bandicoot rat	Bandicota indioca (Bechstein)

Insects and Mites

Insect pests of stored grain are the descendent of earlier insects which came on the scene of existence 250-500 years ago. They belong to phylum Arthropoda, class-Insecta, having 3 pairs of legs and outer skin called cuticle, body divided into three parts head, thorax and abdomen. The head bears one pair of antennae, the thorax carries three pairs of legs one or two pairs of wings. Insects are responsible for enormous spoilage in storage. They feed on grain, bore the kernel, and destroy the germ portion, cause heating and deterioration in stored produce. The insects are not generated denovo in the grain. All insects lay eggs which mostly escape visual detection. The insects pass through egg, larval, pupal stages before becoming adults. As per the Panse Committee Report, the storage losses due to insect pests of stored grain are 2.55%. In addition they are also responsible for the qualitative losses in foodgrains. Insect infestation is normally present along with the grain. Following are the main sources of insect infestation.

Field Infestation

Family Bruchidae contains several important field and stored crop pests (FAO, 2009). The seeds of nearly all legume species are vulnerable to attack by seed beetles, both in the field and in storage. Common insect pests of stored pulses are *Acanthoscelids obtectus* (Say), *Callosobruchus analis* (CA) (Fabricus), *C. chinensis* (Linnaeus), *C. maculatus* (Fabricus), *Zabrotes subfasciatus* (Boh) and *Sitophilus oryzae* (Atwal and Dhaliwal, 2008). Pulse beetle, *C. chinensis* causes substantial economic loss during the storage/preservation of pulses (Shu *et al.*, 1996). It is a notorious pest of green gram (*V. radiata*), Bengal gram (*C. arietinum*), red gram (*V. unguiculata*) and math (*Phaseolus aconitifolius*). Severe infestation reduces quality and germinability to under 20% after 4 months only (Seck *et al.*, 1996) and as much as 90% loss within 6 months of storage (Mendki *et al.*, 2001). Insects like bruchids or *Sitotroga* spp. fly from stores to fields and lay eggs upon maturing grain/pods. These eggs hatch out to larvae in favourable conditions when grains reach the stores. This is termed as "hidden infestation".

Cracks and cervices

When storage godowns/containers are not white wash/plaster, clean properly/thoroughly done. Some grains continue to lie in cracks and cervices. Insect present in the stores keep on feeding and multiplying during this period also. If fresh stocks are kept in such stores the same get infested. Insects from old stocks can crawl/fly to fesh stocks and infest them. This process is called "cross infestation".

Table 2 : Storage insect pests of economic importance

S. No.	Common name	Scientific name	Order	Family	Food infested	Damage
1.	Angoumois grain moth	*Sitotroga cerealella* (Olivier)	Lepidoptera	Gelechiidae	Wheat, maize, sorghum, barley, oats, etc.	Damage is caused by larvae which bores into the grains and feeds on its contents
2.	Indian meal moth	*Plodia interpunctell a* (Hubner)	Lepidoptera	Pyralidae	Grains, soybean, dried fruits, nuts, seeds, etc.	Damage is caused by larva which feed covered under a silken web
3.	Rice moth	*Corcyra cephalonica* (Stainton)	Lepidoptera	Pyralidae	Rice, gram, sorghum, maize, groundnut and cotton	Damage is done by larvae alone by feeding on the grains

Contd...

4.	Khapra beetle	*Trogoderma granarium* (Everts)	Coleoptera	Dermesti dae	Wheat, rice, barley, sorghum, pulses, maize	Damage is caused by grubs which eat the grain near the embryo
5.	Rice weevil	*Sitophilus oryzae* (Linnaeus)	Coleoptera	Curculio nidae	Rice, what, maize and other grains	The damage is caused by both the grubs and adults by breaking and hollowing the grains
6.	Rust-red flour beetle	*Tribolium castaneum* (Herbst)	Coleoptera	Tenebrio nidae	Wheat flour also found in what, cereal products, pulses, dried fruits, etc.	It feeds on powdered product produced by other pests after feeding the grain
7.	Lesser grain borer	*Rhyzopertha dominica (Fabricius)*	Coleoptera	Bostrichi dae	Wheat, rice, sorghum, maize, barley, biscuits, etc.	The grubs eat upon the kernel leaving the husk behind
8.	Almond moth	*Ephestia cautella (Walker)*	Lepidoptera	Pyralidae	Dried almonds, walnut, dates, dried apples, dates, berries, figs, etc.	The damage is done by the caterpillars which tunnel into food material
9.	Saw-toothed grain beetle	*Oryzaephilus surinamensis (Linnaeus)*	Coleoptera	Cucujidae	Maize, sorghum, gram, etc.	Damage is caused by both adult and grubs which attack seeds damaged by other insects
10.	Gram dhora	*Callosobruchus chinensis (Linn)*	Coleoptera	Bruchidae	Gram, mung, peas, cowpeas, sorghum, maize, etc.	Damage is caused by larvae which feed inside the grain
11.	Mung dhora	*Callosobruchus analis (Fabricius)*	Coleoptera	Bruchidae	Mung, mash, peas, cowpea and other pulses	Larvae cause the damage by feeding and breeding inside the grain, consuming the entire contents

Containers / bags

It has been a general practice, especially with the farmers, that they keep without cleaning the empty foodgrains bags for use in the next season. Eggs and larvae of insects remain hidden in the seams and mesh of the bags. During off season they feed on the grain fragments etc. while sticking to the bags. Wherever the grains are filled up in such bags, infestation reappears.

Trunks, trollies and bullock carts

Foodgrains are harvested, threshed in the fields and them transported to the godowns by Trunks, trollies and bullock carts etc. these are very good sources of cross infestation. Insects which may be present in the joints or corners of the carriers migrate to the foodgrain lots and cause infestation.

Table 3 : Important mite pests of stored products

S. No.	Common name	Scientific name	Order	Family	Food infested	Damage
1.	Grain mite/flour mite	*Acarus siro* (Linn.)	Acarina	Tyroglyphidae	Grain and grain products	Grains with seed coats damaged over the embryo are most suitable for feeding
2.	Black legged mite	*Aleuroglyphus ovatus* (Troup)	Acarina	Tyroglyphidae	Grains	Damage is caused by eating the embryo of the grain
3.	Common hairy mite	*Glycyphagus destructor* (Ouds)	Acarina	Glycyphagidae	Grain and groat	-do-
4.	Brown flour mite	*Gohieria fusca*	-do-	-do-	Embryo of the grain is damaged	Wheat, de-husked rice and other grains

Control of Insect-Pests

Various strategies have been used to control pre- and post-harvest insect pests since times immemorial. As per the latest reports, chemicals worth US $ 26 billion are applied to crops all over the world, both on and off the field (Thakore, 2006). France, USA, Germany, Great Britain and Switzerland together produce almost 3/4th of this amount. Though, India is one of the largest producer of synthetic pesticides, usage of them is approximately 0.5 Kg/ha compared to about 3 Kg/ha in USA (Wahab, 2009). Pesticides worth Rs. 27 billion approx. were consumed in India in the year 2005 with cotton alone accounting for the major share (Wahab, 2009). Amongst the present methods of insect control, following are the important methods which can be help in safe storage of foodgrains particularly at farmer's level :

1. Preventive measures; 2. Curative measures

1. Preventive measures

"Prevention is better than cure". Hence the following preventive measures are recommended:

a. Hygiene or Sanitation

1. Threshing floor/yard should be clean, free from insect infestation and away from the Vicinity of villages/granaries.
2. Clean the harvesting and the threshing machines before their use.
3. Trucks, trollies or bullock carts which are used for transportation of food grains should be made free from insect infestation.
4. Clean the storage structure/godowns before storage of newly harvested crop.
5. All dirt, rubbish, sweepings and webbings etc. should be removed from the stores and dumped/destroyed.
6. All the cracks, cervices, holes existing in the floors, walls, ceiling should be plastered with mud or cement permanently.
7. All the rat burrows should be closed with the mixture of broken glass pieces and mud and then plastered with mud/cement.
8. White washes the stores before storage of foodgrains.
9. Foodgrains should be kept in stores which are rat and moisture proof.
10. Proper stacking of bags also helps in grain protection.

b. Disinfestation of Stores / Receptacles

Before the use of receptacles/store they should be disinfested with approved residual insecticides preferably by spraying malathion 50%EC, with dilution of 1:100 and applied at the rate of 3 lit/100 m^2.

2. Legal method

Entry of an insect which is not found in the particular area can be prevented by the Destructive Insect Pests Act 1914.

B. Curative measures

The infestation of stored grain insect pests can be controlled by the following methods.

a) Non chemical control measures
b) Chemical control measures

a) Non Chemical Control Measures

The measures where chemicals are not used for the control on insects of stored grains are:

i) Ecological Control Measures

The safety of stored grains from insect pests largely depends on proper management of three factors viz.

a) Temperature
b) Moisture content of grain
c) Availability of oxygen

All these factors are required for rapid development and multiplication of insects. Hence, they have to be properly manipulated through design and construction of storage structure/godowns and storage practices so as to create ecological conditions unfavourable for attack by insects.

Temperature : Temperature ranging from 20°C to 40°C accelerates the development of insects but above 42°C and below 15°C retards reproduction and development, while 10°C may kill the insects. Heating of grains to 50°C will be lethal to insects but it is not advisable, because the grains are affected and lose their viability (Fig. 1).

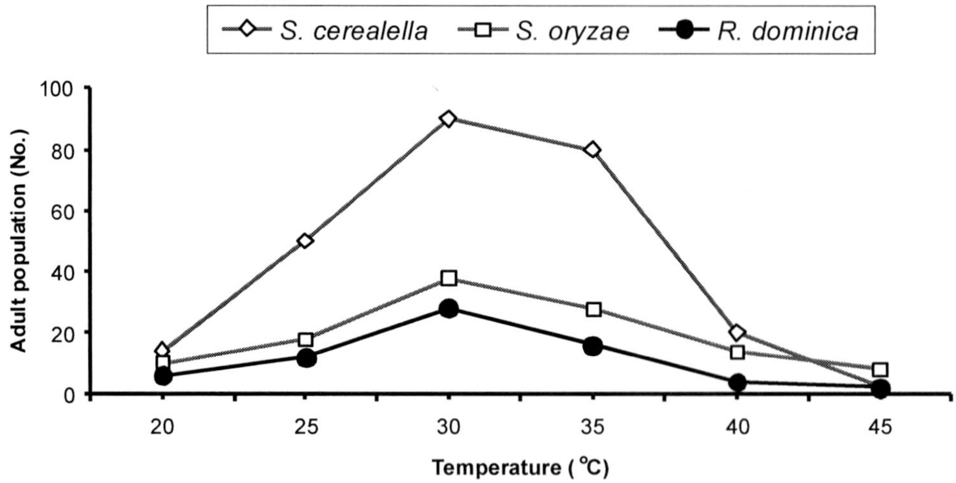

Fig. 1 : Population of grain borers at different levels of temperature (Moisture: 18%)

b) Moisture Content

Moisture is the critical factor in safe storage of foodgrains. Grains stored at around 10% moisture content escapes from the attack of insects (except khapra beetle) (Fig. 2).

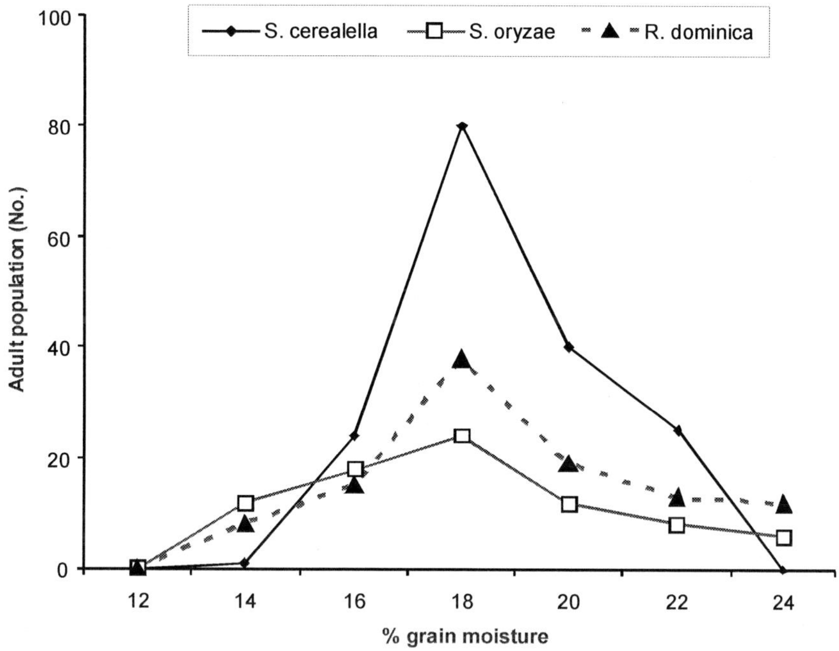

Fig. 2 : Population of grain borers at different levels of grain moisture (Temperature: 28 ± 1°C).

c) Oxygen

In storage oxygen is consumed by grains and insects during respiration and CO_2 is produced. Insects respire at the rate of 20,000 to 1,30,000 times than that of the same weight of the grains. Thus, O_2 level will reduce below 1% and CO_2 level will automatically increase which will be lethal to all the stages of insects.

ii) Mechanical Control Measures

Amongst other methods mechanical methods are quite practicable.

Screening of grains : Broken and cracked grains promote the attack of stored grains pests. Hence, screening/sieving out of such grains reduce the insect infestation and gives a better appearance to the lot. Of course the grains do not become insect free. Following point help in making the screening more effective.

1. Screening should be done regularly and away from the stores to avoid any re-infestation.
2. Immediate destruction of screenings
3. Bags used for carrying the screenings should not be used again unless disinfested.

iii) Physical Control Measures

Tribolium castaneum (Herbst), a species with a well developed chemosensory system (Barrer, 1983), is able to differentiate changes in the physical environment such as temperature (Saxena *et al.*, 1992; Dowdy, 1999), humidity (Evans, 1983), carbon dioxide tension (Soderstrom *et al.*, 1992), and even different hues immediately around it in the environment.

1. *Heat treatment* : Most of the stored grain insects die at 55-60°C within a period of 10-20 minutes. Therefore flour mills and other type of food processing plants have arrangement of super heating by infra red heaters for the control of insects.

2. *Controlled atmosphere* : It involves alternation of the concentrations of the normal atmosphere gases processing gases present in a storage space as to obtain an artificial atmosphere that prevents multiplication of insects, mould growth and quality deterioration of foodgrains. The storage atmosphere can be modified by:

   - Low oxygen atmosphere (2-4%) generated by purging the nitrogen.
   - High CO_2 atmosphere developed by the addition of CO_2. It has been found that at about 9.0-9.5% CO_2 in air is lethal to all insects.

3. *Mixing of inert dust* : This method is used by many farmers in villages. The inert dusts are sand, clay, ash etc. Mixing of such dusts with the grain makes the entry of insects in grains a difficult task and cause physical injuries; block mouth parts and lodges in joints of the insects. Some of the dust made from silica aerosols, activated carbons operate by abrasive action on waxy layer over the insect cuticle, resulting in desiccation and death of insect. Foodgrains stored with inert dust are to be cleaned and washed before consumption.

4. *Management based on hue response* : Sheribha *et al.* (2010) assessed the possibilities for the management of *Tribolium castaneum* (Herbst) on stored products using coloured lighting systems. Adult beetles were exposed in a choice chamber to red, blue, green, and clear tungsten light. *T. castaneum* adults preferred tungsten, blue, and green lights, and avoided red light. For each colour, beetles were separately reared for 3 generations. It is evident that red light is not preferred by *T. castaneum* adults; they always moved away from environments exposed to red light. Thus, if storage areas are lit red, *T. castaneum* beetles could be managed without the use of chemical pesticides. Previous studies reported the presence of colour preference or avoidance in *T. castaneum* grubs and adults (Ramos *et al.*, 1983; Viswanathan *et al.*, 1996; Khan *et al.*, 1998).

5. *Use of edible oils* : Coconut, groundnut and mustard oil at the rate of 0.25-0.5% are effective for the protection of pulses against pulse beetles upto 6 months (Table 4).

iv) Cultural Control Measures

Splitting of pulses: Pulse beetles prefer to attack whole pulses and not split ones. Split pulses escape from the attack and more such most suitable for storage than whole pulses.

v) Engineering Control Measures

a) *Hermetic or airtight storage* : Foodgrains kept in airtight sealed structures remain insect free, reason being that during storage the grain respires and liberates CO_2. Excessive CO_2 accumulation and depletion of O_2 do not allow the insects to survive.

b) *Drying of grain* : In grain having moisture content below 10%, most of the insect sp. does not survive/multiply. In villages the grains are dried by spreading a thin layer of grains in the sun or by dryers. However, it should be kept in mind that too rapid drying or over drying may damage seeds, develop wrinkles, cause cracking of seed coat and destroy vitamins and seed viability.

Table 4 : Efficacy of different vegetable oils against pulse beetle in cowpea

Treatments	No. of eggs laid	No. of adults emerged	% grain damage
Groundnut oil	18.2	2.6	5.2
Sesame oil	21.7	1.9	4.0
Soybean oil	13.8	1.1	2.7
Mustard oil	15.2	1.1	2.2
Coconut oil	12.9	1.2	2.3
Neem oil	8.9	0.0	0.0
Untreated	41.0	33.7	66.5

(*Source :* Bhatnagar *et al.*, 2001)

vi) Use Botanicals

Botanicals are chemicals which repel approaching insects, deter feeding and oviposition on the plants, disrupt behaviour and physiology of insects in various ways (Table 5). Various products of plants have been tried recently with good degree of success as protectants against a number of stored grain insect pests (Pandey *et al.*, 1986; Yadav and Bhatnagar, 1987; Dixit and Saxena, 1990; Varma and Dubey, 1999; shukla *et al.*, 2007, Srinivasan, 2008).

Table 5 : Effect of plant materials and their doses on mortality and adult emergence of *Sitophilus oryzae*

Plant matter (powder)	Per cent adult mortality (days after treatment)			No. of adults emerged (days after treatment)
	2	7	14	40
Curry leaf	26.7	56.7	62.2	8.33
Turmeric	46.7	76.7	86.7	6.33
Mango ginger	37.8	88.9	97.6	4.77
Madar	34.4	66.9	82.2	10.55
Ipomea	23.3	58.9	81.1	8.33
Garlic	37.8	70.0	82.2	8.66
Chilli	43.3	77.0	87.7	7.66
Mustard	27.8	54.0	81.3	7.66
Neem	35.6	63.3	78.9	13.1
Untreated (control)	20.0	23.3	36.6	18.00

(*Source :* Patel *et al.*, 2004)

3. *Chemical Control Measures*

Amongst the present methods of insect control, chemical control is the most popular and perhaps most effective one. The term insecticide may be applied to all those chemicals which are used for both types of treatments.

a) Prophylactic treatment; b) Curative treatment

a) *Prophylactic treatment* : To prevent the insect infestation and cross infestation by spraying the insecticides. These insecticides should not, however, be not sprayed directly on food grain. Some of the insecticides and concentrations recommended for prophylactic treatments of stored grain are given below in table:

Name of insecticides	Concentration of spray solution (%)	Preparation and Dosage	Frequency of treatment
Malathion 50% EC	0.50	1:100 @ 3lit./100M2	15 days
Pirimiphos methyl 50% EC	0.50	1:100 @ 3lit./100M2	15 days
Pyrethrum with 2.0 % Pyrethrin EC	0.02	1:100 @ 3lit./100M2	15 days

b) *Curative treatment*: In spite of regular prophylactic treatment, infection develops due to hidden infestation or cross infestation etc. It can be controlled by the following insecticides and fumigants given below in Table.

Ethylene Dibromide (EDB)

EDB is one of the most toxic chemicals to insects damaging stored foodgrains. The use of EDB, filled in ampoules has been found suitable for fumigation stored in household structures, holding small quantities, which can be made reasonably airtight by sealing it with mud plaster, polythene sheet, etc. leaving the opening at the top for insertion of ampoules.

In small quantities say upto 1 quintal, the ampoule is broken gently just below the top surface of the grain with a piece of stone. The opening is immediately sealed.

In large quantities say upto 10 quintal of food grains, the ampoules are to be distributed evenly in the grain. The ampoules are tied to a stick at (60) cms apart and are broken just before insertion of ampoules into grain mass along with the stick and the container is sealed. Such stocks are left undisturbed for 7 days. Before consumption, the grains are to be aerated.

Dosage : 3ml/quintal for wheat and pulses but 5 ml/quintal for rice and paddy. Ampoules are available in size of 3ml, 6ml and 10 ml.

Ethylene Dibromide Plus (EDB+)

It is a mixture of EDB and carbon tetreachloride in the ratio of 1:8 w/w. It has a greater penetration power and hence can be used for bagged storage under gas proof cover also. It is equally effective in rural storage structures. Ampoules are also available in size of 11ml and 22 ml.

Dosage : 11ml/quintal for wheat and pulses but 15 ml/quintal for rice and paddy.

Ethylene Dichloride Carbon Tetrachloride (EDCT)

It is a mixture of two chemicals (Ethylene dichloride and Carbon tetrachloride) in the ratio of 3:1 v/v. It has been used for large scale food grains fumigation. It is available in the market in tin containers of 500 ml, 1 litre, 5 litres etc.

Dosage : for small scale storage 55ml/quintal. For large scale storage 30-40 litre/100 cu.m.

Aluminium Phosphide

This is a solid fumigant available in the market in the form of tablets of 3 gm each packed in sealed tubes. Of course, it is one of the very potential fumigants and can be used practically on all foodgrains, milled products etc. The active ingredient is phosphine gas which is an acute toxic poison. When the tablet comes in the contact with the moisture of the atmosphere, it liberates ammonia, carbondioxide and phosphine.

Dosage : It is applied at the rate 2 tablets (3gm each)/tonne.

Chemical Control

Name of insecticides	Dose of pesticides	Method of application	Against	Remark
Malathion 50% EC	In ratio of 1:100 with water @ 3lit./100M^2 for flying insects@ 1lit./ cu. meter	Spray on Grain bag, walls floors etc. should not be sprayed on food grains directly	As Prophylactic treatment Against insect pests	Fortnightly interval
EDB	3 ml/ quintal or 22 gms/ cu.mt	the ampoules are inserted n storage after breaking & making the structure airtight	As fumigant against insect pests except for oilseed/ flour)	7days Exposure period
EDCT mixture	30.40 kg per 100 cu.mt. in large scale storage. For small scale storage is 55 ml. per quintal	Fumigation in airtight condition on bags/in bulk	For insect control	
Aluminum phosphide	(a) 2-3 tabs. of 3 gms per metric tonne (b)2 tabs. of 0.5 gms per rat live burrow	Fumigation in airtight condition on bags/in bulk 2 tabs. Of 0.5 gms each are inserted in living burrow & the burrows are sealed with mud	For insect control For field rat control	7days Exposure period ------

Rat control

Name of Insecticide	Dose of pesticides	Method of application	Against	Remark
Zinc phosphide	Preparation of bait Pure Zinc phosphide: 2 parts Foodgrain as base : 96parts Any edible oil : 2 parts	Mix them thoroughly and placed on rats visiting place	For house/ field rat control	Applicatio n after pre-baiting
Anticoagulants (Rodafarin, warfarin, ratafin etc,)	I part rodenticide + 19 parts bait material for 500 gms bait Preparation of bait (i) Flour : 450 gms (ii) Any edible oil : 10gms (iii) Sugar/ Jaggary : 15 gms (iv) anticoagulant : 25 gms	Mix them thoroughly and placed on rats visiting place	For house rat control	Continuous baiting for 10-15 days

General Precautions

Pesticides by their nature are harmful to all living organisms. Hence they must be used with utmost care. Important precautions are listed below;

1. Always read the label carefully and follow the manufacturer's instructions.
2. Keep pesticides in labeled containers only.
3. Store pesticides under lock and key and beyond the reach of children, irresponsible persons and pets.
4. Never store insecticides near foodstuffs. Store them in cool places.
5. Do not use empty containers for any other purpose except for insecticides. Waste containers should be destroyed and dumped.
6. Wash hands with soap and water after using pesticides.
7. Do not use mouth for cleaning nozzle etc. of sprayers.
8. Do not use pesticides on farm animals with out consulting veterinary doctor.
9. Aviod swallowing, inhalation or contact with skin as for as possible.
10. Keep a first aid box ready along with universal antidote which can be prepared as follows:
    - i. Activated charcoal - 2 parts
    - ii. Tannic acid - 1 part
    - iii. Magnesium oxide - 1 part

    15 gms in half tumbler of water is given
11. In case of signs of illness, e.g. giddiness, nausea, headache, blurred vision, vomiting, depression, difficulty in respiration etc. a physician should be called or the patent be rushed to a nearby hospital. The doctor should also be told about pesticide the patient had handled.

References

Agrawal, A., Lal, S., and Gupta, K.C. 1988. Natural products as protectants of pulses against pulse beetles. *Bulletin of Grain Technology*, 26: 154 -164.

Alam, M.Z. 1971. Pests of stored grain and other stored products and their control. *Publ. information Service*, 3, R.K.Mission Road, Dhaka, Bangladesh. 361p

Atwal, A.S. and Dhaliwal, G.S. 2008. Agricultural pests of South Asia and their management. Kalyani Publishers, New Delhi.

Barrer, T.M. 1983. A field demonstration of odour-based, host-food finding behaviour in several species of stored grain insects. *J. Stored Prod. Res.*, 19: 105-111.

Bhatnagar, A., Bhadauria, N.S. and Jakhmola, S.S. 2001. Efficacy of vegetable oils against pulse beetle, *Callosobruchus maculates* in cowpea. *Indian Journal of Entomology*, 63(3): 237-239.

Dixit, O.P. and Saxena, R.C. 1990. Insecticidal action of *Premina integrifolia* against *Callosobruchus chinensis*. *Pesticides*, 24: 29-31.

Dowdy, A.K. 1999. Mortality of red flour beetle, *Tribolium castaneum* (Herbst) (Coleoptera: Tenebrionidae) exposed to high temperature and diatomaceous earth combinations. *J. Stored Prod. Res.*, 35: 175-182.

Evans, D.E. 1983. The influence of relative humidity and thermal acclimation on the survival of adult grain beetles in cooled grain. *J. Stored Prod. Res.*, 19: 173-180.

FAO, 2009. Major insect pests of stored foods Webpage *http:// www.fao.org/inpho/content/ compend/text/ch02-01.htm.*

IGSI, 1990. Save grain. Govt. of India, Ministry of Food, Hapur. 1-61pp.

Khan, A.A., Rath, S. and Shankar, U. 1998. Effect of coloured light on larval growth of rust red flour beetle, *Tribolium castaneum* (Herbst). *Shashpa*, 5: 181-184.

Lal, S. 1988. Saving Gain aftear Harvest. In: G. Kasturi (ed.) The Hindu Survey of Indian Agriculture-1988. National Press, Madras, India, pp. 246-248.

Mendki, P.S., Maheshwari, V.L. and Kothari, R.M. 2001. Flyash as a post-harvest preservative for five commonly utilized pulses. *Crop Protect.*, 20: 241-245.

Pandey, N.D., Mathur, K.K., Pandey, S. and Tripathi, R.A. 1986. Effect of some plant extracts against pulse beetle *Callosobruchus chinensis*. *Indian Journal of Entomology*, 38: 110-113.

Patel, Y., Jakhmola, S. S. and Bhadauria, N.S. 2004. Effect of plant materials on rice weevil *Sitophilus oryzae* in wheat. *Indian Journal of Entomology*, 66(2): 99-101.

Ramos, E., De Conconi, J., Conconi, J., Elorduy, C., Oxley, T. and Barry, S. 1983. Laser light as a new potential method for pest control in preserved foods. *Biodeterioration. Proceedings of the 5th International Biodeterioration Symposium*, Aberdeen, UK. September 1981. Oxley, T.A. and Barry, S., eds. pp. 592-608.

Saxena, B.P., Sharma, P.R., Thappa, P.K. and Tikku, K. 1992. Temperature induced sterilization for control of three stored grain beetles. *J. Stored Prod. Res.*, 28: 67-70.

Seck, D., Lognay, G., Haubruge, E., Marlier, M. and Gaspar, C. 1996. Alternative protection of cowpea seeds against *Callosobruchus maculatus* (F.) (Coleoptera: Bruchidae) using hermetic storage alone or in combination with *Boscia senegalensis* (Pers.) Lam ex Poir. *J. Stored Prods Res.*, 32: 39-44.

Sheribha, P.R.B., Jinham, A.P., Das, S.S.M. and Jasmine, K.R. 2010. Management of *Tribolium castaneum* (Herbst) based on hue response. *Turk J Zool*, 34: 367-375.

Shu, S., Koepnick, W.L., Mbata, G.N., Cork, A. and Ramaswamy, S. 1996. Sex pheromone production in *Callosobruchus maculatus* (Coleoptera: Bruchidae): Electro-antenographic and behavioural responses. *J. Stored Prods. Res.*, 32: 21-30.

Shukla, R., Srivastava, B., Kumar, R. and Dubey, N.K. 2007. Potential of some botanical powders in reducing infestation of chickpea by *Callosobruchus chinensis* L. (Coleoptera : Bruchidae). *Journal of Agricultural Technology*, 3(1): 11-19.

Soderstrom, E.L., Brondl, D.G. and Mackey, B. 1992. High temperature combined with carbon oxide enriched or reduced oxygen atmospheres for control of *T. castaneum* (Herbst) (Coleoptera: Tenebrionidae). *J. Stored Prod. Res.*, 28: 235-238.

Srinivasan, G. 2008. Efficacy of certain plant oils as seed protectant against pulse beetle, *Callosobruchus chinensis* L. on pigeonpea. *Pesticide Research Journal*, 20(1): 13-15.

Thakore, Y. 2006. The Biopesticide market for global agriculture use- Industrial report., The New Biopesticide Market, BCC Research, Code: CHM029B.

Verma, J. and Dubey, N.K. 1999. Prospectives of botanical and microbial products as pesticides of tomorrow. *Current Science*, 76(2): 172-179.

Viswanathan, G., Rao, S.J. and Chakravarthy, C. 1996. Response of *Tribolium castaneum* to different colours of light during night and change in their intensities. *Insect Environment*, 2: 94.

Wahab, S. 2009. Biotechnological approaches in the management of plant pests, diseases and weeds for sustainable agriculture. In: *Souvenir of the 5th International Conference on Biopesticides: Stakeholders' Perspectives*. April 2009. pp 25-47.

Yadav, R.S. and Bhatnagar, K.N. 1987. A preliminary study on the protection of stored cowpea grains against pulse beetle by indigenous plant products. *Pesticides*, 21: 25-29.

❑❑❑

Section

VI

FUTURE STRATEGIES AND POLICY ISSUES

Ecologically Based Integrated Pest Management, 2012
© D.P. Abrol and Uma Shankar (eds.), pp. 841-852
New India Publishing Agency, New Delhi (India)
e-mail : info@nipabooks.com; website : www.nipabooks.com

42

Availability of Inputs and Associated Problems in Integrated Pest Management

R.R. JAT

Introduction

According to an estimate by World Health Organization (WHO), about one million cases of pesticides poisoning and 20, 000 deaths occur every year due to pesticide residues in food chain including cereals, pulses, vegetables, fruits, oils, milk and milk products (including mother's milk), fishes, poultry, meat and water. At global level, chemical pesticide worth US$ 30 billion is being used annually for the management of diseases and insect pests. In spite of this, epidemics of diseases, infestations of insect pests and colonization by weeds results in significant losses in agriculture production worldwide. As a result of crop improvement programs to incorporate resistance to pests, devastations epidemics and infestations are rare. Nevertheless, pests continue to exert a heavy toll in terms of yield losses. The last global estimate of yield losses due to pests, made crops like rice, wheat, maize, cotton, soybean, barley and coffee were equivalent to approximately US $ 250 billion annually during the period the period 1989-1990 (Table 1). Since these eight crops occupy only half of the global cropped area, yield losses due to all pests on all crops would be much high. The estimates indicate that 42 per cent of the total global attainable production of the eight major crops is lost due to pests, of which 16, 13 and 13 per cent losses are due to insects, diseases and weeds, respectively. Post harvest losses are estimated to cause an additional 10 per cent loss. Developing countries incur higher losses than industrial countries. (Table 2).

Table 1 : Yield losses in eight major crops due to pests

Crop	Percentage loss	Equivalent US (in billion)
Rice	51	113
Wheat	37	39
Maize	38	28
Cotton	38	15
Soybean	32	12
Coffee	40	8
Barley	29	6

Source : CABI 2003

Table 2 : Losses due to pests in developed and developing countries

Country	Percentage loss	Equivalent US (in billion)
Africa	49	13
Asia	47	145
Soviet Union	41	22
Latin America	41	22
Europe	28	17
North America	31	23
Oceana	37	2

Source : CABI 2003

In India, crop losses due to pest and disease amount to about US $ 18 billion (Parliamentary Sub-Committee, December 2002) where losses in various crops are about 15 30% and post harvest losses are 18-20%. New pests and diseases are enlarging every day causing heavy monitory losses to the farmers. Suicides by farmers in some states explain the severity of losses they bear.

Now, the challenges are quite different than were in pre-green revolution era or even pre World Trade organization (WTO) regime. Now, we are concerned about food safety and quality that too without sacrificing food security. Extensive use of chemical pesticides is not compatible with present day concept of safe food. Since there is a lot of concern about the ill effects of pesticides, utmost importance is being given to unpolluted environment: soil, water and air, conservation of biodiversity, organically produced food and sustainable agriculture. One of the solutions to these problems lies in integrated pest management (IPM) with emphasis on use of non-chemical methods particularly biopesticides and resistant varieties. The Government of India and some state governments have given lot of impetus and thrust to adopt the use of biopesticides. However, adoption of biopesticides by farmers is still in infancy. Till date, only about 5000 villages and around 0.1 million farmers have reaped the knowledge and benefits of biopesticides. Establishment of "organic cotton", "organic tea", "organic vegetables", "organic basmati" and "organic farming" units has now been initiated. State of Uttranachal has adopted "organic agriculture" as its main agriculture policy particularly for hill.

Important Inputs of IPM and their Availability

IPM modules for different crops

IPM is a knowledge-based technology. It involves integration of different methods of diseases and pest management to manage an important disease or pest of crop or to manage all the important diseases and pests in a particular cropping system. Under IPM, use of chemical pesticides is discouraged if not totally banned. IPM modules are area specific. They take into consideration not only diseases and insect pests but also availability of inputs. Due to extensive efforts of National Center for Integrated Pests Management (NCIPM), ICAR institutes, and a number of state agricultural universities, now IPM modules are available for a number of crops in different areas. Directorate of Plant Protection and Quarantine has published IPM modules for several crops. These modules can be adopted in new areas with little modifications. However, adoption of IPM modules by the farmers is still poor. The major reasons are: i) lack of knowledge at the level of not only farmers but also extension workers, ii) poor extension, iii) non-availability of quality inputs and iv) complex nature of modules. Most of the modules include too many practices, which usually discourage the farmers. They are not ready to invest their entire energy in pest management alone. Therefore, it is important to develop modules, which involve only few critical interventions. For example: reduced application of nitrogenous fertilizers, which can reduce incidence of a large number of diseases and insect pests in crop like rice; and application of biocontrol agents, *Trichoderma* and *Pseudomonas* through compost in pits, which not only effectively suppress soil and seed borne diseases but also improve quality of compost and crop growth. Incorporation of *Pseudomonas / Trichoderma* in compost (FYM, chicken manure, vermicompost, etc.) should be promoted as a general agricultural practice, irrespective of the crop, season or locality. This method is also cost and labour efficient.

Use of resistant varieties

Wherever available, use of resistant varieties should be preferred. They are most important input of any IPM module.

Pesticides

The use of pesticide should be avoided as much as possible. However, need based and judicious use of environmentally safe pesticides is advocated under IPM.

Biocontrol agents

Biological control can be achieved either by introducing biocontrol agents directly into natural ecosystem or by adopting practices, which favour population build-up of biocontrol agents under natural condition. Combination of both the approaches is probably the best solution. In recent years, there has been a tremendous progress in this area. Various kinds of biocontrol agents available in India are listed in Table 3.

Table 3. Biocontrol agents and biorationals available in India

Botanical pesticides	Microbial pesticides	Biocontrol insects	Biorationals
Anona	*Bacillus sphaericus (2)*	*Bracon sp.*	*Pheromone traps,*
Azadorachta (443) [b]	*B. subtilis,* *B. Thuringiensis (11)*	*Chrysoperia carnea*	*pheromone lures**
Chrysanthemum (290)	*Beauveria bassiana (2)*	*Cryptolaemus*	*& sticky traps for:*
Cymbopogan (1)	*Gliocladium* sp.	*Montrouzieri*	*Bactocera* sp.
Derris	*Hirsutella* sp.	*Crytobagous salviniae*	*Chilo* sp.
Nicotianha	*Metarrhizium anisopliae*	*Encarsia peprniciosi*	*Dacus* sp.
Pongamia	*Nomuraea rileyi*	*Epiricania melanoleuca*	*Earias vittella*
Vitex	*Pseudomonas fluorescens*[d,] *Trichoderma* sp. (10)	*Goniozus nephantidis* *Neochatina bruchi*	*Helicoverpa armigera* *Leucinodes orbonalis*
	Verticillium lecanii	*Trichgramma achaeae*	*Pectinophora gossypiella*
	Granulosis virus Nulcear	*T. brasilliensis*	*Plutella xylostella*
	Polyhedrosis virus (3)	*T. chilonis* *T. exiguum*	*Scirpophaga incertulas,*
		Spodoptera litura *T. japonicum* *T. pretiosum*	White grub
		Zygoramma biocolorata	Mating disruptant: *Pectinophora Gossypiella* PB rope (1)

Biocontrol agents like *Trichoderma* and *Pseudomonas* have multiple activities like antifungal, anti-nematode, symbiotic, plant growth promontory, plant defense inducing (induced systemic resistance), composing phosphorus solubilization and probably anti-weed. *Trichoderma* is reported to suppress *Phalaris* minor also. Plant Growth Promoting Rhizobacteria (PGPR) like *Pseudomonas* with biocontrol ability play important role in controlling soil and seed borne and sometimes foliar diseases. They are extremely useful in high humidity and waterlogged condition and in alkaline soils, where fungal biocontrol agents like *Trichoderma* spp. are less effective. They are also useful for forestry, pulses and vegetable crops, particularly for nurseries and therefore are now well accepted in "organic farming".

Bio-pesticides Market in India

At present, the biopesticides share approximately 2.5 per cent of total pesticides in market compared to 1 per cent in 2001. This increase is due to utilization of biopesticides mainly in vegetable crops, sugarcane and paddy. The projected share of the biopesticides by the end of 2005 is 12-15 per cent of total pesticides in market. They are not integral part of "organic farming" and IPM and are available in private market in most of the states like Tamil Nadu, Karnataka, Andhra Pradesh, Gujarat, Kerala, Uttar Pradesh, Uttaranchal, Haryana, Punjab, etc. The estimates of some of the pesticides available in India are given in Table 4.

Table 4 : Estimated biopesticides availability in India

Product	Availability
Azadirachtin formulations	2000 t
Bacillus thuringiensis	500 t
Trichoderma species	500 t
Fungal entomopathogens	100 t
Pseudomonas species	50 t
Nuclear polyhedrosis viruses	0.5 million LE
Trichogramma cards	1.5 million
Pheromone traps	5 million
Pheromone lures	15 million

Indian biopesticides industry is the hub of various biopesticides products available in the country. Many products are being exported now. There are over 400 biopesticide production units with 8-10 major players. There is also availability of surplus manufacturing capacities, data for registration and samples for demonstrations and trials. Technologies for joint ventures in foreign countries are also available.

Constraints in wide adoption of biocontrol agents biopesticides in IPM

In spite of strong research base in country, biopesticides have not become popular with the farmers. Some of the reasons are listed below:

Technological

- Often the performance of biopesticides in field in inconsistent. Unlike chemical pesticides, biopesticides need support even after their application to get established in targeted niche. There is need to develop a delivery system, which not only places biocontrol gent at the required site but also help to get it established there. Recently, this has been achieved to some extent by suing cow dung/FYM/chicken manure/ vermicompost, prer-colonized by *Trichoderma harzianum, Pseudomonas fluorescens* or their mixed formulation (Zaidi and Singh, 2004*ab*).
- Performance is dependent on environmental conditions like temperature, humidity and pH.
- Biopesticides products are often of poor quality. The standard methods for quality analysis are not available and most of the laboratories are not well equipped for quality analysis. The quality of fungal and bacterial antagonists is based on the population of organism in the formulation. There is no consideration of the efficiency of the organism; for example: for testing the antagonism, screening is made against mycelial growth and no consideration is given into sclerotial colonization. Several important plant pathogens like *Rhizoctonia, Sclerotium,* and *Sclerotinia* form sclerotia as resting structures. These sclerotia are more important in diseases cycle than hyphae.
- There is lack of knowledge among farmers about biological control and IPM. Many have misconceptions associated with biocontrol agents.
- Biocontrol agents have short shelf-life.
- There is lack of simple illustrated biointensive IPM modules and well executed demonstrations on them.
- Non-adoption of IPM practices in whole area.
- Inconsistent and unplanned demands.
- Inappropriate application technologies and equipments.
- Dealer's interest in chemical pesticides rather than in biocontrol agents.
- Lack of interest in multinational companies in biopesticides.

The most practical way for using biocontrol agents is as component of IPM rather than the only treatment. IPM or ICM modules must be developed using biocontrol agents as an important component.

Policy

The plans of central government thrust on biopesticides and facilitate regulatory measures for biopesticides. Although no one can dispute the requirement of toxicity data for registration but they can be made less stringent considering the fact that most of the bioagents are natural in occurrence. The central government should reduce excise duty and provide incentives to industries, engaged in the production of biopesticides.

The state government plans the implementation of IPM schemes. However, at times they are not able to contribute to meet the requirement of 25 per cent funding to obtain the central funding of 75 per cent. The state governments charge varying rates of sales tax, (up to 9.8 per cent) and procure bio-inputs often untimely or at times at the lowest rates, irrespective of the quality. It often lays stress mainly on IPM kits. Moreover, the Pollution Control Board still considers biocontrol agents hazardous like the conventional pesticides.

The state governments also promote research at state agricultural universities and issue package of practices without emphasis on biopesticides. Although it provides extension but it is poor, inadequate and inappropriate. There is very little feedback on the efficacy of biopesticides supplied under various schemes. The state governments face various constraints in IPM programs and functioning of biocontrol laboratories and pesticide testing laboratories for testing the biopesticides. The promotion of IPM is through '*Kisan melas*' and workshops, but emphasis and expenditure are mainly on travel. There are only a few states that implement well the IPM programs and ensure quality procurements.

Very few enterprises initiate industry and marketing related to biopesticides. Most entrepreneurs set up units with no quality consciousness; and only a few seriously manufacture quality products, follow regulatory procedures and contribute significantly to proper extension down to end-users. However, still the manufacturing capacity remains under utilized. There are about 220 manufacturing units in the country, out of which only 8-10 are operational.

The dealer, who is the final guide to the farmers, mainly works for chemical inputs and gives advice to the farmers without much knowledge. These people care little about quality of the product and therefore work for products, which provide high margins. Farmers depend on the advice of dealer and have developed habit of using chemical inputs. They lack confidence in biopesticides as they expect instant results and consider them costly. Rarely a farmer gets proper information on biopesticides and demonstrations in local languages that match his level of understanding.

Suggestive measures for accelerated adoption of bio-pesticides

Some of the suggestive measures that may accelerate the adoption of biopespticides are listed below:

Policy

- Biopesticides should be classified as separate from chemical pesticides.
- At least 50 per cent of the plant protection budget should be reserved for eco-safe biocontrol agents, biopesticides, botanical pesticides and pheromones.
- Custom and excise duties on biological products should be reviewed.
- Nodal biopesticides quality testing laboratories should be set up and the pesticide-testing laboratories be upgraded for testing biopesticides.
- Efficacy and quality based purchase of bio-inputs should be mandatory.
- The manufacture of good quality biopesticides should be listed as "accredited/preferred manufacturers".
- In "package of practices", biopesticides should be included with details of products, utility and methodology for use.
- The seed corporations should replace chemical treatment of seeds with biopesticides.
- The sale of pesticides should be against prescriptions from plant protection officials/scientists/industry staff only.
- Training and programmes for dealers and distributors (hub of the activity) on bio-control and biopespticides in collaboration with research organizations and industry should be organized periodically.
- Qualifying tests for dealers/and distributors should be introduced.
- Education to farmer through programs on television channels should be encouraged.
- Incentives should be given to biopesticides/biocontrol industries for ensured availability of quality bioagents to the farmers.

Research

The issues that need extensive research are:

- Quality control standards for biopesticides.
- Improvement of shelf life without affecting parasite fitness.

- Development of microbial consortium involving fungal and bacterial antagonists along with entomopathogens and biofertilizers.
- Development of value added strains of biopesticides with characteristics like phosphorus solubilization, straw decomposition, weed control, etc.
- Development of strains of bioagent suitable for adverse climatic conditions; like acidic and *usar* and dry soils, cold, upland, etc.
- Factors affecting the efficacy of biocontrol agents (inoculum load, pH, moisture, etc.).
- Formulation strategies for foliar application and rain protection.
- Development of inexpensive and effective methods of application/ delivery of biocontrol agents (seed biopriming, compost colonization, FYM, green manure, etc.).
- Development of phagostimulants for faster feeding and accelerated impact of biopesticides.
- Studies on performance of biopesticides on newly emerging biotypes and species, e.g. *Helicoverpa.*
- Revision of policy related matters involving central and state governments, industries, dealers and farmers.

For biopesticides practioners

- Should consult industry/research scientists.
- Practice of deep ploughing and soil solarization.
- Use organic manures and optimum irrigation.
- Invariably add *Trichoderma* and/or *Pseudomonas* in compost pits.
- Use resistant or tolerant cultivars, wherever available.
- Practice seed and nursery treatments with biocontrol agents. Prefer seed biopriming.
- Practice proper crop hygiene.
- Practice surveillance and monitoring.
- Practice mixed cropping and traps crops.
- Use the advised biopesticides at right time and right site with proper equipment.
- Procure biopepsticides from a reliable source.

Convergence and synergies

A consortium of partners involving research scientists, state agriculture universities, extension departments, farmer's associations, non governmental organizations, seeds and bio-input industries, and product buyers should be set up to demonstrate, implement, train and promote integrated crop management (ICM) at farmer's fields. Such synergies would benefit all the players. Farmers can have latest upgraded technology and also an assured availability of input products at their fields. Seed and bio-input companies will be benefited by having an increased market penetration and assured sale. In return buyers of agricultural produce will have timely availability of high quality produces for their outlets at competitive prices. Scientists will be exposed to actual problems faced by various parties and will have effective outlet for their research outputs with respect to products and technologies.

IPM in chickpea

The most important pest of chickpea is pod borer (*Helicoverpa armigera*). It is followed by root rots and wilt complex caused by *Rhizoctonia solani, Sclerotium rolfsii* and/or *Fusarium oxysporum* f sp. *Ciceri. Botrytis* grey mould caused by *Botrytis cineria* is prevalent in central and northern parts of India and *Ascochyta blight* caused by *Ascochyta rabie* is also a disease of economic significance, particularly in cooler regions of the country like Punjab. Although a number of resistant/moderately resistant (MR) varieties are available against chickpea wilt, availability of their seeds remains a serious problem. Resistant varieties available for wilt and root rot are: GCP 105 (MR); BG 1053 (root rot); L 551: KAK-2 (MR); Pusa 391 (MR); JAKI 9218 and Pant G 186; and for Botrytis grey mould: Pusa 209; ICCV-2; and Pant G 186. No resistance is available against pod borer. Scientists at ICRISAT have developed transgenic chickpea, which is resistant to pod borer. However, it is yet to be tested under field conditions and if found resistant, it might take another decade before it can be available to the farmers.

IPM module developed for chickpea at GBPUAT, Pantnagar

An IPM module developed for chickpea and validated at GBPUA&T Pantnagar is given below:

- Use *Trichoderma harzianum* (TH) colonized compost, if available.
- Use wilt and root rot resistant/tolerant varieties.
- Seed treatment with *Trichoderma harzianum* (TH) or *Pseudomonas fluorescens* (PsF) or mixed formulation (*Trichoderma harzianum*+ *Pseudomonas fluorescens*) (@ 10 g kg^1 seed). Suspend the formulation in 1% of jaggery (gur) solution in water to ensure proper adhesion. After this treatment, coat the seeds with *Rhizobium* culture @ 5g kg^1 seed).
- Remove wilted plants.

- In case *Ascochyta* blight or *Botrytis* gray mould are observed, spray mixed formulation of TH + PsF (@ 5 g l^{-1}) or carbendazim + PsF (@ 1 + 4 g l^{-1}).
- Practices for the management of pod borers:
  - Use pheromone traps (@ 20 ha^{-1}) at equal distance (20mX20m) at the time of flowering. Use 1 mg lure Z-9 Hexadecanal + Z11-Hexadecanal in the ratio of 3:97 (v/v) and change at 15 days intervals. If change is required at the time of heavy infestation, use additional traps and remove old traps after three days of installing new traps.
  - Alternate spray of H*a*NPV (1.5×10^{12} POB ha-1) and *neem* seed kernel extract (5%). Soak 12.5 kg *neem* seed kernel overnight in 25 l water, filter, add 500 g Nirma soap and suspend in 250 l water for spray in 1 ha area.
  - If unable to manage by above means, spray insecticides like endosulfan (35 EC, 750 ml ha^{-1}) or chlorpyriphos (20 EC, 625 ml ha^{-1}).
- Store properly dried seeds (moisture content approx. 8 to 9%) and fumigate with aluminium phosphide.

Conclusion

Integrated pest management (IPM) approach has been globally accepted for achieving sustainability in agriculture. Though IPM approach is not new, its need is being felt more these days because of our concern for food and environmental safety. IPM modules are available for most of the crops. However, for successful implementation of IPM, timely availability of quality inputs is the basic necessity. Non-availability of inputs is major bottleneck for large-scale adoption of IPM modules. Even if the IPM modules not implanted *in toto*, because of various reasons, certain practices like addition of *Trichodrma* or *Pseudomonas* through compost and seed treatment with biocontrol agents in lieu of chemicals can go as general recommendation and they should be part of all IPM modules. Setting up of an assemblage of partners involving research scientists, state agriculture universities, extension departments, farmer's associations, non-governmental organizations, industries and farmers for demonstration, implementation and trainings to successfully promote integrated crop management at farmer's fields is an ut5most need of the present time.

References

CABI, 2003. Crop Protection Compendium. CAB International, UK. (CD).

Singhal, V. 2003. Biopepsticides: Production, utilization and demand. In: Recent Advances in Integrated Pest Management-Proceedings & Recommendations. NATP Interactive Workshop on IPM, NCIPM, New Delhi, pp 258-264.

Zaidi, N.W. and Singh, U.S. 2004*a*. Use of farmyard manure for mass multiplication and delivery of biocontrol agents, *Trichoderma harzianum* and *Pseudomonas fluorescens*. *Asian Agri-History*, 8.

Zaidi, N.W. and Singh, U.S. 2004*b*. Use of cow dung as a substrate for mass multiplication of *Trichoderma* spp. *Indian Phytopathology*, 57.

❑❑❑

Ecologically Based Integrated Pest Management, 2012
© D.P. Abrol and Uma Shankar (eds.), pp. 853-876
New India Publishing Agency, New Delhi (India)
e-mail : info@nipabooks.com; website : www.nipabooks.com

43

Methodologies for Dissemination of Integrated Pest Management Technologies and Their Impact

RAJINDER PESHIN AND RAKESH SHARMA

Introduction

Diffusion is the process by which an innovation is communicated through certain channels over time among the members of a social system (Rogers, 1962, 1983, 1995, 2003). It is a broader term, which encompasses unplanned as well as planned and directed spread of an innovation. Dissemination is planned and directed diffusion of an innovation. In case of Integrated pest management which is a complex sort of decision making process, many technology transfer approaches were/are tried for dissemination of IPM technologies to farmers. These include radio, television, print media, training and visit, field days, demonstrations, extension field visits and participatory learning approaches like participatory technology development (PTD) and farmers field school (FFS). While Training and Visit (T&V) system failed, Farmers Field School (FFS) approach of educating farmers followed by 'Community IPM' activities and 'Farmers to Farmer' approach registered success in IPM implementation in developing countries (Matteson, 2000).

The dissemination and adoption of IPM has been generally slow in both the developed and the developing world, despite some successes. Pesticides are still the main strategy of many IPM programmes (Peshin *et al.,* 2009a). Once the IPM message is right, the challenge is to take it to as many people as possible. There are basically two types of extension methods: Face to face and mass media. Mass media is less intense method that provides a basic level of awareness knowledge but train a relatively large number of farmers per dollar spent. Face to face method namely FFS is very intensive method that provides a deep understanding but trains a relatively small number of farmers per dollar spent on the programme. Face to face methods is the personal way of reaching many people. Going public is another face to face method for a mass audience. An extensionist goes to a market or another crowed place, and delivers a short message, and then repeats it. It is especially well suited to rather simple messages that must show something (e.g. a disease symptom, a new insect). Going public has been used in Bolivia, Bangladesh, Uganda and elsewhere (Bentley *et al.,* 2003).

In this article we first discuss the different extension methods adopted for dissemination of IPM technology, other than farmer field school. The second part of the article deals with dissemination of IPM in the developed countries and in section three the classical farmer field school model is discussed. In the last section an overview of the evaluation model and impact of IPM programme is presented.

Categorization of Extension Methods

While IPM practices and technologies are categorized based on complexity, IPM dissemination methods can also be categorized by training intensity. At one end of this intensity continuum are less intense methods such as mass media, e.g. radio broadcast. A broadcast lasts for a short time and farmers are instructed how to use a specific IPM technology or practice. Less intense methods help create awareness of IPM that may lead to farmers seeking further information from other sources. A type of method that is relatively intense than a mass media radio broadcast is a field day. Field days may last an entire day or for several consecutive days. The use of extension personnel, trained in IPM, who visit farmers for several hours at a time, is a different type of method that could be classified as a moderately intense training method. Farmer field schools (FFS) are an example of training methods that can be classified as highly intense. FFS generally take place for several hours a day, once a week for ten to fifteen weeks during a growing season. There is a tradeoff between highly-intensive participatory training programme such as FFS that train and empower a relatively small number of farmers to make informed decisions about IPM and less-intensive IPM programme such as field days which train a greater number of farmers but provide perhaps a shallower level of understanding per farmer trained (Norton *et al.*, 2005).

Mass Media

Mass media methods include print media such as newspapers, leaflets and bulletins that provide IPM information to farmers. These printed methods give farmers awareness knowledge of IPM practices. These are inexpensive to produce and transmit but do not provide higher levels of cognition and analytical skills. Mass media can also transmit information to farmers quickly where other more intensive methods may take longer for messages to spread. Methods such as radio broadcasts and television programme are more intense than printed media. Radio broadcasts and television programming with agricultural themes are broadcast/telecast nearly every day for an hour or two. These methods also have the potential to reach a wide audience and can cover topics in more depth then print media.

While mass media can potentially reach many people in a short time, one of its limitations is its inability to reach the entire population. Print media is limited in the number of people it can reach because of illiteratacy. The difficulty in reaching people with television and radio is that while it can reach illiterate people, there are many areas particularly in rural areas without electricity making access to television difficult and impossible. Despite these limitations, mass media is still a viable method for reaching a large segment of the population quickly and with a relatively low average cost per farmer reached. Radio has been used recently in Vietnam to teach people to avoid insecticide abuse in rice. Radio together with leaflets, T.V. broadcasts led to a 53% reduction in insecticide use and no loss in production (Bently *et al.*, 2005).

Field Day Demonstrations

Field day method is more intense than mass media. The typical field day demonstration lasts for one day or several days. During an IPM field day, experts come to a village and put on demonstrations to introduce new IPM technologies or practices to farmers. Several hundred farmers may attend a field day demonstration. The benefit of field days is that a large number of farmers can receive information about an IPM technology or practice at the same time. The field day provides more depth of understanding about specific practices for farmers than does mass media, because participants are able to observe the IPM innovations first hand. The costs associated with field days include salaries of the experts that administer the event, the opportunity cost of the trainers' time, transportation for the experts to the event, and cost of food and refreshment for participants and the opportunity cost of participants' time.

Extension Visits

Extension agents living in villages and visiting local farms individually and in group meetings is perhaps the oldest and most common extension method in the world. The extensionist job is to provide farmers with information and training about new agricultural technologies and practices. The transactions cost of having extension agents visit farmers is very low because the agents live in close proximity to the farmers with whom they work. The costs of the agent visits include paying the agents' salaries and providing them with training along some food and transportation costs. The drawbacks of this system are that it is insufficiently funded so it is impossible for extension personnel to interact with every farmer. Extension personnel are over extended in many parts of the country. Provide them with incentives to actually work with the farmers. Furthermore, the quality of extension agents varies as does their ability to connect with farmers and influence them to adopt new practices and technologies. All the extension workers are not well trained in IPM.

IPM Clubs

Farmer field schools will have a sustainable impact of spreading IPM only if those farmers who attend FFS continue to use the IPM knowledge they receive after the formal training ends. For this reason the attempts were made that each of their FFS turn into a self sustaining IPM club. The IPM club is less intense than a FFS and serves as a forum for farmers to meet and discuss pest problems and to create further awareness about IPM by encouraging farmers who did not attend the FFS to join the club. The concept of an IPM club has many benefits, such as requiring little financial support from the government and encouraging farmers other than the original 25 FFS trainees to join the club and adopt IPM practices. The problems with IPM clubs are that without any substantial funding, it is difficult for the club to establish itself and purchase necessities such as materials and supplies, and create a clubhouse for holding meetings. The success of the IPM club also depends entirely on the initiative of the farmers in the community. Finally because it is a club, the organization will always remain exclusive in some way and it will be unlikely that everyone in the village will join the club and learn about IPM.

Training and Visit Method

The training and visit method of extension emphasizes regular training for extension field workers and contact farmers, with better linkages between research and extension. The training for extension workers is given once a fortnight, while on the other days during the fortnight the extension worker are supposed to extend the messages from the training to the contact farmers.

Contact farmers from the group receive recommendation for the forthcoming from the extension worker and then spreads the messages to other group members. But all attempts to disseminate IPM technology by this method did not yield desired outcome.

Picture Songs

It is one of the most innovative techniques used by Non Governmental Organization (NGOs) in Bangladesh to teach appropriate rice technology to thousands of people, especially about natural enemies and using organic fertilizer. As a performer sings out the message (and dances) the rest of the troupe accompanies her with music, and rolls out the illustrations on the scroll. Hundreds of the people can see each memorable performance at one sitting and as of 2005 and about 25,000 people had seen and heard the message. (Bentley *et al.,* 2005).

Video Movies

Videos have been used in combination with farmer participatory research and community meetings. Researchers develop appropriate technologies with farmers' participation and then they make videos where farmers speak on camera. Their honest words were convincing to other farmers, who could identify with them. Extension agents showed the videos in communities, and then answered questions from the audience, which allowed many people to be trained at once, in a relatively short time. (Van Mele *et al.,* 2005).

IPM Websites

The most comprehensive and easy to use IPM information for arthropods and diseases infesting wheat is available on centralized individual IPM websites for Oklahoma and Texas in the United States of America. Information on these sites can be accessed by area/regional and county Cooperative Extension Service personnel, but also by producers and other stakeholders. The Oklahoma IPM website (Oklahoma IPM 2007) is a central resource area organized primarily by commodity allowing access to printable fact sheets, newsletters and computer decision support models. The Texas IPM website (Texas IPM 2007) is quite unique in that the just topic on the outline provides a link to the overview of IPM (Giles and Walker 2009). In India, Punjab Agriculturale University (PAU) has also created 'Cotton IPM Website'. But till date no study has been conducted on its impact. In the developing countries, where overall literacy is low all villages are not having electricity, this method does not work.

Plant Health Clinics

There are new extension method being implemented in Nicaragua, Bolivia, Uganda, Bangladesh and elsewhere, pioneered by the Global Plant Clinic. They started in Bolivia in the 1990s, so farmers from distant areas could bring plant samples and get advice about plant health problems. Most of the clinics are 'mobile'. The plant clinics provide a place for personalized consultations between farmers and agronomists. The plant clinics can be easily combined with other methods like fact sheets, radio, and short courses. (Danielson *et al.*, 2006). In Indian Punjab, the PAU has plant clinic for diagnosis of insect pest, disease and other plant problems. Its search is limited as it is not 'mobile'.

Written Material

It includes fact sheets, journals and newspapers which are also useful, especially when written for farmers and validated by farmers before distribution. In India, coffee farmers in Karnataka have journals & magazines in their homes. *Adika Pathrika* is a magazine that started as a newsletter in the 1980s and rapidly expanded (Padre and Tripp, 2003).

Street Plays

Street plays can play an important role in creating awareness about the harmful effects of pesticides and the IPM philosophy. Street play *"Sambal Samaj Kar Hee Keet Nashak Ka Upyog: Ek Mukti Marg"* (Judicious use of pesticide: a way out of pesticide treadmill) was used in IPM based insecticide resistance management programme in Punjab (India) (Fig. 1). The street play went a long way in creating awareness about the IPM programme (Peshin, 2005).

Fig. 1 : Street play at IRM-IPM village (*Source :* Peshin, 2005)

Dissemination of IPM Technology in the Developed Countries

In the developed countries like United States of America (USA), the Cooperative Extension Service (CES) continues to play the major role in disseminating IPM programme information. Individual states in the USA have IPM coordinators who are associated with land grant universities and work with a multidisciplinary group of faculty and varying structures of regional and county extension educators to deliver IPM to stakeholders. Additionally, United States Department of Agriculture (USDA), Agricultural Research Service (ARS) scientists often cooperate with university personnel to discover develop and evaluate IPM technologies. Stakeholders are identified primarily as growers, but also include commodity groups, private businesses with crop advisors and larger agricultural companies who interact with universities and ARS personnel together to evaluate and disseminate IPM technologies (Fuchs, 2007). The CES approach includes presentations at field days, growers meetings and pesticide certification meetings. CES personnel at the state levels interact with individuals within areas/regions and individually at county offices to develop and disseminate IPM information. Published documents in the form of fact sheets of production manuals that incorporate IPM recommendations' in the context of crop production goals. These are developed and provided to stakeholders by CES and/or other practitioners. IPM information for corn in USA has historically been disseminated by CES personnels and other stakeholders to producers (Giles and Walker, 2009). A number of extension methodologies are used to dissemination of IPM technologies to the farmers. These include training about pest and natural enemies and field scouting, on farm result demonstrations, field days, use of print and electronic media regarding information on the periodic/current status of pest and natural enemies, economic thresholds, computerized pest forecasting, newsletter, etc. In some European countries like Denmark, Netherlands, Norway, Sweden and Switzerland "IPM innovation system" approach is the cornerstone for high adoption of IPM practices by the growers (Peshin *et al.*, 2009a). The European Parliament develops regulatory framework for taxes on pesticides. Different strategies are adopted for implementation of IPM. For example pest warning system, legislation, providing information about IPM to growers, employment of pest control advisors by large farmers and growers associations, providing incentives for low pesticide produce and support of market in encouraging sale of low pesticide use products. Different initiatives are taken in Europe for implementation of IPM e.g. in Belgium there is fruit growers association to promote IPM and hazardous pesticides have been prohibited as per IOBC norms. Following are selected cases from the Europe:

Denmark

The first Pesticide Action Plan 1986–1997 targeted a 25% reduction in total pesticide consumption by 1992 and 50% by 1997. It also included measures to encourage the use of less hazardous pesticides and educating farmers to improve their knowledge and skills. The second Plan 1997–2003 introduced the indicator treatment frequency index. The target was to reach a treatment frequency of less than 2.0 before 2003 and establish 20,000 ha of pesticide-free zones along key watercourses and lakes. The objective of the third Pesticide Action Plan 2003–2009, is to lower the treatment frequency below 1.7 by 2009, to promote pesticide-free cultivation and establish 25,000 ha pesticide-free zones along watercourses and lakes. This plan includes the fruits and vegetables sector for the first time. Danish agriculture advisory service educates farmers about IPM, government provides incentives to encourage IPM, and to discourage pesticide use it has imposed pesticide taxes.

Italy

Environmental NGO promoting pesticide free fruit and vegetables cultivation in Italy, and NGO provides guidelines to farmers on IPM. Labeling of IPM produce as LAIQ.

Netherlands

IPM for crop protection introduced by the cabinet decision in the Netherlands. New initiatives based on multi-stakeholders launched in 2003 with Euro 14 million for integrated crop management (ICM). Experimental advisory service for low pesticide farming methods, development of environmental impact cards with indicators, development of best practice protocols for IPM in major crops and market support to ICM. Farmers adopting ICM in apple, strawberry, cabbage, lettuce etc. offered premium by the market.

United Kingdom

In the UK, the UK cooperative group one of the largest consumer's cooperative in the world manages 10000 ha of cooperative owned land and 20000 ha of farmland owned by land owners. Farmers are provided guidelines on Integrated Farm Management and have prohibited use of 23 and restricted use of 32 pesticides which is aimed to reduce pesticide use by 50%. They ensure the adoption of biological and mechanical crop protection ahead of pesticides.

Similarly in Norway and Sweden the government has levied tax on pesticide in order to encourage the farmers to adopt IPM. There are active advisory services which disseminate IPM to farmers by forcasting, demonstrations, trails and training. In Switzerland farmers are provided incentives for adopting IPM. The farmers are provided with pest warning services, testing of spray equipments and other facilities.

Dissemination of IPM Technology in the Developing Countries

In developing countries particularly in Asia, IPM programme was implemented after the Indonesian government's rice self sufficiency programme failed, in most part due to brown plant hopper attack during the 1970s and 1980s. The IPM programme implementation modeled after some pioneer projects, introduced a new plant protection approach, mainly by involving farmers as early as possible in decision making, and reducing the use of pesticides. In the beginning, many farmers got confused as they had been accustomed to chemical based approach and they were not ready to accept the new approach and lost their confidence in extension service. The Indonesian experts then created a training-cum-extension programme called farmers field school (FFS) on rice and for about five years (1988-1993) trained about one million rice farmers. The training was later extended for other crops.

Basics of Farmer Field Schools

Farmer Field Schools (FFSs) were first conceptualized and adopted on a broad scale in Indonesia in 1989 to reduce reliance on the use of pesticides in rice cultivation by enhancing farmers' understanding of crop ecology (Kenmore, 1996). The principles supporting the development of the FFSs grew out of the traditions of literacy education and village-level basic health care. The FFS approach focuses on farmers' self-development and skill building. Farmers become experts in adapting their farming practices to local conditions by testing and adopting innovative solutions. Through weekly field observations and analysis followed by management action, farmers gain expertise and confidence in decision making. The FFS approach therefore reverses the system of top-down, research driven extension and pays particular attention to the conventional wisdom of farmers. Farmers are no longer positioned as receivers of already developed technological packages but as field experts who collaborate with the government extension staff to find solutions relevant to local realities. FFS outcomes and impacts range from technical achievements to gains in social and human capital. Basic concepts and essential elements are common to IPM FFS programmes in several countries such as: the practice of agro-ecosystem analysis (AESA) as the decision-making tool for field management; field experimentation to strengthen farmers' analytical skills; and adult non-formal education to enable the learning of all participating farmers. Good FFS require technically strong facilitators and a season-long curriculum linked to the crop phenology. So far, FFS programmes on IPM as well as other topics such as soil, water and livestock management have been implemented in 74 countries (Braun *et al.*, 2006). Farmer field schools are currently being promoted by many development organizations, such as the Food and Agricultural Organization (FAO), as the most appropriate method for disseminating IPM technologies and information. The first FFS began in the 1980s in Java Indonesia, teaching farmers about rice IPM. The principal component of any FFS is that it emphasizes experiential learning, with a participatory approach. Hands-on training is important to attract both literate and illiterate farmers and to keep them interested in learning about IPM. India, adopted the FFS model of IPM implementation since 1993. FFS curriculum calls for establishing IPM trials in farmers' fields and demonstration plots to show the benefits of IPM through observation and experimentation (Feder *et al.*, 2004a). Farmer field schools are run by facilitators rather than instructors in order to create a group learning environment rather than a classroom setting with a teacher giving instructions. Farmer Field Schools are based on the premise that farmers already posses sizeable knowledge about farming so the FFS curriculum strives to incorporate principles of IPM into farmers understanding and create an awareness for farmers of the agro-ecological system that exists on their farms (Den Tex, 2004). The FFS system is an intense IPM training method. The benefits of farmer field schools are that they provide participants with a great depth of understanding about IPM. The argument against farmer field schools is that while the farmers who attend FFS training may receive a deep and substantial understanding of IPM, training only 25 farmers in a village is not enough to have a significant impact. More farmers might be reached with the same amount of money if a less intensive method is employed. FFSs are now active in Asia (including East, South-East, South, Central and Middle East), Africa (Western, Southern, Eastern and Central), Latin America (South and Central America), the Caribbean and Eastern Europe. FFS-type activities are conducted in Australia through Rice Check programme and in the USA on fruit trees. Braun *et al.* (2006).

Principles of IPM

The implementation of IPM is based on sound scientific foundation and the following four principles of the IPM- FFS should always be observed:

Grow a healthy crop

Healthy crop allows plants to recover better from abiotic and biotic stress. It provides defense against pest injury and promotes natural defences toe many insects and diseases. Proper crop and plant management methods are used which include primary and secondary cultural practices.

Conserve natural enemies

Conservation of natural enemies namely predators, parasites etc. provide free biological control of insect pests and diseases. Parasites, predators and pathogens have long been recognized to control pest insects, but recent research shows microbial antagonists, and competitors of plant diseases are also important. Conservation usually implies avoiding inappropriate pesticide applications or improving soil organic matter necessary for beneficial soil micro-organisms. Natural enemy habitat protection and development are more active methods of conserving natural enemies. Inoculation or inundation of reared natural enemies may be possible under special circumstances but usually only after conservation efforts have already been implemented.

Observe crops regularly

Crops need to be observed regularly. It means informed decision making for appropriate interventions to be made quickly for water, soil, and plant management. Inputs used are based on an ecological and economic assessment.

Farmers become experts

Farmers experimenting, observing, and analyzing in their own fields is crucial for long term management of soils, pests and crops. Expertise implies a basic understanding of the agro-ecological system, and decision making processes. Simple rules and directives may provide short term benefits but cannot sustain long term local developments.

Elements of IPM Education

In general, Farmer Field Schools (FFS) consist of groups of people with a common interest, who get together on a regular basis to study the "how and why" of a particular topic. The topics covered can vary considerabley from IPM, organic agriculture, animal husbandry, and soil husbandry, to income-generating activities such as handicrafts (Ghallager, 2003). A diverse and evolving pest complex requires a cadre of trained individuals with enhanced management skills that ensure human health and environmental protection. It is important for practitioners to acquire new skills to implement targeted IPM strategies using new technologies, including genetic engineering, reduced risk pesticides, cultural practices, and biocontrols. The elements that commonly appear in successful FFS programmes as listed by Ghallanger (2003) are:

The group

A group of people with a common interest form the core of the FFS. The group may be mixed with men and women together, or separated, depending on culture and topic. The group could be an established one, such as a self-help, women's, or youth group. Participatory technology groups, for example, sometimes undertake a season of study in FFSs before starting their research. The FFS tends to strengthen existing groups or may lead to the formation of new groups.

The field

FFSs are about practical, hands-on topics. In the FFS, the field is the laboratory and teacher, and it provides most of the training materials like plants, pests and real problems. Any new "language" learned in the course of study can be applied directly to real objects, and local names can be used and agreed on. Farmers are usually much more comfortable in field situations than in classrooms. In most cases, communities can provide a study site with a shaded area for follow-up discussions.

The facilitator

Each FFS needs a technically competent facilitator to lead members through the hands-on exercises. There is no lecturing involved, so the facilitator can be an extension officer or a farmer field school graduate. All facilitators need season long training to (re)learn facilitation skills, learn to grow crops with their own hands, and develop management skills such as fund-raising and development of local programmes. Computer literacy is often included in the training of facilitators, especially for preparing local training materials, budgets and project proposals. Once the facilitators have completed their training and are leading the FFS process, it is easy to identify capable farmers who are interested in becoming facilitators. Farmer Field School graduates are usually given special farmer facilitator training (10-14 days) to improve technical, facilitation and organizational skills.

The curriculum

The FFS curriculum follows the natural cycle of its subject, be it crop, animal, soil or handicrafts. For example, rice transplanting in the FFS takes place at the same time as farmers are transplanting their own crops - the lessons learned can be applied directly. One key factor in the success of the FFSs has been that there are no lectures – all activities are based on experiential (learning by doing), participatory, hands-on work. This builds on adult learning theory and practice. Each activity has a procedure for action, observation, analysis and decision making. The emphasis is not only on "how" but also on "why". Experience has shown that structured, hands-on activities provide a sound basis for continued innovation and local adaptation, after the FFS itself has been completed. It is also one of the main reasons that farmer facilitators can easily run FFSs - once they know how to facilitate an activity, the outcomes become obvious from the exercise itself. Activities are sometimes season-long experiments especially those related to soils or plant physiology.

Other activities in the curriculum include 30-120 minutes for specific topics. Icebreakers, energizers, and team/organization building exercises are also included in each session. The curriculum of many FFSs is combined with other topics.

The programme leader

Most FFS programmes exist within a larger programme, run by government or a civil society organization. It is essential to have a good programme leader who can support the training of facilitators, get materials organized for the field, solve problems in participatory ways and nurture field staff facilitators. This person needs to keep a close watch on the FFSs for potential technical or human relations problems. They are also the person likely to be responsible for monitoring and evaluation. The programme leader must be a good leader and an empowering person. He or she is the key to successful programme development and needs support and training to develop the necessary skills.

Methodology for Evaluation of IPM Programmes

Evaluation is an integral part of any development programme and it should be planned at the programme designing stage to ensure that the evaluation will be useful for programme management, improvement and accountability. This scientific feedback needs to be utilized for further improvement of programmes and for decision making by stakeholders and policy makers. The desired and undesired consequences of new technologies need to be measured as was not done during the green revolution period that resulted in un-sustainability and over exploitation of natural resources (Peshin *et al.*, 2009b). The worth of IPM technologies can not only be measured in terms of adoption of IPM practices, reduction in pesticide use and increase in yields but it should consider all aspects related to socio-economic development, health, sustainability, environment conservation etc. that represent the true worth of IPM programmes. IPM programmes have been, and are being implemented throughout the world since last four decades and the primary purpose of these programmes is to reduce the use of pesticide by adopting alternate pest management practices. The purpose of IPM programme evaluation is to seek the information needed for programme improvement and weighing the results and consequences. The two major types of evaluation are formative evaluation and summative evaluation. Formative evaluation is conducted to assess the programme process for its improvement. Summative evaluation is used to access the programme results for accountability. The input-process impact (IPOI model) based logic model developed by Peshin *et al.*, 2009b should be used to design and conduct formative and summative evaluation of IPM programmes (Table 1). The input in terms of the resources (available time, quality of trainers and adoptability/adaptability of IPM practices), processes (training interventions, implementation of IPM programme, development and delivery of educational materials, whether farmers are provided experimental training in agro-ecosystem analysis, use of sampling methods for determining economic threshold levels (ETL) of pests for taking pest management decisions). Process evaluation during programme implementation is generally formative in nature but can also be summative, e.g., if the stakeholders need to know about the actual delivery of the programme and its application by farmers for justification of programme spending (Peshin, 2005). Outcome evaluation is conducted once the programme is settled or immediately at the end of the programme to identify and measure the benefits of the programme. Outcome evaluation is summative in nature and it is conducted at the end of programme to measure immediate impact and over time to measure long term impact. The detailed framework IPOI model of IPM programme evaluation at farm level is given in Table 1.

Table 1 : Theory based logic model for IPM programme evaluation

System Approach for evaluation of IPM programmes based on logic model				
	INPUTS →	ACTIVITIES OR PROCESS →	OUTCOME →	IMPACT
Method of evaluation	Input Evaluation	Process Evaluation	Outcome Evaluation	Impact Evaluation
Type of evaluati on	Formative	Formative	Summative	Summative
Research de-sign	Non-experimental research design based on IPM theory and participatory evaluation model	Quasi-experimental research and non-experimental research designs, in-depth observation and interviewing	Quasi-experimental and non-experimental survey design (Critical multiplism model)	Quasi-experimental design for evaluating immediate impact and ex-post facto design (longitudinal studies) for evaluating long-term impact
	With IPM	With/Without, Before/During IPM	With/Without, Before/After IPM	With/Without,, Before/After IPM
Time dimensi on	T1 Before the IPM programme	T2 During the implementation of IPM programme	T3 At the end of IPM programme	T4 Over a period after IPM programme

contd...

Table 1. contd...

Evaluation indicators	1. Evaluation of IPM technology	Implementation of IPM programme	Outcome of IPM programme	Impact of IPM programme
	To find out whether IPM practices fit the farmer by conducting pilot studies	Active participation of farmers and their reactions regarding IPM pro-gram	Gain in cognition and analytical skills	Continued adoption of IPM practices
	Adoptability of IPM practices in terms of technological attributes	Experiential learning of farmers, e.g. agro-ecosystem analysis, sampling for ETL	Change in attitude	Reduction in number of pesti-cide application and use (technical grade material)
	Adaptability of IPM practices	Demonstration of IPM technology validated during input evaluation	Adoption of IPM practices which have been identified at input evaluation stage	Reduction in pesticide expenditure
	2. Evaluation of implemen-tation (extension) methodology	Whether IPM intervention yields desirable changes?		Increased population of natural en-emies and reduction in population of insect pests
	3. Input evaluation in terms of human resources, available budget, etc.	Causal relationship		Reduction in insecticide resistance in insect pests
				Higher input/output ratio
				Multiplier effect (farmer to farmer diffusion)

Source : Peshin *et al.* (2009b)

Indicators for IPM programme evaluation

An indicator can be defined as a variable (Indicators for the evaluation of the EU's Rural Development programmes, 2005). According to Powell *et al.* (1996), "Indicators are observable evidence of accomplishments, changes made or progress achieved." These indicators are measured to quantify the changes made, inputs utilized or outcomes achieved. Evaluation indicators can be categorized into the following four groups:

a) Indicators for measuring programme inputs
b) Indicators for measuring programme outputs
c) Indicators for measuring farmer participation
d) Impact indicators for measuring the programme outcomes and impacts

Impact / Outcome of IPM Programme in Developing Countries

IPM-FFS programme was first started in Indonesia in 1989, after the banning of 57 broad spectrum pesticides in 1986. IPM-FFS programmes were carried out in 12 Asian countries after observing its success in Indonesia. Later on IPM-FFS were implemented in vegetable, cotton and other crops. The programme spread to Africa, Latin America, the Middle East and Eastern Europe (van den Berg and Jiggens, 2007). FFS programmes are being implemented in 78 countries and four million farmers have been trained under this programme, with 91% of these from Bangladesh, China, India, Indonesia, the Philippines and Vietnam (Braun al.,2006). The coverage of IPM-FFS was just 1-5% of all households in Asia (1989-2004). By 2002, the FAO Intercountry programme (ICP) had spent US 45 million dollars on training activities in Bangladesh, India, Indonesia, Loas, Malaysia, Nepal, the Philippines, SriLanka, Thailand and Vietnam. Preliminary pooled average results from seven studies on cotton IPM in five Asian countries indicate that FFS graduates increases their income by 31% in year after training, increases yield by 10% and reduces pesticide expenditure by 39% in relation to control (non IPM) farmers (FAO, 2004). The studies on impact evaluation of IPM-FFS in Asia conducted by van den Berg (2004) and Miller (2004) have shown that the insecticide use in these countries decreases and the corresponding yield in majority of these Asian countries increases (Table 2).

Table 2 : Outcome of IPM-FFS programme in Asia

Country	Crop	Insecticide Use	Yield
China	Cotton	Decreased (6.3 to 3.1 application)	-
Bangladesh	Brinjal	Decreased (7.0-1.4 application)	Increased
Cambodia	Rice	Decreased (64%)	-
Vietnam	Rice	Decreased (1.7 to 0.3 application)	-
Srilanka	Rice	Decreased (2.2 to 0.4 application)	Increased (23%)
Indonesia	Rice	Decreased (65%)	Increased (15%)
Thailand	Rice	Decreased (60%)	-
Vietnam	Tea	Decreased (50-70%)	Increased

Source : van den Berg (2004), Miller (2004)

Case study of outcome evaluation

The Insecticide Resistance Management based Integrated Pest Management (IRM-IPM) programme is being implemented in 28 districts of 10 cotton growing states of India since 2002. The objectives of the programme are: to help rationalize and reduce the use of insecticides and to overcome the development of resistance in cotton pests, especially *Helicoverpa armigera.* An evaluative study was conducted to measure the outcome of the programme at farmers' level. The evaluation indicators were: i) gain in knowledge and analytical skills of the farmers, ii) adoption of IPM practices iii) adoptability of IPM practices, iv) reduction in pesticide use and v) economic benefits. The IRM-IPM programme resulted in a reduction in insecticide consumption (technical grade material) by 30%, and it reduced the number of sprays by 15%. The majority of IRM farmers avoided using highly toxic insecticides as they gained significant knowledge out of it but the adoption of IPM techniques was low. However it did not result in a significant change in productivity (Table 3).

Table 3 : Outcome of IRM programme in Punjab

Evaluation indicators	Before IRM programme			After IRM programme		
	IRM villages	Non-IRM villages	Difference	IRM villages	Non-IRM villages	Difference
Average number of insecticide applications	15.34	14.93	0.41NS	13.07	15.43	2.36*
Technical grade insecticide use (kg/ha)	NA	NA	–	5.602	8.032	2.430*
Tank mixtures of insecticide (No.)	NA	NA	–	3.1	5.1	2.0*
Insecticide expenditure ($ US/ha)	229.00	247.67	18.67NS	128.73	152.78	24.05**
Insecticide expenditure to total cost of cultivation (%)	39.70	42.73	3.03	30.26	34.76	4.50
Average seed cotton yield (kg/ha)	1749	1704	045NS	2300	2209	091NS

contd...

Table 3. Contd...

Average net income ($ US)[a]	414.00	406.44	7.56	431.10	388.00	43.11
Output/input ratio	NA	NA	-	1.86	1.77	5.08
Adoption of sampling for ETL (% farmers)b	0	0	-	34	0	34

NS = Non-Significant, * Significant at p = .05, ** Significant at p = 0.01 a. 1$ = Rs. 45
Source : After Peshin *et al* (2009c)

Many studies have been conducted by using before/after, with/without and both before/after and with/without methods of evaluation. A study conducted by Feder *et al.* (2004b) for the World Bank using time series data in Indonesia founds no significant difference in change in yields or pesticide use when comparing FFS participants with non participants. They reported that FFS trained farmers had a greater knowledge of IPM than non FFS farmers, but that knowledge of IPM is not spreading to farmers in villages with FFS who did not attend the training. These findings are significant. If FFS graduates do not share their knowledge of IPM with their neighbours, then there is no diffusion of IPM in the farming system. The high cost of training farmers through FFS has also been questioned by Feder *et al.* (2004b) for cost-effectiveness and sustainability in diffusing IPM technology at national level. Whereas, van den Berg and Jiggins (2007) in the synthesis of 25 studies reported that there is substantial immediate and developmental benefits of participation in Farmer Field Schools (see Table 4).

Table 4 : Relative change in insecticide use frequency and crop yield

Country	Crop	Insecticide use %	Yield
Longitudinal (Before/After)			
Indonesia[1]	Rice	Decreased (61%)	-
Vietnam[2]	Rice	Decreased (82%)	Increased (7%)
Vietnam[3]	Tea	Decreased (61%)	-
India[4]	Rice	Decreased (23%)	-
Latitudinal (With/Without)			
Bangladesh[5]	Rice	Decreased (92%)	Increased (9%)
Bangladesh[6]	Eggplant	Decreased (80%)	Increased (25%)
Cambodia[7]	Rice	Decreased (43%)	Increased (4%)

contd...

Indonesia[8]	Rice	Decreased (35%)	Increased (8%)
Srilanka[9]	Rice	Decreased (81%)	Increased (23%)
Srilanka[10]	Rice	Decreased (65%)	-
c. Longitudinal + Latitudinal (Before/After, With/Without)			
China[11]	Cotton	Decreased 52% in FFS farmers field & 8% in non IPM farmer field	Increased 16% in FFS farmers field & 2% in non-IPM farmers field
India[12]	Cotton	Decreased 78% in FFS farmers field & 12% in non IPM farmer field	Increased 20% in FFS farmers field & 15% in non-IPM farmers field
Indonesia[13]	Rice	Increased 81% in FFS farmers field & 169% in non-IPM farmers field	Decreased 11% in FFS farmers field & 15 % in non IPM farmer field
Pakistan[14]	Cotton	Decreased 13% in FFS farmers field & increases 21% in non IPM farmer field	Decreased 30% in FFS farmers field &39% in non IPM farmer field
Thailand [14]	Rice	Decreased 58% in FFS farmers field & 10% in non IPM farmer field	-

Sources : FAO (1993)[1], Pincus (1999)[2], Le Toan (2002)[3], Peshin and Kalra (1998)[4], Larsen *et al.* (2002a)[5], Larsen *et al.* (2002b)[6] ,van Duuren (2003)[7] ,SEARCA (1999)[8],van den Berg *et al.* (2002)[9] ,Tripp *et al.* (2005)[10] NAESC (2003)[11], Mancini, *et al.* (2006)[12], Feder *et al.* (2004a)[13],Khan *et al.* (2005)[14] Modified after van den Berg and Jiggins (2007).

Impact / Outcome of IPM Programmes in Developed Countries

Pesticides remain the mainstay of many IPM programmes throughout the globe. In developed nations like USA and European Union, there is government legislation and mechanisms for implementing and evaluation of IPM programmes, especially in Europe, where IPM innovation systems involving the government, researchers, farmers, advisory agencies and market forces are part of a system to reduce pesticide use. The European countries provide incentives to the growers for compliance with IPM tactics to reduce pesticide use. The European commission considered levying taxes on plant protection products to encourage pesticide free or low pesticide farming. Norway, Denmark and Sweden have levied taxes on pesticides which results in 67% reduction in pesticide use (Table 5).

Table 5 : Impact of IPM in Europe

Country	Crop	Insecticide use % (a.i)	Farmers adopted %
Austria	Pome fruits	-	51%
Belgium	Pome fruits	-	31%
Denmark	All fruits	Pesticide use frequency reduces from 3.1 to 2.1	-
Germany	Pome fruits	-	17%
Netherland	All major crops	Decreases (57%)	-
Norway	-	Decreases (54%)	-
Switzerland	Pome fruits	-	39%
Switzerland	Cereals, rapeseed, potato, fruits	Decreases (40%)	18000 farmers members of IP-SUISSE and 3000 out of 4000 fruit producers grow IP-SUISSE fruits Apple-92%, Strawberrie-85%, Raspberries-70%
UK	Carrot, potato, cauliflower, mushroom, avocado, pineapple	50% reduction in pesticide use	-

Source : Cross *et al.*, 1995, PAN Europe, 2004, 2007, Neumeister, 2007, IP SUISSE, 20005, 2006, Cannell, 2007.

An overview of impact/outcome of IPM programme in the USA

The synthesis of the published impact studies of different IPM programmes in the USA were presented by Fernandez-Cornejo at the 5th National IPM Symposium in the USA in 2006 (Table 6). These studies used economic indicators for measuring the impact of IPM by employing ex-post facto design.

Table 6 : Impacts of IPM on Pesticide Use, Yields and Net Returns in U.S.A

Crop	Pesticide use	Range%	Yield per acre	Net return per acre
Cotton	Increased	-64 to +92	Increased	Increased
Cotton	Decreased	-98 to +34	Increased	Increased
Soybeans	Decreased	-21 to +83	Increased	Increased
Soybeans	Decreased	-100 to +85	NA	Increased
Corn	Increased	+15 to +47	Increased	Increased
Corn	Decreased	-15 to +67	Increased	NA
Peanuts	Decreased	-81 to +177	Increased	Increased
Fruits/Nuts	Decreased	-43 to +24	Increased	Increased
Fruits/Nuts	Decreased	-41 to +37	Same or increased	Same or increased
Vegetables	Decreased	-67 to +13	Same	Increased

Source : Presentation by J. Fernandez-Cornejo at the 5Th National IPM Symposium, St Louis, Missouri, April 4, 2006.

Conclusion

The IPM tactics have been developed in the last five decades but the uptake of IPM at the farm level has been slow. There are different extension methodologies for dissemination of the IPM technology. To disseminate IPM to as many farmers as possible, at an affordable cost, a combination of extension methods need to be used, besides the classical FFS model in developing countries. The Cooperative Extension Services in the USA through participatory extension, websites and on demand print documents is successfully disseminating the IPM technology (Giles and Walker, 2009). Besides, government policies and regulations (as in Denmark, Switzerland and USA) are necessary to provide impetus to success of the IPM uptake. The stakeholders (biological scientists) of IPM programmes must incorporate evaluation in their programmes by the scientists well versed with the evaluation methodology to find out the outcome and impact of IPM (Peshin *et al.*, 2009b). The robust methodology for determining the impact of IPM programmes is Longitudinal + Latitudinal (with/without and before/after) design–double diffusion model.

References

Bentley, J.W., Boa, E., Van Mele, P., Almanza, J., Vasquez, D. and Eguino, S. 2003. Going public: A New extension method. *International Journal of Agricultural Sustainability*, 1 (2):108–123.

Bentley, J.W., Nuruzzaman, M., Nawaz, Q.W. and Haque, M.R. 2005. Picture songs. In Van Mele,P., Salahuddin, A. Magor, N.P. pp. 115–123 (Eds.) *Innovations in Rural Extension: Case Studies from Bangladesh*. CABI and IRRI, Wallingford, UK.

Braun, A., Jiggins, J., Roling, N., van den Berg, H. and Snijders, P. 2006. *A Global Survey and Review of Farmer Field School Experiences*. Report Prepared for the International Livestock Research Institute (ILRI), Wageningen, The Netherlands.

Cannell, E. 2007. European farmers plough ahead: Pesticide use reduction. Pesticides News, 78, December 2007. *http://www.pan-Europe.info*.

Cross, J.B., Bonauer, A., Bondio, V., Clemente, J., Denis, J., Grauslund, J., Huguet, C., Jorge, E., Koning, S., Kvale, A., Malavota, C., Marcelle, R., Morandell, I., Oberhofer, H., Pontalti, M., Polesny, F., Rossini, M., Schenk, A., de Schaetzen, C. and Vilajeliu, M. 1995. *The current status of integrated pome fruit production in western Europe and its achievements*, pp. 1–10. In: Polesny, F., Muller, W. and Olszal, R. W. (eds.) Proceedings of the International Conference on Integrated Fruit Production. Cedzyna, Poland, August 28–September 2, 1995.

Danielsen, S., Boa, E. and Bentley, J.W. 2006. Puesto para Plantas: A Clinic Where You Can Bring Your Sick Plants. Funica, Managua. (available on-line at *http://www.funica.org.ni/docs/* Book-PPP-Eng.pdf, accessed 6 February 2008).

Den Tex, N. 2004. *Amplifying the IPM Message through CARE'S Small Holder Pest Management Programme in Vegetables*. Manilla, Philippines.

FAO, 1993. The impact of IPM training on farmers' behavior. A summary of results from the second field school cycle. *IPM National Programme*. Jakarta: FAO.

FAO, 2004. Benefits from Farmer Education: Impact of IPM-FFS in Cotton in Asia. Cotton IPM Newsletter No. 5, August 2004. FAO, Bangkok, Thailand.

Feder, G., Murgai, R. and Quizon, J.B. 2004a. Sending farmers back to school: The impact of farmer field schools in Indonesia. *Review of Agricultural Economics* 26: 45-62.

Feder, G., Murgai, R and Quizon, J. 2004b. The acquisition and diffusion of knowledge: The case of pest management training in farmer field schools, Indonesia. *Journal of Agricultural Economics* 55: 217-239.

Fuchs, T.W. 2007. Diffusion of IPM programmemes in commercial agriculture: concepts and constraints. In: Koul, O. and Cuperus, G. (Eds). *Ecologically Based Pest Management*. CAB International, Oxfordshire, UK: 432–444.

Ghallanger, K. 2003. Fundamental elements of a farmer field school, *LEISA* March, 2003: 5-7.

Giles, K.L. and Walker, N.R. 2009. *Dissemination and Impact of IPM programmes in US Agriculture*. 29 In: Peshin, R and Dhawan, A.K. (eds) *Integrated Pest Management: Dissemination & Impact*, Vol.2, Springer Verlag.

IP Suisse, 2005. Medienmitteilung, Richtpreise knapp erreicht, Erl¨os jedoch tiefer als imVorjahr, IP SUISSE, 6.6.2005, Zollikofen, Switzerland.

IP Suisse, 2006. Medienmitteilung, IP-SUISSE Labelkartoffeln im Aufwind, IP SUISSE, 30.3.2006, Zollikofen, Switzerland.

Kenmore, P. 1996. Integrated Pest Management in Rice. pp. 76-97. In: Persley, G. (Ed.), *Biotechnology and Integrated Pest Management*. CAB International, Wallingford.

Khan, M.A., Ahmad, I. and Walter-Echols, G. 2005. Impact of an FFS-based IPM approach on farmer capacity, production practices and income: evidence from Pakistan. pp. 45–60. In: Ooi, P. A. C. Praneetvatakul, S. Waibel, H. and Walter-Echols, G. (Eds.) *The impact of the FAO-EU IPM programme for cotton in Asia. Pesticide Policy Project Publication Series*, Special issue no. 9. Hanover, Germany.

Larsen, E.W., Haider, M.L., Roy, M. and Ahamed, F. 2002a. Impact, sustainability and lateral spread of integrated pest management in rice in Bangladesh. *Report SPPS 73*. Dhaka: Department of Agricultural Extension and DANIDA.

Larsen, E.W., Haider, M.L., Roy, M. and Ahamed, F. 2002b. Impact, sustainability and lateral spread of integrated pest management in vegetables in Bangladesh. *Report SPPS 74*. Dhaka: Department of Agricultural Extension and DANIDA.

Le Toan, 2002. IPM impact evaluation of training and development of IPM in tea. *Report. Phu Tho (Vietnam)*: Plant Protection Department.

Mancini, F., Termorshuizen, A.J., and van Bruggen, A.H.C. 2006. Impact of integrated pest management farmer field schools on pesticide use and farmers'ecological knowledge on cotton farming in India. pp. 27–39 In Mancini, F. (Ed.) Impact of integrated pest management farmer field schools on health, farming systems, the environment, and livelihoods of cotton growers in Southern India. *Published doctoral dissertation*, Wageningen University, the Netherlands.

Matteson, P.C. 2000. Insect pest management in Tropical Asian Irrigated Rice. *Annual Review of Entomology*, 45:549–574.

Miller, G.T. 2004. Sustaining the earth, 6th ed. Thompson learning, Inc. Pacific Grove, CA, pp. 211-216.

Mancini, F., Termorshuizen, A.J. and van Bruggen, A.H.C. 2006. Impact of integrated pest management farmer field schools on pesticide use and farmers'ecological knowledge on cotton farming in India. In F. Mancini (Ed.), Impact of integrated pest management farmer field schools on health, farming systems, the environment, and livelihoods of cotton growers in Southern India (pp. 27–39). *Published Doctoral dissertation*, Wageningen University, the Netherlands.

NAESC, 2003. *Report on impact assessment of China/EU/FAO cotton IPM programme in Shandong Province,PR China.* Report prepared for the National Agro-technical Extension and Service Center, Ministry of Agriculture, Beijing.

Neumeister, L. 2007. Pesticide Use Reduction: Strategies in Europe Six Case Studies. PAN Europe, Pesticide Action Network Europe, London, UK. *http://www.pan- Europe.info*

Norton, G.W., Moore, K., Quishpe, D., Barrera, V., Debass, T., Moyo, S. and Taylor, D. 2005. Evaluating socio-economic impacts of IPM. pp. 225–244 In *Globalizing Integrated Pest Management*, Eds. G. W. Norton, E. A. Heinrichs, G. C. Luther, and M. E. Irwin. Blackwell, Ames, IA.

Oklahoma IPM, 2007. IPM: integrated pest management. Oklahoma State University Cooperative Extension. Available at: *http://www.ento.okstate.edu/ipm/.*

Padre, S., Sudarshana, P and Tripp, R. 2003. Reforming farm journalism: The experience of *Adike Pathrike* in India. Overseas Development Institute, London, *Agricultural Research and Extension Network* Paper No. 128.

PAN Europe, 2004. Pesticide Use Reduction is Working: An Assessment of National Reduction Strategies in Denmark, Sweden, the Netherlands and Norway. PAN Europe, London, February 2004, 19pp. Available at *http://www.pan Europe, info/publications/index. html.*

PAN Europe, 2007. Pesticide Use Reduction Strategies in Europe: Six Case Studies. *http://www.pan- Europe. info/Media/PR/080623.html*

Peshin, R. 2005. Evaluation of the dissemination of insecticide resistance management programme in cotton in Punjab. *Ph.D Dissertation, Punjab Agricultural University*, Ludhiana, India.

Peshin, R. and Kalra, R. 1998. Integrated pest management at farmer's level. *Man and Development* 22:137–141.

Peshin, R., Vasanthakumar, J and Kalra, R. 2009a. Diffusion of innovation theory and integrated pest management, pp1-29 In: Peshin, R and Dhawan, A.K.(eds) *Integrated Pest Management: Dissemination & Impact,* Vol.2 Springer, Verlag.

Peshin, R., Jayaratne, K.S.U. and Singh, G. 2009b. Evaluation research: methodologies for evaluation of IPM programmes, In: Peshin, R and Dhawan, A.K.(Eds) *Integrated Pest Management: Dissemination and Impact,* Vol.2 Springer, Verlag.

Peshin, R., Dhawan A.K., Kranthi, K.R. and Singh, K. 2009c. Evaluation of the benefits of an insecticide resistance management programme in Punjab in India, *International Journal of Pest Management*, 55(3):207-220.

Pincus, J. 1999. The impact of IPM farmer field schools on farmers' cultivation practices in their own fields. *Report prepared for the FAO Programme for Community IPM in Asia.*

Powell, E.T., Steel, S. and Douglah, M. 1996. Planning a Programme Evaluation: Programme Development and Evaluation. University of Wisconsin-Extension. *http://learningstore.uwex.edu/pdf/G3658-1.PDF.*

Quizon, J., Feder, G. and Murgai, R. 2001. Fiscal sustainability of agricultural extension: the case of farmer field school approach. *Journal of International Agricultural and Extension Education,* 8: 13-24.

Rogers, E.M. 1962, 1983, 1995, 2003. *Diffusion of Innovation*. The Free Press, New York, USA.

SEARCA, 1999. Integrated pest management training project, World Bank loan 3586-IND: Impact evaluation study. *Report prepared for the SEAMEO Regional Center for Graduate Study and Research in Agriculture*, Laguna, Philippines.

Texas, IPM. 2007. Texas IPM programme. Texas Cooperative Extension, Texas A&M University. Available at: *http://ipm.tamu.edu/.*

Tripp, R., Wijeratne, M., and Piyadasa, V.H. 2005.What should we expect from farmer field schools? *A Sri Lanka case study. World Development*, 33:1705–1720.

Van den Berg. H. 2004. IPM Farmer Field Schools: A Synthesis of 25 Impact Evaluations. Wageningen University, January 2004. Prepared for Global IPM Faculty.

van den Berg, H. and Jiggins, J. 2007. Investing in farmers–the impacts of farmer field schools in relation to integrated pest management, *World Development, 35*:663–686.

van den Berg, H., Senerath, H., and Amarasinghe, L. 2002. Participatory IPM in Sri Lanka: a broad scale and an in-depth impact analysis. *Report prepared for the FAO Programme for Community IPM in Asia, Jakarta.* Summarized in: van den Berg, H., Senerath, H., and Amarasinghe, L., 2003. Farmer field schools in Sri Lanka: Assessing the impact. *Pesticide News*, 61: 14–16.

Van Mele, P., Zakaria, A.K.M. and Bentley, J.W. 2005. Watch and learn: Video education for appropriate technology. Van Mele, P., Salahuddin, A. and Magor, N.P. pp. 77–88. (Eds.) *Innovations in Rural Extension: Case Studies from Bangladesh*. CABI and IRRI, Wallingford, UK.

❑❑❑

Ecologically Based Integrated Pest Management, 2012
© D.P. Abrol and Uma Shankar (eds.), pp. 877-889
New India Publishing Agency, New Delhi (India)
e-mail : info@nipabooks.com; website : www.nipabooks.com

44

Methodological Aspects of IPM Evaluation

S.K. KHER, R. PESHIN AND P.S. SLATHIA

Introduction

Plant protection in India is mainly based on use of pesticidal chemicals. Chemical control is one of the effective and quicker methods in reducing pest population where farmer gets spectacular results within a short time. However, indiscriminate use of pesticides results in a series of problems in the agricultural ecosystem mainly the development of resistance in insects to insecticides, resurgence of treated population, environmental contamination and residue hazards. Destruction of natural enemies of insect-pests, expenses on pesticides etc. contributed to a new way of thinking concerning pest control practices i.e. integrated approach of pest control. IPM was first proposed by Stern and his collegues for integration of biological and chemical control measures. This is not altogether a new concept; it was practiced before the advent of modern chemicals. The date of sowing the crop were carefully studied to ensure that crop was not being planted when it would encounter severe pest problems, cultural practices such as ploughing after harvest, timely weed control, well timed irrigation and reduced use of fertilizers all contributed to reduce pest population. Most of these methods were curtailed when modern pesticides become available it was thought that these chemicals alone could control pests, but now we know that this is not possible and a single method of approach in pest control is not feasible. An integrated approach of pest management (IPM) was devised on the principles of managing the pest rather than eradicating them. IPM is a broad ecological approach for managing pest problem encompassing available methods and techniques such as cultural, biological, mechanical and chemical in a compatible manner to reduce and maintain the pest problem below the economic injury level.

Aims of IPM

- Reduce the use of synthetic pesticides
- Provide consumable safe food
- Minimize risk to human health
- Strengthening agro-eco system

Principles of IPM

- Identification of key pests and beneficial organisms
- Defining the management unit, the agro-eco system
- Development of management strategy
- Establishment of economic threshholds
- Development of assessment techniques
- Evolving description of predictive pest models

Late 1960's – Smith, R.F., Stern, V.M., Bosch, R.V.D. and Harold, T.R. of the University of California introduce the IPM concept as a pest population management system that uses all suitable techniques. These include attractants for monitoring and in baits, sterile male insect release, introduction of natural enemies, resistant species and varieties, and selected pesticides when and as needed.

Late 1980's – Evaluation of IPM programs in 15 states documents benefits and indicates IPM users increase profits by more than $575 million annually compared to nonusers. Growers use IPM successfully on cotton, soybeans, corn, vegetables, hay, and on external parasites of livestock.

In many countries IPM is considered as a reliable alternative to sole dependence on pesticides for the management of pests of different crops. Integrated Pest Management has brought significant improvements in enhancing in crop yield, savings in pesticides cost. The major tool of the IPM is Farmers Field School (FFS). FFS was started in the year 1989 in Indonesia to reduce farmer's reliance on pesticides in rice. FFS is basically a community based hands-on practically oriented field study programme. FFS provides opportunity to the farmers to learn together using interesting hands-on methods of discovery learning. FFS works on the principles

- Conserve natural enemies
- Grow healthy crop
- Conduct regular field observations

IPM programmes have been implemented throughout the world since the last 4 decades. The primary purpose of the programme is to reduce the use of pesticides by adopting alternate management practices. As such, the evaluation of IPM programmes is vital for making proper programmatic decisions. The purpose of IPM programme evaluation is to seek information needed for programme improvement and weighing the results and consequences.

Evaluation of IPM

A regular evaluation of IPM is essential to determine the success of pest management strategies, to assess its productivity, effectiveness and efficiencies and to evaluate its achievements. Evaluation is a method for determining how far an activity has progressed and how much further it should be carried to accomplish objectives. (Sandu, 2003).

"Evaluation is the systematic assessment of the worth or merit of some object or value of things" (Scriven, 1967).

Steps in Evaluation

Although each definition and in fact each evaluation is slightly different, there are several different steps that are usually followed in any evaluation. An overview of the steps of a typical evaluation follows:

Benefits of Evaluation

a. Evaluation gives direction to continued improvement in programming
b. It provides fresh data regarding situations essential for improving programming functions.
c. It helps to determine the degree to which specific objectives are being achieved and also helps to clarify these objectives.
d. It provides realistic information to report to the public, parliament and legislative bodies.
e. It enhances our knowledge.

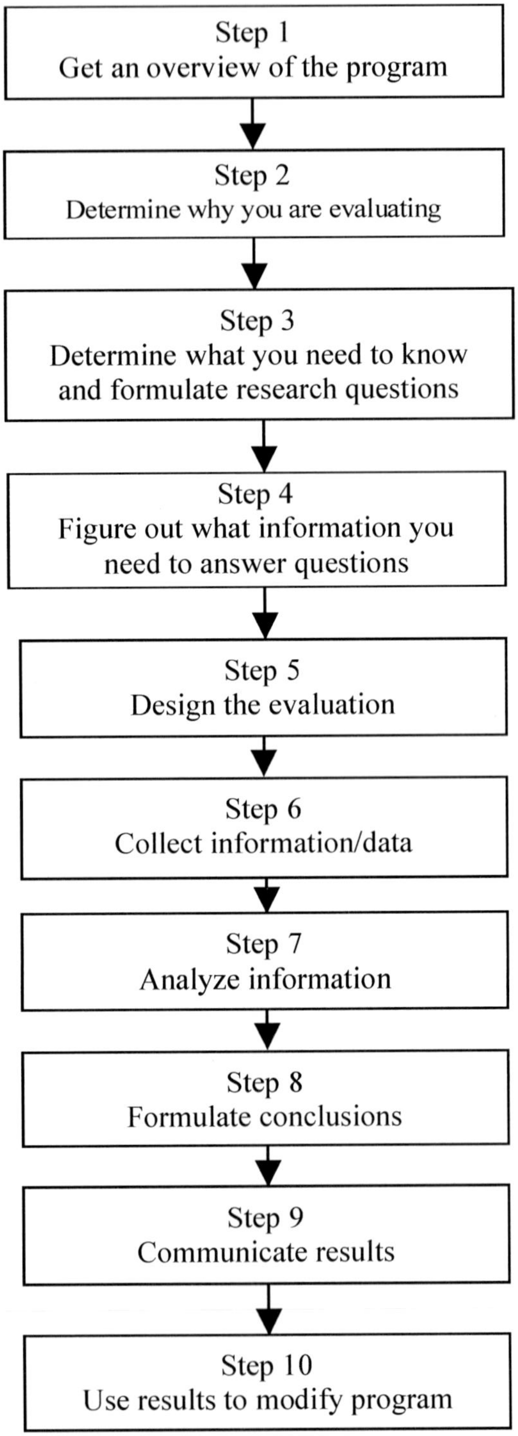

Source : http://www.ed.gov/pubs/ed tech guide/whatseval.html.

f. It helps to determine the short term and long term impact of extension programmes in terms of social and economic dimensions (Sandhu, 2003).

Types of Evaluation

There are different types of evaluations depending on the object being evaluated and the purpose of the evaluation. Perhaps the most basic distinction in evaluation type is that between formative and summative evaluation.

Formative evaluation is conducted to provide program staff with judgments useful in improving the program whereas summative evaluation is generally conducted after completion of the program (or when a program has stabilized) and for the benefit of some external audience or decision maker. The findings from summative evaluation could be used to decide whether to continue the program or not. The main difference is that the aim of summative evaluation is to report on the program whereas a formative evaluation reports to the program.

Formative Evaluation includes Several Evaluation Types

- Needs assessment determines who needs the program, how great the need is, and what might work to meet the need.
- Evaluability assessment determines whether an evaluation is feasible and how stakeholders can help shape its usefulness.
- Structured conceptualization helps stakeholders define the program or technology, the target population, and the possible outcomes.
- Implementation evaluation monitors the fidelity of the program or technology delivery.
- Process evaluation investigates the process of delivering the program or technology, including alternative delivery procedures.

Summative Evaluation can also be Subdivided

- Outcome evaluations investigate whether the program or technology caused demonstrable effects on specifically defined target outcomes.
- Impact evaluation is broader and assesses the overall or net effects – intended or unintended – of the program or technology as a whole.
- Cost-effectiveness and cost-benefit analysis address questions of efficiency by standardizing outcomes in terms of their dollar costs and values.
- Secondary analysis reexamines existing data to address new questions or use methods not previously employed.

- Meta-analysis integrates the outcome estimates from multiple studies to arrive at an overall or summary judgment on an evaluation question. (Trochim, 2006).

Methods of Evaluation

Methods of evaluation as described by Dart and Petheram (1998) are given as under:

1. *Case Study Evaluation*: A method at the opposite end of the spectrum to the large scale survey. It substitutes rich details of a few cases for evaluative conclusions. Can provide valuable insight, especially when used in combination with survey data.
2. *Experimental Evaluation Design*: Involves pre-and-post test designs. Involves randomly assigning program participants to groups to minimize bias so that statistical analysis can be conducted.
3. *Quasi experimental designs*: Used when not possible to randomly select the sample population.
4. *Delphi technique*: A form of interactive (postal) surveying that utilizes iterative questionnaires and feedback approach, and provides participants with an opportunity to revise earlier views based on the response of other participants until some desired level of consensus is reached. Valuable when participants cannot meet face to face.
5. *Focussed Groups*: gather people together to discuss a topic, usually in the presence of a facilitator to pose questions to get the ball rolling and to help the discussion stay on track.
6. *Goal Attainment Scaling*: Involves the development of outcome scales specifically tailored to the individual, group, agency or system whose program is to be measured.
7. *Participant Observation*: A form of inquiry in which the inquirer (the observer) plays two roles-firstly as an observer and secondly as a genuine participant, having a stake in the group's activities.
8. *RRA (Rapid Rural Appraisal):* generally a repertoire of techniques aimed at collecting information efficiently and involving use of innovative and visual techniques.
9. *PRA (Participatory Rural Appraisal):* uses similar methods to RRA, but the participants become more involved in the inquiry. A main difference is in the behaviour of the researchers. In PRA the researchers aims to facilitate for participants to elaborate their own research and hands over the stick. The contention is that this process can be empowering to the participants.

10. *PLA (Participatory Learning and Action):* A concept that evolved from PRA. It differs from PRA in that there is greater emphasis on action-on the basis that it is not enough to simply appraise a rural situation but it is also necessary to facilitate action.
11. *Pre-post Design*: Used when a pre-treatment measure can supply useful information. Many evaluation designs use pre-tests as benchmarks to report the change that has occured from before the program to its conclusion. The danger is that, unless a control group is considered, there is no means of differentiating between change that occurs in the group naturally and as a a result of the program (i.e. through maturation, other learning and intervening events).
12. *Qualitative methods in General*: Include naturalistic inquiry, open ended questions, participant observation.
13. *Questionnaires*: Used in formal surveys, or in interviews, or as 'interview guides' to informal surveys. Can be developed to explore or measure attitudes, opinions, behaviour, level of knowledge, life circumstances or other issues.
14. *Semi-structured interviews*: Used in informal guided interviews where only some questions are predetermined and new questions are developed from discussion. An interview guide is used to ensure that the same format and topics are covered with each respondent.

Economic Evaluation of IPM Programmes

- *IPM evaluation at farm level:* Intervention employed at farm level are based on the focus to improve profitability or risk reduction of a particular farm operation which fits well for profit maximization and risk reduction from farming. Owing to this reason most of the evaluation efforts have focussed largely on assessing profits and risk on the farms concerning with IPM programmes. Impact on profits has largely been measured with budgeting techniques. Many studies have measured profit to the difference between gross revenue and cost of adopting IPM intervention. Napit *et al.*, 1988 evaluated the economic benefits of extension integrated IPM programmes by using budgeting techniques for different crops. They used net returns as a measure of profits and used T-test for measuring significant difference to management per unit of land and costs of pesticides. Trumble *et al.*, 1997also used the budgeting technique for economic and environmental impact analysis of IPM programmes. Net profits for different treatments were estimated by using partial budgets. The evaluation established that IPM programme caused significant reduction in pest population and rise in average yields and net profits. A rise in net profit by more than US $410 per ha was achieved by the programme with 40% less use of conventional pesticides.

- *IPM evaluation at aggregate level:* There are two broad approaches of impact evaluation of IPM programmes at aggregate level. These are economic surplus approach and econ-metric approach. Economic surplus approach is widely used in both ex-ante and ex-post evaluation of economic benefits. The supply seems to be shift up as a result of adoption of IPM programme producing surplus for both producer and consumer. The approach tends to measure such surplus resulting from adoption. Economic surplus further tends to measure the welfare impact of the programme.

Models of Evaluation

- *Bennett's hierarchy*: A framework that sets out the hierarchical goals of agricultural extension. It can be a very useful tool for clarifying what aspects of a program are to be evaluated.
- *CIPP*: A systems based framework used in planning evaluation, rather than a method. It is based on the assumption that it is better to improve than to prove. The 4 approaches to evaluation (context, input, process and product) bear some resemblance to owen's 5 form of evaluation.
- *Logical Frameworks*: are used for both planning projects and monitoring activities. Frequently used in social and natural resource projects in developing countries.
    - Owen's five forms of program evaluation: A meta model of approaches to evaluation useful in planning. The 5 categories of evaluation are:
    - Evaluation for program development,
    - Evaluation for program design clarification,
    - Process evaluation
    - Evaluation in program management and Impact evaluation (Owen, 1993)
- Synder Model: A systems-based approach to program evaluation. A main feature is that outputs are divided into 3 time perspectives:
- Immediate effects of project activities
- Targets (present goals)
- Vision or ultimate goal

Indicators for IPM Evaluation

Some of the important indicators to evaluate IPM program are:

a. Knowledge gain of the farmers
b. Extent and level of adoption of IPM
c. Reduction in pesticide use
d. Reduction in pesticide expenditure
e. Economic benefits of IPM programmes
f. Attitude of farmers towards IPM
g. Constraint in adoption of IPM

Table : Evaluation indicators of IPM programmes

Area	Design	Indicators	Reference
Indonesia	Case study with/without and before after	Reaction of FFS participants, knowledge gain, reduction in pesticide use, increased yield	Van de Fliert, 1993
Indonesia	Sustainable livelihood analysis and participatory research process	Organizational indicator: Participation, Cohesiveness, common goal, time management, plan to follow, leadership	Jiggins, 2002
India (A.P.)	Before/After with control farmers and control village	Change in agronomic practices, input use and farmer's attitude towards pest management, knowledge, decision making	Mancini et. al, 2007
Asia	With/Without	1. Knowledge about ecology 2. More sustainable production 3. Experimentation skills improved livelihoods 4. Improved crop management ability to deal with risks, opportunities 5. Pesticide reduction innovation 6. Yield increase more cost-effective production 7. Risk reduction improved biodiversity 8. Reduced pesticide poisoning	Van den Berg and Jiggins 2007
India	With/without	Natural capital, Human capital, social capital, financial capital, physical capital	Mancini F. 2006

contd...

contd...

India	Before/ After, With/ without	Knowledge gain and analytical skills, adoption, adoptability, reduction in pesticide use, pesticide expenditure and number of sprays.	Peshin, R.2005
Asia	Cost benefit analysis at farm level and at program level	Farmers awareness, knowledge, efficient farm input use, increase in yield, lower pesticide use and net profit of farmers	Feder and Quizon 1999
Pakistan	Single difference comparison (between trained and untrained) and difference in difference model (FFS farmers, exposed and unexposed farmers)	Decision making skills, field experiments, observed biodiversity, attitude environment, social recognition, input use, revenue and gross margin	Khan et. al, 2005

Impact of IPM

Birthal and Kumar (2005) in their economic evolution of IPM in cotton concluded that IPM appears to be an effective alternative to chemical pest control. They measured the Technical potential of IPM in terms of reduction in pest use, efficacy of pest suppression and conservation of natural enemies' of insect pests. The cumulative effect of these is reflected in the additional yield of IPM farms over non-IPM farms. The mean values of these parameters are presented in the table.

The IPM CRSP Phillipines program evaluate the impact of IPM program in onions which results that yields of onions have increased and production costs reduced in those areas that adopted IPM technology. The onion IPM program reduced pesticide use for adopters resulting in health related, environmental and economic benefits.

Economic impact estimates for onion IPM indicate that the adoption of IPM technologies results in a 30% reduction in pesticide sprays and input cost and a 50% decrease in herbicide use and hand weeding activities along with an estimated average yield increase of 10%. Farmers who have adopted the practices have increased their net income upto 25%.

Table 1 : Pesticide use, yield and incidence of pests and their natural enemies under IPM and non-IPM situations.

Items	Non-IPM	IPM
• Pesticid use (a.i.kg/ha)e	3.16	0.014
• Crop yield (kg/ha)	880	1090
• Pest infestation (no's per 3 leaves)[1]		
Sucking pests		
Jassids	3.5	2.1
Aphids	21.6	6.8
Thirps	1.0	0.9
Whitefly	0.1	0.1
Bollworms		
American	0.23	0.13
Spotted	0.14	0.09
Pink	0.30	0.17
• Natural enemies (no's per 3 leaves)[1]		
Coccinellid grubs	0.1	0.3
Coccinellid adults	0.7	2.7
Chrysopa eggs	0.2	0.5
Chrysopa adults	0.0	0.1

1-Observations are from 10 plots each from IPM and non-IPM villages. [Birthal and Kumar (2005)]

Table 2 : Economic impact of onion IPM program in Phillipines

Impact Indicators	Decrease %	Increase %
1. Reduction in pesticide sprays and input cost	30	-
2. Herbicide use and hand weeding	50	-
3. Increase in yield	-	10
4. Increase in income	-	25

Conclusion

Evaluation is an integral part of any developmental programme and it should be planned at the program design stage to ensure that evaluation will be useful for programme management and accountability. Evaluation envisages desired as well as undesired consequences of new technologies. The worth of IPM technologies not only measured in terms of adoption of IPM practices, reduction in pesticide use and increase in yields but it also should consider all the aspects related to socio-economic development, health, sustainability, environment conservation etc. which will represent the true wealth of IPM programmes. Further, evaluation of IPM programmes provides feedback to stakeholders for necessary improvements if required. Evaluation can be utilized to make IPM practices compatible to fit the farming system.

References

Birthal, P.S. and Kumar, S. 2005. Economic evaluation of integrated pest management in cotton. In: Joshi, P. K., Birthal, P. S. and Bantilam, H. C. S. (Eds.) Impact of Agricultural Research: Post Green Revolution evidence from India, National Centre for Agriculture Economics and Policy Research (NCAP), New Delhi, India.

Dart, J. and Petheram, R.J. 1998. Review of evaluation in agriculture extension, RIRDC.

Feder, G., Quizon, J.B. 1998. Integrated pest management in Asia: Are there real returns to IPM and its diffusion? In Waibel, H., Fleischer, G., Kenmore, P.E., and Feder, G. (eds.) Evaluation of IPM programmes-Concepts and methodologies. A publication of pesticide policy project Hannover, April 1999. Publication series No. 8. Publication No. 98/136, Australia.

Jiggins, J.L.S. 2002. New approaches to evaluation. International learning workshop on farmer field schools: Emerging issues and challenges, 21-25, Oct. 2002, Yogyakarta.

Khan, M.A. and Ahmad, I. 2005. Impact of FFS based IPM knowledge and practices on rural poverty reduction: Evidence from Pakistan. In: Ooi, P.A.C., Praneetvatakul, S., Waibel, H. and Echols, G. W.(eds.) Imapct of FAO-EU IPM program for cotton in Asia. A publication of the pesticide policy project Hannover, Special issue publication series, No.9.

Mancini, F., Van Bruggen, A.H.C. and Jiggins, J.L.S. 2006. Evaluating cotton integrated pest management (IPM) Farmer Field School noutcomes using the sustainable livelihood approach in India. *Experimental Agriculture* 43:97-112.

Napit, K.B., Norton, G.W., Kazmierczak, R.F. Jr. and Rajjot, E.G. 1988. Economic impacts of extension integrated pest management programs in several states. *Journal of Economic Entomology* 81:251-56.

Owen, J.M. 1993. Program Evaluation, Forms and Approaches, Allen and Unwin, St. Leonards, NSW, Australia.

Peshin, R., Kalra, R, Dhawan, A.K. and Tripat, K. 2007. Evaluation of insecticide resistance management program. *AI and Society* 21: 357-381.

Sandhu, A.S. 2003. Extension programme planning. Oxford and IBH Publishing Co. Pvt. Ltd., New Delhi.

Scriven, M. 1967. The methodology of evaluation. In stake, L. (eds.) *Curricullum Evaluation.* Rand McNally, Chicago, USA.

Scriven, M. 1991. Evaluation Thesaurus (4th edition), Newbury park, Sage.

Trumble, J.T., Carson, W.G. and Kund, G.S. 1997. Economic and environmental impact of sustainable integrated pest management program in Celery. *Journal of Economic Entomology* 90(1):139-146.

Van den Berg, H. Senerath, H. and Amarasinghe, L. 2002. Participatory IPM in Sri Lanka. Brofor community IPM in Asia. Summary published as: Farmer Field School in Sri Lanka: Assessing the impact. *Pesticide News* 61 (2003), 14-16 ad. Scale and in-depth imapct analysis, FAO programme.

Vandeman, A., Fernandez Cornejo, J., Jans, S. and Lin, B. 1994. Adoption of integrated pest management in US Agriculture, United States, Department of Agriculture Economic Research Service. Agriculture Information Bulletin 707. Washington, D.C. USA.

Vatta, K., Dhawan, A. and Peshin, R. 2009. Economic evaluation of integrated pest management programmes. In: Integrated pest management: *Dissemination and Impact*. Vol 2, 103-117.

□□□

Ecologically Based Integrated Pest Management, 2012
© D.P. Abrol and Uma Shankar (eds.), pp. 891-901
New India Publishing Agency, New Delhi (India)
e-mail : info@nipabooks.com; website : www.nipabooks.com

45

IPR Issues in Agriculture

SUDHAKAR DWIVEDI

The World Trade Organization (WTO) is an international organization designed to supervise and liberalize international trade. It was established on 1st January 2005 and is responsible for making and enforcing rules for trade between nations. WTO marks a major change in global trade rules. As an organization, it replaces the General Agreement on Tariffs and Trade (GATT), which had been in existence since 1947. The Eighth Round of Multilateral Trade Negotiations under GATT, which started in Uruguay in 1986, was concluded in 1994, leading to the creation of WTO as the new permanent international trade organization. The role of WTO is much more extensive than that of GATT, which dealt with trade in goods. Apart from goods, the two other broad areas that WTO covers are services and intellectual property, which previously belonged to the domestic domain. Accordingly, WTO administers not only the Multilateral Trade Agreements (MTAs) in goods but also the General Agreement on Trade in Services (GATS) and the Agreement on Trade Related Aspects of Intellectual Property Rights (TRIPS), which came into existence with WTO. All the agreements annexed to the Agreement establishing the WTO were signed as part of a package deal. Member countries did not have the option of choosing some and rejecting others. Another important difference with the erstwhile GATT is that WTO has a stronger compliance mechanism than did GATT-a member's failure to meet the obligations can invoke retaliation across agreements and sectors.

GATT	Uruguay Round	WTO
(1947)	(1986-94)	(1995)

As one of the WTO agreements, TRIPS is binding on all member countries of WTO. TRIPS aims at establishing strong minimum standards for intellectual property rights (IPRs). Apart from patents, intellectual property includes copyrights, trademarks, geographical indications, industrial designs, integrated circuits and trade secrets. The protection of IPRs is binding and legally enforceable.

All the agreements of WTO, including TRIPS, came into force on January 01, 1995. But Article 65.2 of TRIPS permits developing countries such as India, a transition period of five years to implement the provisions of TRIPS. In addition, if a country did not provide product patent protection in any field when TRIPS came into force, then under Article 65.4, she gets another five years (in addition to the five years permissible under Article 65.2) to introduce such protection (Anonymous, 1998).

Intellectual property (IP) is a product of the mind. Intellectual property is intangible in contrast to real property (land) or physical property, which one can see, feel and use. With any type of property there are property rights. When IPs are expressed in a tangible form, they can also be protected. Intellectual property rights (IPRs) have been created to protect the right of individuals to enjoy their creations and discoveries. In fact, IPRs can be traced back to the fourteenth century, when European monarchs granted proprietary rights to writers for their literary works (Menon, Usha).

IPRs have been created to ensure protection against unfair trade practice. Owners of IP are granted protection by a state and/or country under varying conditions and periods of time. This protection includes the right to: (i) defend their rights to the property they have created; (ii) prevent others from taking advantage of their ingenuity; (iii) encourage their continuing innovativeness and creativity; and (iv) assure the world a flow of useful, informative and intellectual works.

Protection of Intellectual Property

Intellectual property is protected and governed by appropriate national legislations. The national legislations specifically describe the inventions, which are the subject matter of protection and those, which are excluded from protection. For example, methods of treatment of the humans or animals by surgery or therapy, inventions whose use would be contrary to law or morality, or inventions which are injurious to public health are excluded from patentability in the Indian legislation.

World Organizations

WTO	*World Trade Organization*
GATT	*General Agreement on Tariffs and Trade*
WIPO	*World Intellectual Property Organization*
UPOV	*International Union for Protection of New Plant varieties*
TRIPs	*Trade Related Aspects of Intellectual Property Rights*

GATT was framed in 1948 and was meant to be a temporary arrangement to settle disputes, amicably among countries regarding who gets what share of the world trade. GATT serves as a code of rules for international trade and a forum to discuss and find solutions to trade problems of the member countries. Agriculture was included for the first time in the 8th round of GATT negotiations in 1986 at Uruguay. The Uruguay round discussions were started in 1986, but successfully concluded on December 15, 1993. The negotiations were signed in the form of an accord on April 15, 1994 at Marrakesh, Morocco, by about 124 countries and lead to the formation of World Trade Organization (WTO) on 1st January 1995.

Under WTO obligations each member country is supposed to provide for a minimum prescribed level of protection to Intellectual Property Rights (IPRs). The purpose behind such a commitment is to facilitate cross border enforcement of such rights. The Agreement on Trade –Related Aspects of Intellectual Property Rights (TRIPS) expects minimum standards to be adopted by the member countries for protection of IPRs.

The Trips agreement is based on the basic principle of the other WTO Agreements, like non-discrimination clauses- National treatment and Most Favoured Nation Treatment and are intended to promote "technological innovation" and "transfer and dissemination" of technology. It also recognizes the special needs of the least developed member countries in respect of providing maximum flexibility in the domestic implementation of laws and regulations.

The following are the IPRs , which are sought to be protected under WTO obligations (Chawla and Singh, 2005).

1. Trademarks
2. Copyrights and related rights
3. Patents
4. Trade secrets
5. Geographical indications
6. Designs
7. Layout designs of integrated circuits

1. Trademarks

A trademark is a distinctive sign which identifies certain goods or services as those produced or provided by a specific person or enterprise. Its origin dates back to ancient times, when craftsmen reproduced their signatures, or "marks" on their artistic or utilitarian products. In research, laboratory equipments bear trademarks that are well known to workers in their field. Trademark law, unlike patent or copyright law, confers a perpetual right. So long as the trademark continues to identify a single source, anyone who uses a very similar mark may be liable for trademark infringement. The perpetual right of trademarks depends on the use. A trademark provides protection to the owner of the mark by ensuring the exclusive right to use it to identify goods or services, or to authorize another to use it in return for a payment.

Trademark rights are so important that multinational companies spend large amount of money to maintain their respective trademarks around the world. Every country has different trademark laws. However, there are agreements to ensure that a company's trademark in one country is protected in another country. India has a Trade Marks Act, 1999. In this category there are certification marks like AGMARK, FPO, ISO etc. service marks viz. LIC, SBI, PNB, etc. and collective marks viz. INTUC, AA, etc.

2. Copyright and Related Rights

Copyright is a right given by the law to creators of literary, dramatic, musical and artistic works and producers of cinematograph films and sound recordings. It is a bundle of rights including, *inter alia,* rights of reproduction, communication to the public, adaptation and translation of the work. There could be slight variations in the composition of the rights depending on the work.

Copyright was created to provide protection to composers, writers, authors and artists to protect their original works against those who copy; that who take and use the form in which the original work was expressed by the author. Computer software/program is another mode of expression. A computer program is produced by one or more human authors but, in its final mode or form of expression it can be understood directly only by a machine (the computer) not by human readers. In India, The Copyright Act 1957 as amended in 1994 is in force. The Copyright protection of computer software is under the Information Technology Act, 2000 (Anonymous, 1993).

3. Patent

A patent is an exclusive right granted by a country to the inventor to make, use, manufacture and market the invention that satisfies the conditions of novelty, innovativeness and usefulness. A patent is a personal property, which can be licensed or sold like any other property. In India, The Indian Patents Act of 1970 amendment 2005 allows now product patent for food, agrochemicals and drugs. The duration of the patent in India is 20 years (Chawla, 2005).

There are three criteria to issue a patent.

i. *Novelty*: The inventor must establish that the invention is new or novel. The novelty requirement refers to the prior existence of an invention. If an invention is identical to an already patented invention, the novelty requirement is not met, so a patent cannot be issued.

ii. *Inventive*: It is invention and not merely discovery. It is non obvious to one skilled in the field. The non-obvious requirement refers to the level of difficulty required to invent the technology. If the invention is so obvious that anyone having an ordinary skill would have thought of it, then it does not meet this requirement.

iii. *Industrial application*: It has a utility or is useful for the society. The useful requirement refers to the practical use of invention. If the invention provides a product that is required or needed in some manner, then it meets this requirement.

In the patent adequate disclosure should be made so that others can also work on it. It should have the features: i) be a written description; ii) enables other persons to follow; iii) adequate and iv) deposit mechanism.

The purpose of a patent is to promote the progress of science and useful arts. The patent law promotes this progress by giving the inventor the right of exclusion. In exchange for this right to exclude others, the inventor must disclose all details describing the invention, so that when the patent period expires, the public may have the opportunity to develop and profit from the use of invention. A patent is enforced in the country, which issues it. For each country a separate application is to be filed in that country where protection is sought.

4. Trade Secret

A Trade Secret or undisclosed information is any information that has been intentionally treated as secret and is capable of commercial application with an economic interest. It protects information that confers a competitive advantage to those who possess such information, provided such information is not readily available with or discernible by the competitors. They include technical data, internal processes, methodologies, survey methods used by professionals, recipes, a new invention for which a patent application has not yet been filed, list of customers, process of manufacture, techniques, formulae, drawings, training material, source code etc.

Trade secrets have an unlimited term, which may be perpetual. The term is as long as it takes the public or a competitor to determine how to make the product and to ascertain the nature and identity of the trade secret. Disclosure of a trade secret and its unauthorized use can be punished by the court and the owner may be allowed compensation.

5. Geographical Indications

Geographical Indications (GIs) of goods are defined as that aspect of intellectual property which refers to the geographical indication referring to a country or to a place situated therein as being the country or place of origin of that product. Typically, such a name conveys an assurance of quality and distinctiveness which is essentially attributable to the fact of its origin in that defined geographical locality, region or country. Products such as Scotch whisky, Champagne, California wines etc. fall under this category. Recent instances of violations of possible geographical indications, is the case of basmati rice of India and Pakistan origin.

GIs can be used, when protected in legal terms, to prevent others from misleading the public or creating unfair competition in trade for the goods that have their origin from a particular territory, region or locality. In the TRIPs there is an important provision that if some one uses a GI as a trademark and it misleads the public as to the true place of origin, then registration of trademark is refused or invalidated *ex officio*.

In India, the Geographical Indications of Goods (Registration and Protection) Act, 1999_came into force with effect from 15th September 2003. This Act seeks to provide for the registration and protection of Geographical Indications relating to goods in India. The Controller General of Patents, Designs and Trade Marks is also the registrar for the Geographical Indications, and the Geographical Indications Registry is located at Chennai. A law to the effect in the country was necessary because other member countries would be obliged to give protection to the goods only if there are laws to protect the goods in the country of origin. The Indian GI Act 1999 means an indication in relation to goods, which identifies such goods (agricultural goods, natural goods or manufactured goods or any goods of handicrafts or of industry, including food stuff) as originating, or manufactured in the territory of country, or a region or locality in that territory as the case may be, where a given quality, reputation or other characteristics of such goods is essentially attributable to its geographical origin. GI cannot be a trademark, the only exception is that trademark held prior to the Act or in good faith. According to GI an indication includes any name (excluding the name of origin country, region or locality), geographical or figurative representation or any combination of them.

6. Designs

Designs are an element of intellectual property. Designs refer to creative activity, which result in the ornamental or formal appearance of a product. Design rights refer to a novel or original design that is accorded to the proprietor of a validly registered design. But it does not include any mode or principle or construction or anything which is in substance a mere mechanical device.

The expression 'Designs' means only the feature of shape, configuration, pattern or ornament applied to any article by any industrial process or means whether manual, mechanical or chemical, separate or combined, which in the finished article appeal to end are judged solely by the eye. Design means the features of shape etc. applied to an article and not the article itself. The features are conceived in the author's intellect. S/he gives those ideas conceived by her/him a material (visual) form as a pictorial illustration or as a specimen, prototype or model. By registration under The Designs Act, 1911 has been amended and now Designs Act, 2000 is in force in India, in which the features are protected as design. The act confers exclusive right to apply a design to any article in any class in which the design is registered. To qualify for registration the design should fulfill the following conditions:

i. it must be a new or original design.

ii. it must not have been previously published in India.

7. Layout Designs of Integrated Circuits

The basis for protecting integrated circuit designs (Topographies) in the TRIPS Agreement is the Washington Treaty on Intellectual Property in Respect of Integrated Circuits, 1989. India is a signatory to this international agreement. In India, the IPRs on the layout designs of integrated circuits are governed by the Semiconductor Integrated Circuits Layout-Design Act, 2000.

Under this Act, a layout-design shall be considered original if it is the result of its creator's own intellectual efforts and is not commonly known to the creators of layout-designs and manufacturers of semiconductor integrated circuits at the time of its creation. But a layout-design, which is not original, or which has been commercially exploited anywhere in India or in a convention country; or which is not inherently distinctive; or which is not inherently capable of being distinguishable from any other registered layout-design, shall not be registered as a layout-design. But if a layout-design which has been commercially exploited for not more than two years from the date on which an application for its registration has been filed either in India or in a convention country shall be treated as not having been commercially exploited.

Integrated circuit topographies are the 3D configurations of electronic circuits embodied in integrated circuits, products or layout designs. Today they are at the heart of modern information, communication, entertainment, manufacturing, medical and space technologies. Such technologies have a direct of indirect impact on agriculture sector. In India layout designs of the Integrated Circuits are covered under the Semiconductor Integrated Circuits Layout Design Act, 2000 (Sahal, 1992).

Plant Variety Protection

Patents are probably the most important IPR today for agricultural goods and services as they provide, wherever these are available, the strongest protection for patentable plants and animals and biotechnological processes for their production. Trade related intellectual property rights (TRIPs), one of the agreements bound to the WTO, mandate the countries signatory to it, to put in place all legal and administrative changes required on IPR domains by January 2000 and 2005 (Rao, 1997). Under this commitment, WTO member countries are required to grant patent to all inventions, whether processes or products, in all fields of technology (including agriculture). TRIPs further demand India, as also any other member country, to grant patent to microorganisms, microbiological and non-biological processes and plants and animals produced by non-biological processes. Thus it implies that member countries should protect plant varieties either by patent, or an effective system of *sui generis* protection, or a combination of these two. In this context India initiated its legislative process to protect plant varieties. Though all the three options i) patent, ii) *sui generis*, and iii) patent and *sui generis* were available, India chose a *sui generis* system for protection of plant varieties. An Act named as Protection of Plant Varieties and Farmers' Rights (PPVFR) Act 2001 has been passed. The PPVFR Act 2001 provides protection to following types of plant varieties.

i. Newly bred varieties
ii. Extant varieties - The varieties which were released under Indian Seed Act, 1966 and have not completed 15 years as on the date of application for their protection.
iii. Farmer's varieties - The varieties which have been traditionally cultivated, including landraces and their wild relatives which are in common knowledge, as well as those evolved by farmers
iv. Essentially derived varieties
v. Transgenic varieties

The PPVFR Act is notable for farmer's rights and integration of the some of the important elements of CBD, such as prior informed consent for the use of Indian genetic resources, the concept of benefit sharing and creation of gene fund for strengthening agro biodiversity conservation. The farmers right in this act provides right to register farmers varieties, right to claim compensation for under performance of a protected variety from the promised level, benefit sharing for use of biodiversity conserved by farming community, protection from legal proceedings of any alleged infringement. According to the concept of benefit sharing, whenever a variety submitted for protection is bred with the possible use of a landrace, extant variety or farmer's variety, a claim can be referred either on behalf of the local community or institution for a share of the royalty (Gupta, 1999).

Convention on Biodiversity

The convention on Biological Diversity (CBD) ensures conservation and sustainable use of biological diversity, besides fair and equitable sharing of benefits from their utilization. In June 1992, 170 countries met in Rio-de-Janerio to discuss the details of a proposed Biodiversity treaty under the Biodiversity convention. Though US has not signed this treaty but this convention on Biodiversity (CBD) dealt with the sovereign rights of the nations over their genetic resources. This CBD has come into force in December 1993 as the first legal mechanism dealing with biological resources (Sahai, 1994). In respect of biological resources, earlier status according to FAO International Undertaking on Plant Genetic Resources, a non-legally binding mechanism that had adhered to since inception in 1983 that 'plant genetic resources are the heritage of mankind' and should be made available without restriction. But CBD which has come into force in 1993 as the first legal mechanism dealing with biological resources, these plant genetic resources are deemed not to be the heritage of mankind, but are the properties of the countries and are therefore tradable commodities. This is in contrast to protection of intellectual property rights, which decrees that natural products are the God's gift to mankind, and hence, cannot be exploited exclusively by any one. The objective of the CBD, to be pursued in accordance with its relevant provisions included, i) the conservation of biological diversity, ii) the sustainable use of its components, and iii) the fair and equitable sharing of the benefits arising out of the utilization of genetic resources. The latter may include appropriate a) access to rightfully owned/ possessed genetic resources, b) transfer of relevant rightfully held technologies, and c) funding.

Countries need to legislate Biodiversity bills (protecting nature's treasures), the basic principle being, that no foreigner or foreign organization, could take away any biological resources for research or commercial use without permission of the country of origin. No local organization would be allowed to transfer even research results on biological resources to any foreigner without permission. The legislation ensures that benefits between conservers of the resources and users are equitably shared. Even though all the signatories had confirmed such approaches for protection of their sovereign rights, most are yet to enact appropriate legislations in this regard. A weak point in the CBD is its ambiguous treatment for the equity. It may be seen that whereas patenting of products of biotechnology is clearly recognized, but there are no effective guidelines and conditions defined to recognize and reward the contributions of indigenous communities and the other informal innovators who have been responsible for nurturing, using and developing biodiversity worldwide.

The key developments in India concerning legislative and regulatory provisions include the enactment of two legislations. The Biological Diversity Act, 2002 for which the rules have also been framed. A National Biodiversity Authority has been created with its headquarters at Chennai (Chawla and Jaiswal, 2004). It has the clear mandate to control the biodiversity and no person shall without the prior approval of authority transfer the results of any research relating to any biological resources occurring in or obtained from India for monetary consideration or otherwise to any person who is not a citizen of India or body corporate which is not registered in India.

References

Anonymous, 1993. Parliamentary Standing Committee on Commerce, Third Report on Draft of Dunkel Proposals, Evidence, Rajya Sabha Secretariat, New Delhi .

Anonymous, 1998. Policy paper on Conservation, Management and use of Agro-biodiversity. National Academy of Agricultural Sciences, India.

Chawla, H.S. 2005. Patenting of Biological Material and Biotechnology. *Journal of Intellectual Property Rights*, 10: 44-51.

Chawla, H.S. and Singh, A.K. 2005. Intellectual Property Rights. Vol II: Copyrights, Trade Marks, Trade Secrets and Geographical Indications, University Press, GBPUAT, Pantnagar, pp-75.

Chawla, H.S. and Jaiswal, J.P. 2004. Intellectual Property Rights, Vol 1: Patents, Protection of Plant Variety and Biodiversity, University Press, GBPUAT, Pantnagar, pp-32.

Chawla, H.S. 2002. Introduction to Plant Biotechnology (2nd edition). Oxford & IBH Publishers, New Delhi, pp- 538.

Gupta, A.K. 1999. Implications of WTO for Indian Agriculture: The Case of Intellectual Property Rights and Emerging Biosafety Protocol, IIMA Working Papers 1999-10-06, Research and Publication Department, Indian Institute of Management, Ahmedabad.

Menon Usha, 1991. Intellectual Property Rights and Agricultural Development, *Economic and Political Weekly*, 6(13): 1660-1667.

Rao, N.C. 1997. Plant Variety Protection and Plant Biotechnology Patents: Options for India, Policy Paper No. 29 for UNDP funded Project LARGE, UNDP, New Delhi.

Sahai, S. 1992. Patenting of Life Forms: What It Implies? *Economic and Political Weekly*, 25: 878-879.

Sahai, S. 1994. Government Legislation on Plant Breeders' Rights', *Economic and Political Weekly*, 25:1573-1574.

❑❑❑

Ecologically Based Integrated Pest Management, 2012
© D.P. Abrol and Uma Shankar (eds.), pp. 903-923
New India Publishing Agency, New Delhi (India)
e-mail : info@nipabooks.com; website : www.nipabooks.com

46

Critical Evaluation of WTO's Regulations and Policies in Indian Agriculture

SUDHAKAR DWIVEDI

Introduction

Impact of WTO or Globalization of Indian agrarian sector claims to extend benefit of trade even to widely spread villages. In India, a sizeable number of rural masses are self-employed and able to produce what is needed just for their livelihood. They cannot be brought into market exchange relationship. So, there is every possibility of their being deprived of expected benefits. Globalization, rather, will widen the disparity between who are associated with market and who are not. Developed countries seem to have held globalization as a powerful weapon in their hands. They are stressing on opening up of national boundaries for movement of their goods and capital to developing countries but they are putting lot of restrictions on technology and labour movement, which can be far more beneficial to developing countries.

An enormous size of population as in India with so much heterogeneity cannot be surrendered to market forces for its subsistence and well-being. In

India, where agriculture is playing very dominant role not merely as an occupation but more as a way of life for two – third of the country's population, complete market – orientation of agriculture is even unthinkable. It is a healthy sign that the realization of this fact to some extent, transpires in the recent budget speech of Union Finance Minister, 2007.

Globalisation could be seen as the common evolutionary tree of mankind, which may end up in positive outcomes or dismal ones depending upon how choices are made at each critical node of decision-making in the 21st century in relation to the main challenges stemming from one global economy and environments.

Broadly speaking, the term 'Globalisation' means integration of economies and societies through cross country flows of information, ideas, technologies, goods, services, capital, finance and people. The current processes of Globalisation of agriculture threaten to under mine all three dimensions of sustainable agriculture. They are undermining ecological security by removing all limits on the concentration of ownership of natural resources (land, water and biodiversity) and by encouraging non-sustainable resource exploitation for short-term profits (Whitewell, G. 1998). Trade liberalization of agriculture has not been guided by the need for providing livelihood security for two thirds of India's people, who are farmers for tile food security of the poorer half of Indians, or for India as a whole. Thus the livelihood base of millions of farmers and food security at the household, regional and national level is severely threatened by the globalization of agriculture.

The new economic policy being implemented since mid 1991 has progressively started posing new challenges in all sectors of the economy and is also opening up new vistas in the global markets. Agriculture cannot remain insulated from the ongoing economic reform process as the indirect impact of liberalisation of the economy in general on agriculture may turn out to be far more important than the possible impact of the liberalisation measures directly affecting this sector. The presumption that the liberalisation of the economy would result in higher investment and growth in agriculture induced by favourable terms of trade for the agriculture sector these challenges come primarily amidst

- Opening up of the economy to the influences of global markets through more free imports and export of goods and services.
- Entry of multinational companies and corporation in agro-based processing industry and input supplies, contract production for processing for domestic and export markets, and
- Pressures for improvement of quality of production so that the products become competitive in the global markets.

Therefore, it is indeed a dire need for developing the techniques of production, processing and market technologies. The emergence of great economic co-operation among various countries and international institution like the International Monetary Fund (IMF), the world Bank and the General Agreement on Trade and Tariffs (GATT) now transformed into World Trade Organisation (WTO) have also contributed to the process of globalisation. It is important to number that WTO come into existence because of the need for such and organization perceived by the developed countries which were important exporters of agricultural commodities.

In India the vast majority of poor people are located in rural areas and engaged primarily in subsistence agriculture for their survival, even today The core problems of widespread poverty, growing inequality rapid population growth and rising unemployment, find their origin in the stagnation and often retrogression of economic life in rural areas. Millions of people have been bypassed by whatever economic "Progress" that has been attained in India before and after independence. If "development" is to take place and becomes self-sustaining really and substantially encompassing especially the poor. It will have to start in the rural areas in general and the agricultural sector in particular. In India terms of trade always worked against agriculture as a result of high protection policy to industry, and also due to an over valued exchange rate. The underlying assumption perhaps was that a rapidly growing manufacturing sector and the urban areas would siphon off excess labour from agricultural sector too, to modernize itself with a more favourable land-man and ratio and raise productivity in agriculture. Creating a charm between rural-urban India and rural India appears to have lagged much behind even with occasional injection of technological breakthrough like green revolution and subsidies.

Indian agriculture has come a long way from taking care of the basic food requirements of its burgeoning population to visualizing the goal of transforming it into a commercial and export oriented commodity to usher in a new era of prosperity Powerful boost provided by agro-industries and agro-exports to the economies of Japan, South Korea, Taiwan and Thailand perhaps promoted India's policy-makers to view agriculture in an entirely new light. India's share in world trade in agricultural commodities is less than one percent. The overall global environment is being considered as conductive to enhance India's exports of foodgrains, fruits, flower and vegetables. The academicians, supporting globalization, enthusiastically foresee that agricultural products alone have the potential of fetching $ 14 billions in export of all categories and are sanguine of its being raised even to the targeted level of $ 50 billion in few years.

WTO and Indian Agriculture: A critical review

The new policy was laissez-faire in spirit as it reinforced the feeling that only liberalisation can remove the shackles that chained the Indian economy. The most crucial issue is that how can agriculture serve as the bigger safety net in the process of liberalisation. The liberalisation policies through structural adjustment measures are designed basically to improve efficiency of allocation of resources through the operation of incentives in a market economy. Infact the challenges lie in generating the surpluses in agricultural sector through increased public and private investment infrastructure, in research and development, new technology and the market infrastructure.

It is evident that agriculture in India has been dis-protected while industry has been heavily protected. As a result, the terms of trade for agriculture have been unfavourable when compared to the situation that would have prevailed in the absence of such protection for industry. Therefore, liberalisation of non-agriculture, and trade would result, on balance, in higher returns to the farmers and hence, in greater flow of investible resources into agriculture. Fortunately, India is now comparatively in advantageous position in respect of certain food grains, especially rice and wheat. In the international market, advantage could be eroded, at least in respect of some crops. There is no risk of compromising with food security, since India has surplus in foodgrains even after improving the policy for raising the effective domestic demand. The export potential of foodgrains will strengthen food security and it would improve the incentives for farmers, raise their resource position and contribute to technological up gradation of agriculture. There is limited scope in the near future for higher budgetary allocations for agricultural subsidies and for that matter any subsidy. Although the level of subsidy in the Indian agricultural sector estimated at about 5 percent of the agricultural product is much below the permissible limit of 10 percent under GATT provisions. According to some, budgetary burden of agricultural sector in India is in fact net-taxed. Yet the budgetary burden of agricultural inputs and output subsidies is becoming progressively unbearable for the national exchequer. The recurrent expenditure on various input subsidies for irrigation, credit and other agricultural inputs has been mounting and the governments at various levels have fewer surpluses for investment. There is need to gradually reduce the subsidies and allocate resources so saved for investment in irrigation systems which have high potential to increase yields. It has also been pointed out that for every rupee of investment in agriculture there is more than a rupee by way of subsidies on inputs in agriculture. These subsidies have led to excessive, unbalanced and, therefore, inefficient use of chemical fertilizer, endangering the intrinsic fertility of land. The excessive use of fertilizer also leads to over-exploitation of water resources (leading to severe subsidence of natural water level) encouraged by subsidies irrigation and electricity charges (Agricultural Statistics at a Glance, 2004). In a situation of resource crunch, the agricultural sector will have to perforce move away from the regime of subsidies. The government could be induced to go easy on administered prices and procurement through support price system for its foodgrain stock. The public distribution system may also not remain in demand to that extent unless pressures for diversification of production patterns away from foodgrains and traditional crop systems to high value products that would be suitable for processing and would be more in demand in domestic and world market in such competitive environment.

The pressure on agriculture land has continuously been persisting with the increase in population in India. There is little justification for doing away with ceiling on the ownership of land at this stage, especially because farmers are able to augment their operational holdings in response to the changing technological and economic factors. Productivity/unit of operated land is only consideration that may be important from the point of view of total national production. It has been realised that small farms would provided for minimum level of income to the farm family that would make them survive and would move the farm-factory on a growth path. It is therefore, essential that Land Ceiling Act and agrarian laws are reviewed to provide the much needed impetus to growth and development in the agriculture sector.

However, experience suggests that freeing the lease market for land may contribute to equity as well as efficiency by bringing into open the lease transactions. The farmers with uneconomic holdings or otherwise unable to efficiently carry out agricultural operations should be able to lease out their land to efficient operators who may be small or marginal farmers. Freeing the lease market may thus be conducive to both efficiency and equity. Land-leasing laws should also, to be changed, not be any bar upon the period and size of leasing out of land as also the ownership of land should not be in danger. This will help the operational farm size to expand and the farm business growth to generate viable enterprises and at the same time avoid unwarranted pressures on small or marginal farms to dispose off their land.

As there was a downward slide in the international prices of agricultural commodities, a large share of manufactures helped in raising net terms of trade for India, from 79 in 1990 to 104 in 1999. As an interesting sideline one may note that the previous agreements between India and developed countries have done little in expanding markets for our agricultural goods in the developed countries. It is quite dear that the developed countries are not adhering to the rules of the game. Under the guise of various "boxes" the subsidies extended to their agriculture producers have increased and on the plea of sanitary and phytosanitary considerations entry of agriculture goods from the developing countries is resisted.

Agriculture employs 60% of the Indian population today, yet it contributes only 16.6% to the GDP in 2007-08 (Anonymous, 2009). Agricultural production fell by 12.6% in 2003, one of the sharpest drops in independent India's history. Agricultural growth slowed from 4.69% in 1991 to 2.6% in 1997-1998 and to 1.1% in 2002-2003 (Agricultural Statistics at a Glance, 2006). This slowdown in agriculture is in contrast to the 6% growth rate of the Indian economy for almost the whole of the past decade. India's agriculture imports were of the order of US$ 1.86 billion in 2000-01, but they increased to $ 2.29 billion in 2001-02. If the stage continues, then the interests of Indian farmers would be seriously affected. Earlier, Indian agricultural prices were lower than international prices mostly. But as a result of the heavy subsidization of agricultural exports by developed countries, the situation under took a dramatic about-turn. The Indian farmers have been put to serious disadvantage. The phenomenon of farmers' suicides and the growing unrest in several states because of the distress of farmers particularly in agricultural commodities and their exports is a very serious human problem. Farmer suicides were 12% of the total suicides in the country in 2000, the highest ever in independent India's history (Hardikar, 2006). (Unofficial estimates put them as high as 100,000 across the country, while government estimates are much lower at 25,000. This is largely because only those who hold the title of land in their names are considered farmers, and this ignores women farmers who rarely hold land titles, and other family members who run the farms (Sainath, 2004). Agricultural wages even today are $1.5 - $2.0 a day, some of the lowest in the world (Issac, 2005). Institutional credit (or regulated credit) accounts for only 20% of credit taken among small and marginal farmers in rural areas, with the remaining being provided by private moneylenders who charge interest rates as high as at the rate of 24% per month (Sainath, 2005). An NSSO (National Sample Survey Organization) survey in 2005 found that 66% of all farm households own less than one hectare of land. It also found that 48.6% of all farmer households are in debt. The same year, a report by the 'Commission of Farmer's Welfare' in Andhra Pradesh concluded that agriculture in the state was in 'an advanced stage of crisis', the most extreme manifestation of which was the rise in suicides among farmers. Given the performance of agriculture and figures of farmer suicides across the country, this can be said to apply to Indian agriculture as a whole.

Commercially no worthwhile or any business enterprise can be sustained without external finance. Agricultural enterprise is as good an industry as any enterprise can be in secondary or tertiary sectors of the economy. The financial sector of the economy has to, therefore, gear itself up to meet this progressively growing credit demand. Any credit poured into marginal farm holdings becomes overdue and goes in default. Concessional institutional credit has benefited rich farmers more than the small and marginal farmers due to institutional imperfections, but has adversely affected the viability of credit institutions and has resulted in high capital intensity in labour surplus Indian agriculture. The recommendations of Narasimhan Committee to reduce the priority sector lending by commercial banks to 10 percent of the total lending is contrary to the bank nationalization objectives. Under the new economic policy subsidies directed to agriculture have been withdrawn and farm finance has become restricted to the market-determined agents or big farmers at market-determined rate. It is essential to ensure the availability of institutional credit to farmers, particularly small and marginal farmers, by ensuring a stipulated share for agriculture, as being done at present.

It certainly is an unequal world, and perhaps the most debasing and demeaning of all the world's inequalities is the manner in which the cattle in the rich countries are pampered at the cost of several hundred million farmers in the developing world. Irrespective of the stark inequalities, the new agreement throws a stronger protective ring around the domestic producers in the richest trading block i.e. Organisation for Economic Cooperation and Development (OECD). Unmindful of the negative consequences inflicted with impunity, the highway robbers are getting ready once again. Cancun provides them a perfect playground to arm-twist the developing countries into submission.

The Worldwide Critics on WTO's Impact on Agriculture

The WTO has already inflicted - through the first phase of trade robbery on the poor and vulnerable ever since the Magna Carta for hunger, food insecurity and destitution was unleashed in January 1995. In the Philippines, agricultural export earnings were expected to increase by billions of pesos a year after 1994, generating 500,000 additional jobs a year in the Philippines. Instead, traditional exports such as coconut, abaca and sugar have lost markets. Corn production suffered significant negative growth between 1994-2000, partly because of cheaper subsidized grains with incomes falling; the agricultural sector had lost an estimated 710,000 jobs, and another 2 million by the year 2000 (Isaac, G.E. and W.A. Kerr, 2003).

Trade liberalization has already exposed developing country farmers to ruinous competition, driving down prices, undermining rural wages and exacerbating unemployment. In the Philippines, opening up of corn market in 1997 reduced corn prices by one-third. At that time, US corn growers were receiving US $ 20,000 a year on average in subsidies, while Filipino farmers in Mindanao had average income levels of US $ 365. Between 1993-2000, cheap corn imports from US into Mexico increased eighteen times, leading to accelerated migration from rural areas to urban centres (Isaac, G.E. and W.A. Kerr, 2003).

In Central America, Colombia, Costa Rica, El Salvador, Guatemala, Honduras and Nicaragua, the price of coffee beans have fallen to just 25 percent of its level in 1960, and the region lost an estimated US $ 713 million in coffee revenues in 2001. In these countries, traditionally dependent upon coffee exports, over 170,000 jobs were lost the same year with the loss in wages computed at US $ 140 million. The negative impact was also felt in sub-Saharan Africa, where Ethiopia and Uganda reported huge losses in export revenues. In 2000-01, Uganda exported roughly the same volume, but it earned the country $ 110 million, a steep drop from $ 433 million that it notched five years earlier in 1994-95. Ethiopia reported the export revenues dropping from US $ 257 million to US $ 149 million between 1999-2000. Ironically, in January 2002, the EU and USAID warned of increased poverty and food insecurity in Ethiopia not realizing that much of the fault rests with their own policies. In Vietnam's Dak Lak province, farmers who were solely dependent upon coffee are now categorized as 'pre-starvation'. In India, coffee plantations have laid off over 25 percent of the workers in the southern provinces of Karnataka and Tamil Nadu. In Brazil, low coffee returns have resulted in increased unemployment and hunger (Sainath, 2005).

In Honduras, such has been the terrible impact that the World Food Programme reported in March 2002 that the coffee crisis, coupled with prevailing drought, had left some 30,000 farmers in the hunger trap, with hundreds of children so malnourished that they needed to be hospitalised. In 2001, the 25,000 US cotton growers received roughly $3.9 billion in subsidy payments, for producing a cotton crop that was worth only US$ 3 billion at world market prices (One Arkansas cotton grower received US$ 6 million, equal to the combined annual earnings of 25,000 cotton farmers in Mali). It's also more than the gross domestic product of several African countries and three times the amount the US spends on aid to half a billion Africans living in poverty. In 2002, direct financial assistance by a number of exporting countries, including China, European Union and the US, to the tune of 73 percent of the world cotton production, destroyed millions of livelihoods in West African countries (Benin, Burkina Faso, Mali, and Chad). India and Pakistan too have been forced to lower import duties, allowing a surge of cotton imports thereby pushing farmers out.

In the dairy sector, the EU subsidized exports have hit the dairy industry in Brazil, Jamaica and India. While the Jamaican dairy producers have time and again spilt milk onto the streets, the Indian dairy industry too complains of export dumping. In 1999-2000, India imported over 130,000 tonnes of EU's highly subsidized skimmed milk powder. This was the result of Euro 5 million export subsidies that were provided, approximately 10,000 times the annual income of a small-scale milk producer (Sidhu, R.S. and Bhullar, A.S. 2004).

Butter export subsidy paid by the EU, for instance, is currently at a five-year high and butter export refunds have risen to an equivalent of 60 percent of the EU market price. Consequently, butter oil import into India has grown at an average rate of 7.7 percent annually. This trend has already had a dampening effect on prices of ghee in the domestic market. Ironically, India is the biggest producer of milk in the world, and does not provide any subsidy for the dairy sector.

Indonesia was rated among the top ten exporters of rice before the WTO came into effect. Three years later, in 1998, Indonesia had emerged as the world's largest importer of rice. In India, the biggest producer of vegetables in the world, the import of vegetables has almost doubled in just one year - from Rs 92.8 million in 2001-02 to Rs 171 million in 2002-03. Far away in Peru, food imports increased dramatically in the wake of liberalization. Food imports now account for 40 percent of the total national food consumption. Wheat imports doubled in the 1990s, imports of maize overtook domestic production, and milk imports rose three times in the first half of the previous decade, playing havoc with Peruvian farmers.

Looks shocking, but this is merely a peep into the destruction wrought by the 'disagreement' on agriculture. Everyday, thousands of farmers and the rural people in the majority world - without land and adequate livelihoods constituting a reservoir of frustration and disaffection, trudge to the cities, their abject poverty contrasting vividly with the affluence of the urban centres. These are the victims - in fact, the first generation of the affected of the great trade robbery. These are the helpless sufferers, who are being fed a daily dose of promises - increase in poverty in the short-run is a price that has to be paid for long-term economic growth (Sainath, P. 2005).

The complete impact on human lives, women and children in particular and the resulting loss in livelihood security and thereby the accelerated march towards hunger and destitution cannot be easily quantified. Surging food imports have hit farm incomes and had severe employment effects in many developing countries. Unable to compete with cheap food imports, and in the absence of any adequate protection measures, income and livelihood losses have hurt women and poor farmers the most.

Disparities in the WTO Laws and their Implementations

Farmers in the developing world have suddenly become the children of a lesser god. They are the neo-poor. Through a variety of instruments, the rich countries have ensured complete protectionism. Trade policies therefore have remained highly discriminatory against the developing country farmers. Such is the extent of protection that the benevolence OECD exhibits through development aid to all countries, totaling US $ 52 billion, dwarfs before the monumental agricultural subsidies of US $ 311 billion that these countries provided to its own agriculture in 2001. In reality, you don't even give by one hand to take back with the other. Rich countries effectively use development aid to convince the domestic audiences of their generosity towards human suffering, in essence using aid as the human face for 'ambitious' one-way trade from the OECD to the rest of the world.

The colourful band of boxes - green box, blue box and amber box - have come in handy for the rich countries to protect its subsidies to agriculture, and at the same time dump the surpluses all over the world. Considering that the world commodity prices are far from adequate anywhere to provide them with a living, these subsidies are actually the cause of excessive supplies in the world markets, and thus resulting in low prevailing world markets. Still further, US is permitted under AoA to provide $ 363 million in export subsidies for wheat and wheat flour, and the EU can limit it to $ 1.4 billion a year. At the same time, the US incurs annually $ 478 million under its Export Enhancement Programme (EEP), which is not being subject to any reduction commitments.

With the availability of all such subsidies, agribusiness companies find it much easier and economical to export. Export credits, used primarily by the US, and not counted as export subsidies, doubled in just one year to reach US $ 5.9 billion in 1998. The export subsidies and credits are therefore cornered by the food exporting companies. In the US, for instance, more than 80 percent of the corn exports is handled by three firms: Cargill, ADM and Zen Noh. The level of dumping by the US alone hovers around 40 percent for wheat, 30 percent for soybeans, 25 to 30 percent for corn and 57 percent for cotton. Further, each tonne of wheat and sugar that the United Kingdom sells on international market is priced 40 to 60 percent lower than the cost of production (Gulati, A. and Kelley, T. 1999).

The shocking levels of food dumping and its little understood but horrendous impact on the farming sector in the developing countries is the result of clever manipulations at the WTO. The US and EU were successful in ensuring that some subsidies -and that included direct payments – have little or no impact on production levels and so have little or no impact on trade. Using sophisticated models and taking advantage of the un-preparedness of the developing country negotiators, they devised a complicated set of rules that termed only 'amber box' subsidies as 'trade distorting' that needs to be cut. As it turned out, these were the type of subsidies that the poor countries were also using.

On the other hand, 'green box' and 'blue box' subsidies categorise the farm support that only the rich countries were providing, and which the developing countries are not in a position to afford. Subsequently, in July 2002, the US proposed significant cuts in 'trade distorting' domestic support for all products and trade partners, with a ceiling of five percent of the value of agricultural production for industrial countries and 10 percent for developing countries. This however does not mean that the US will make any major cuts in its farm subsidy support, despite the US Farm Security and Rural Investment Act 2002, which provides for US $ 180 billion in subsidies to agriculture for the next ten years, with more than a third coming in the first three years (Berkeley, Kumar Venkat, 2003).

New EU Common Agricultural Policy reform proposals that have been announced prior to the Cancun WTO Ministerial have also made no attempt to make radical changes in reduction commitments. Moving on US lines, it has shifted most of the 'blue box' subsidies to 'green box'. European agriculture will continue to be subsidised to the tune of Euro 43 billion for another decade, and that amount will increase further when the new members join in. Like a magician, both the US and EU have managed to juggle the farm support from one box to another without making any significant commitments. The magical trick is now being used to create an illusion of sincerity of the rich towards 'free' trade.

EU Trade Commissioner, Pascal Lamy, had made the 'hidden' intentions abundantly clear during his quick visit to India soon after the Doha Ministerial. "Who gave you the impression that we are going to reduce agriculture subsidies? Let me make it clear that EU will not be reducing farm subsidies in the years to come except for what has been agreed upon. I am committed to keep my seven million farmers on the farm," he told a group of civil society representatives. "And how will India gain if I reduce subsidies as a result of which the number of farmers in Europe climbs down to three million?" he asked.

When asked how India will protect the livelihood of its 600 million farmers if EU does not eliminate farm subsidies, he quipped: "That's for your government to decide".

As if the massive subsidies are not enough, developed countries have used high tariffs to successfully block imports from developing countries. They have used Special Safe Guards (SSG), used only by 38 rich countries so far, to restrict imports from developing countries. Developed countries took advantage of this flexibility by reserving the right to use the SSG for a large number of products: Canada reserves the right to use SSG for 150 tariff lines, the EU for 539 tariff lines, Japan for 121 tariff lines, the US for 189 tariff lines, and Switzerland for 961 tariff lines. On the other hand, only 22 developing countries can use SSG. A majority of the developing countries, whose trade in agricultural products takes place under a tariff only regime, have no access to these instruments (Chakraborty, D. 2004).

Interestingly, there is a talk of phasing out the SSG provisions in the next 5 to 7 years. By that time, the developing countries would have been forced to open up their markets still further with devastating impacts. The markets of the developed countries will however remain protected for the next seven years.

At the same time, these countries have managed to fulfill the technical requirements for tariff cuts under AoA without any meaningful reductions. Technically speaking the reductions in tariff cuts are in place, but in reality they have defied the letter and spirit of the agreement. Although the US, EU, Japan and Canada maintain tariff peaks of 350 to 900 percent on food products such as sugar, rice, dairy products, meat, fruits, vegetables and fish the thrust of the ongoing negotiations remain on pierce opening the developing country markets to more subsidised exports.

The United States, for instance, is in a neck-to-neck race with the European Union on retaining the supremacy over agriculture trade. While steadily expanding foreign demand brought on by income gains, trade liberalization, and changes in global market structures have helped American exports double over the past 15 years to $53.5 billion estimated for 2002, its market share had dropped from 24 percent of global agricultural trade in 1981 to 18 percent in 2001. The EU, on the other hand, has increased its performance from 13.5 percent to 17 percent, in the consecutive period.

"Losing six points over 20 years may not sound like much, but every percentage point loss of market share amounts to $3 billion in lost export sales and a reduction of $750 million in agricultural income. But, the good news is that every percentage point we can recover will add $3 billion in export sales and $750 million to agricultural income each year," Mattie Sharpless, the then Acting Administrator, Foreign Agriculture Service of the US Department of Agriculture had told the Senate Agriculture Committee in 2002.

"Dollar for dollar, America export more meat than steel, more corn than cosmetics, more wheat than coal, more bakery products than motorboats, and more fruits and vegetables than household appliances," Sharpless had said, adding that agriculture is one of the few sectors of the US economy that consistently contributes a surplus to its trade balance. In fact, the US projections for 2002-03 were that 53 percent of its wheat crop, 47 percent of cotton, 42 percent of rice, 35 percent of soybeans, and 21 percent of corn to be exported (Anderson, K. 2003). This has only been made possible by the heavy subsidies and the removal of trade barriers by the developing countries.

The US, therefore, has adopted an aggressive posture. After ensuring that the developing countries are made to conform to the WTO obligations of phasing out or lifting of quantitative restrictions that allow easy penetration of the American farm commodities and the processed products, it is now preparing for the final assault. The new policy is directed at the 600 million "new consumers" in Asia and Southeast Asia and another 400 million in Latin America and Central America. It also meets "an eye for the eye" with the EU's Common Agricultural Policy. And in this 'clash of civilizations' the battle is primarily between the developed and the developing countries, between industrial agriculture and food security, between value-added functional foods and growing hunger.

Keeping the economic interests of the developed countries in mind, the chair of the agricultural negotiations, Stuart Harbinson, had proposed a compromise formula that suggested creation of two new instruments: 'special products' and 'special safeguard mechanisms'. For crops which are crucial for food security needs, the proposal is to put them in the SP category for which the tariff reduction should on an average be of 10 percent with a minimum reduction rate per tariff line of five percent. For the remaining products, tariff reduction should be between 25 to 40 percent. The US-EU proposal however does not make a mention of the special products. The general feeling is that developing countries do require special safeguard measures that act as a partial protection rather than they will be allowed a more permanent feature of resorting to 'special products'.

For all practical purposes, the concept of 'strategic products' or 'special products' is merely a proxy for the 'development box', a proposal that will eventually turn out to be more damming if implemented. Moreover, the negotiations are going to be centered around the number of 'special products' that a country can claim. In other words, the debate will very conveniently shift from the more contentious issue of agricultural subsidies in the west. The AoA does not realize that production of crops and its imports into developing countries cannot be equated with industrial production. What the developing countries need is a trading system that recognises their specific needs of food security and rural development based on the principle of production by the masses rather than production for the masses.

The hypocrisy of the developed countries has been echoed by the World Bank Chief Economist Nicholas Stern, while traveling through India recently, denounced subsidies paid by rich countries to their farmers as "sin ...on a very big scale" but warned India against any attempts to resist opening its markets. "Developing countries must remove their trade barriers regardless of what is happening in the developed countries." No wonder, while the negotiation continues and the developing countries are kept busy with diversionary tactics like 'special products'; agricultural exports from the OECD countries continue to rise. Between 1970 and 2000, France increased its share from 5.7 percent to 8.1 percent, Germany from 2.6 percent to 5.9 percent and United Kingdom from 2.7 percent to 4.1 percent.

A true reform in agriculture is only possible when the global community accepts the guiding principle that food for all is an international obligation. It can only be achieved when the need for national food self-sufficiency becomes the cornerstone of the AoA. It can only be put into practice when the developed and the developing countries refrain from a battle of food supremacy to reorient efforts to bring equality, justice and human compassion in addressing the mankind's biggest scourge - chronic hunger and acute malnutrition.

Developing countries therefore cannot afford to be a silent spectator. If the rich industrialized countries can protect their agriculture, and take an aggressive approach, developing countries should not feel shy in doing the same. Instead of succumbing to pressure tactics that comes along with olive branch of a 'development box' or 'special products' that helps in partially protecting agriculture, the entire effort should be to demand the abolition of agricultural subsidies in the OECD countries. A collective stand based on the following three planks is the only route for developing countries to protect agriculture, the mainstay of their economies.

Zero tolerance on agricultural subsidies: Developing countries should make it categorically clear that the negotiations will move ahead only when the subsidies (under all 'boxes') are removed. The Agreement on Agriculture should wait till the subsidies in the west are grounded. Any agreement without the subsidies being removed will play havoc with developing country agriculture.

Restoration of Quantitative Restrictions: Developing countries should demand restoration of quantitative restrictions (and special safeguard measures for those countries which did not follow the QR route). In fact, the removal of subsidies should be linked with the removal of quantitative restrictions. That alone will provide the necessary safeguard for developing country's agriculture and food security.

Among the new issues to be introduced at Cancun, the developing countries need to strive for the inclusion of a Multilateral Agreement against Hunger. This should be based on the guiding principle of the right to food and should form the basis for all future negotiations. Such a multilateral agreement would ensure that countries will have the right to take adequate safeguard measures if their commitment towards the WTO obligations leads to more hunger and poverty.

The Debt Trap and the Role of Liberalization

The biggest input for farmers is seeds. Before liberalisation, farmers across the country had access to seeds from state government institutions. For example, AP's APSSDC3 produced its own seeds, was responsible for their quality and price, and had a statutory duty to ensure seeds were supplied to all regions in the state, no matter how remote. The seed market was well regulated, and this ensured quality in privately sold seeds too (Anonymous, 2009). With liberalization, India's seed market was opened up to global agribusinesses like Monsanto, Cargill and Syn Genta. Also, following the deregulation guidelines of the IMF, 14 of the 24 units of the APSSDC's seed processing units were closed down in 2003, with similar closures in other states. This hit farmer doubly hard: in an unregulated market, seed prices shot up, and fake seeds made an appearance in a big way. Seed cost per acre in 1991 was Rs. 70 but in 2005, after the dismantling of APSSDC and other similar organizations, the price jumped to Rs. 1000, a hike of 14.28%, with the cost of genetically modified pest resistant seeds like Monsanto's BT Cotton costing Rs. 3200 or more per acre, a hike of 35.55% (Sainath, P. 2005). BT Cotton is cotton seed that is genetically modified to resist pests, the success of which is disputed: farmers in Andhra Pradesh and Maharashtra now claim that yields are far lower than promised by Monsanto, and there are fears that pests are developing resistance to the seeds. (Menon & Jayaraman, 2002) Expecting high yields, farmers invest heavily in such seeds. Also BT Cotton and other new seeds guarantee a much lower germination rate of 65% as opposed to a 90% rate of state certified seeds. Hence 35% of the farmer's investment in seeds is a waste (Sainath, P. 2004). Output is not commensurate with the heavy investment in the seeds, and farmers are pushed into debt. The abundant availability of spurious seeds is another problem which leads to crop failures. Either tempted by their lower price, or unable to discern the difference, farmers invest heavily in these seeds, and again, low output pushes them into debt. Earlier, farmers could save a part of the harvest and use the seeds for the next cultivation, but some genetically modified seeds, known as Terminator, prevent harvested seeds from germinating, hence forcing the farmers to invest in them every season.

Fertilizer and Pesticide

One measure of the liberalisation policy which had an immediate adverse effect on farmers was the devaluation of the Indian Rupee in 1991 by 25% (an explicit condition of the IMF loan). Indian crops became very cheap and attractive in the global market, and led to an export drive. Farmers were encouraged to shift from growing a mixture of traditional crops to export oriented 'cash crops' like chilli, cotton and tobacco (Anonymous, 2005). These need far more inputs of pesticide, fertilizer and water than traditional crops.

Liberalisation policies reduced pesticide subsidy (another explicit condition of the IMF agreement) by two thirds by 2000. Farmers in Maharashtra who spent Rs. 90 an acre now spend between Rs. 1000 and 3000 representing a hike of 1000% to 3333%. Fertilizer prices have increased 300% (Sainath, 2005) Electricity tariffs have also been increased: in Andhra Pradesh tariff was increased 5 times between 1998 and 2003 (Seeds of ruin, 2005). Pre-liberalisation, subsidised electricity was a success, allowing farmers to keep costs of production low. These costs increased dramatically when farmers turned to cultivation of cash crops, needing more water, hence more water pumps and higher consumption of electricity. Andhra Pradesh being traditionally drought prone worsened the situation. This caused huge, unsustainable losses for the Andhra Pradesh State Electricity Board, which increased tariffs. (This was initiated by Chandrababu Naidu in partnership with Britain's DFID4 and the World Bank). Also, the fact that only 39% of India's cultivable land is irrigated makes cultivation of cash crops largely unviable, but export oriented liberalisation policies and seed companies looking for profits continue to push farmers in that direction. (Isaac, R. 2005).

Low Price of Output

With a view to open India's markets, the liberalization reforms also withdrew tariffs and duties on imports, which protect and encourage domestic industry. By 2001, India completely removed restrictions on imports of almost 1,500 items including food (Anonymous, 2005). As a result, cheap imports flooded the market, pushing prices of crops like cotton and pepper down. Import tariffs on cotton now stand between 0 – 10%, encouraging imports into the country. This excess supply of cotton in the market led cotton prices to crash more than 60% since 1995. As a result, most of the farmer suicides in Maharashtra were concentrated in the cotton belt till 2003 (after which paddy farmers followed the suicide trend). (Hardikar, 2006). Similarly, Kerala, which is world renowned for pepper, has suffered as a result of 0% duty on imports of pepper from SAARC5 countries. Pepper, which sold at Rs. 27,000 a quintal in 1998, crashed to Rs. 5000 in 2004, a decline of 81%. As a result, Indian exports of pepper fell 31% in 2003 from the previous year (Sainath, P. 2005). Combined with this, drought and crop failure has hit the pepper farmers of Kerala hard, and have forced them into a debt trap. Close to 50% of suicides among Kerala's farmers have been in pepper producing districts (Mohankumar & Sharma, 2006).

Lack of Credit Facilities and Dependence on Private Money Lenders

In 1969, major Indian banks were nationalized, and priority was given to agrarian credit which was hitherto severely neglected. However, with liberalisation, efficiency being of utmost importance, such lending was deemed as being low-profit and inefficient, and credit extended to farmers was reduced dramatically, falling to 10.3% in 2001 against a recommended target of 18% (Sainath, P. 2004). A lack of rural infrastructure deters private banks from setting up rural branches, with the responsibility falling on the government, which has reduced rural spending as a result of its liberalisation policies. Rural development expenditure, which averaged 14.5% of GDP during 1985 – 1990 was reduced to 8% by 1998, and further to 6% since then. This at a time when agriculture was going through a crisis proved disastrous for farmers, who turned to private money lenders who charge exorbitant rates of interest, sometimes up to 24% a month (Sainath, P. 2004). With input costs and output prices being what they are, coupled with crop failures and drought, they are pushed into debt which is impossible to repay. 12 out of India's 28 states have 50% and higher indebtedness among farm households. Andhra Pradesh has the highest percentage of indebted farm households-82%. 64.4% of Kerala's farm households and 54.8% of Maharashtra's farm households are indebted (NSSO, 2003) Indebtedness has been identified as the single major cause of suicides in Andhra Pradesh, Kerala and Maharashtra. (Analysis of Farmer Suicides in Kerala, 2006, Who's suicide is it anyway, 2005, Saving small farmers, 2005).

Liberalisation and How it Failed

Branco Milanovic, a World Bank economist describes how he believes liberalisation helps developing countries achieve growth: 'when a country lowers trade barriers, reduces government intervention in the market in order to allow market forces to operate freely, increases competition and attracts foreign investment, it will increase productivity and reduce inefficiency, which will lead to economic growth, and in a few generations, if not less, the poor will become rich, illiteracy will disappear, and poor countries will catch up with the rich (Anonymous, 2005). This argument is an economic rationalist one, which views government intervention with profound suspicion, and has equally profound faith in unfettered market forces (Whitewell, G. 1998). What Mr. Milanovic neglects to mention, though, is that rich countries which now preach liberalisation protected their 'infant industries' at the time they began to industrialize, till they were strong enough to compete globally. The US government, for example, had a protectionist trade policy in the late nineteenth century to help US companies become competitive in the world. Besides, apart from wool, the US, Germany, Britain and France were all almost self-sufficient in the raw materials that they needed for industrialization, and took off from that platform, a luxury that India and other developing countries do not have (Issac, R. 2005). As German economist Friedrich List says, the adoption of these values (of liberalisation) assumes that all countries are at the same starting place, which (as we have seen above) is clearly not the case (Issac, R. 2005). In fact, it is this very reason that has brought about the crisis that Indian agriculture is facing today. Most farmers in India were already in a position of minimum security, with no education system, credit facilities, access to alternative employment, or efficient technology. Their only support was government subsidy and regulation. Liberalisation policies came in and dismantled their only support structure. It halted the sharp reduction in rural poverty from 55% in the 1970s to 34% in the 1980s. Not only has the incidence of poverty in rural areas not gone lower than 34% in the 1990s, it has gone to higher levels of 42% in individual years.

The second most popular argument of the economic rationalists in favour of liberalisation is that competition will weed out the inefficient, and in the growth that ensues, employment will be provided in other areas of the economy, thus lifting the poor out of poverty. This argument however assumes that the poor will be able to take advantage of the opportunities presented to them. As Robert Issac says in 'The Globalization Gap', "Globalization encourages the well positioned to use tools of economics and politics to exploit market opportunities, boost technical productivity, and maximize short-term material interests." This is compounded in India, where the gap between one who is 'well positioned' and one who is not can be extreme. With a lack of investment, chances of generation of rural employment are slim. Unemployment and underemployment are chronic problems in India, with the rate of unemployment being close to 10% in 2004 (Sainath, R. 2005). Primary education in rural areas is mismanaged and bad quality, and there is no system which helps agricultural workers find alternate employment, or develop alternate skills. In the face of such obstacles, it is nearly impossible to expect agricultural workers to shift to alternate fields. Coming back to AP, the IT Revolution spearheaded by Chandrababu Naidu attracted companies like Google, Amazon, Microsoft and Dell and created thousands of jobs. However, given the skills and education of most farmers, it is obvious that none of this translated into job opportunities for them. The final argument that supporters of globalization have is the much touted 10% reduction in poverty (60 million decline in poor) in India in the year 2000. However, this figure was challenged by experts. Poverty is defined according to how many people consume less than the nutritional minimum prescribed. (2400 calories for rural areas, and 2100 for urban areas) Major changes in survey design in 1999-2000 not only made the resultant estimates incomparable to previous years' estimates, but an over-estimation of consumption (meaning people were getting enough food, hence were not considered poor) meant a sharp reduction in poverty figures. After experts challenged it, the Planning Commission of India accepted that the figure was inaccurate, and could not be compared to previous years' estimates; hence the 10% drop in poverty is incorrect. With adjusted figures, experts have determined that the decrease in poverty was a mere 2.3%, and that the number of poor increased by nine million in 2002 as compared to 1999 (Agricultural Statistics at a Glance, 2004).

The contribution of knowledge as a factor of production is beginning to acquire dominant role in future trade, investment and technological change in agriculture as well as other sectors of economy. The management of knowledge not just in farms and firms but also in non-farm sector will, thus, become crucial. But the production and reproduction of knowledge will no more be governed by the conventional norms of public space, scrutiny and substantive needs. It is the tension between public need and private control that will mount the first challenge. The conflict between chemical intensive agriculture (despite declining productivity of inputs) and the non-chemical sustainable technological innovations generated by farmers as well as firms (national or international) will pose second challenge. The increasing trend towards larger areas under fewer varieties and the need for food security through diversified biological systems will be the third source of conflicts.

The strategy proposed is aimed at making Indian agriculture and biodiversity based livelihood strategies of millions of disadvantaged communities and individuals not only globally more competitive but also domestically more progressive by using *knowledge* as a strategic resource. The major contention is that India should not view the challenges posed by WTO as if it will remain always an importing country and that it has no substantive intellectual property to offer to world market. The critical NGOs and other colleagues who criticize the concept of intellectual property rights have perhaps not been exposed to the inventive potential of Indian society. Honey Bee network has demonstrated over last ten years through its data base having about ten thousand entries of innovations and outstanding examples of traditional knowledge, innovations and practices, the immense contribution that grassroots innovator can make towards this cause. Add to this the potential that Indian scientists have and one would know why TRIPS under WTO can indeed make R and D in formal and informal sector as the pivot of socio-economic transformation of our society. It is true that India must negotiate changes in TRIPS to suit our requirements. But we can lobby for these changes because we are part of WTO.

As expected, the United States and the European Union have arrived at a new accord, just ahead of the fifth WTO Ministerial at Cancun, which in letter and spirit lays out a detailed road map for what can be called as the second phase of the great trade robbery.

The new framework - a "common vision" rather than a detailed plan - is aimed at further destroying whatever remains of the strong foundations of food self-sufficiency in some of the developing countries already wilting under the compound impact of the Agreement on Agriculture (AoA). For the small farmers and giant agribusiness in North America, Europe and the Pacific, it will however be business as usual. Rich countries subsidies agribusiness by allowing them to buy very cheap, with the government then making up some of the differences with direct payment to farmers.

So much so that the recipients of the US agricultural subsidies in 2001 also included Ted Turner and David Rockefeller - no wonder, the CNN continues to stifle the voice of the developing countries farmers. The richest man in the United Kingdom, the Duke of Westminster, who owns about 55,000 hectares of farm estates, receives an average subsidy of 300,000 pound sterling's as direct payments, and in addition gets 350,000 pounds a year for the 1,200 dairy cows he owns. As expected, the United States and the European Union have arrived at a new accord, just ahead of the fifth WTO Ministerial at Cancun, which in letter and spirit lays out a detailed road map for what can be called as the second phase of the great trade robbery.

Conclusion

It certainly is an unequal world, and perhaps the most debasing and demeaning of all the world's inequalities is the manner in which the cattle in the rich countries are pampered at the cost of several hundred million farmers in the developing world. In comparison of the life of the western cow with that of the Third World farmer, it didn't realize that this would hit the sensibilities of at least some of the economists and policy makers. It has now been worked out that the EU provides a daily subsidy of US $ 2.7 per cow, and Japan provides three times more at US $ 8, whereas half of India's 1000 million people live on less than $ 2 a day. Irrespective of the stark inequalities, the new agreement throws a stronger protective ring around the domestic producers in the richest trading block - Organisation for Economic Cooperation and Development (OECD). Unmindful of the negative consequences inflicted with impunity, the highway robbers are getting ready once again. Cancun provides them a perfect playground to arm-twist the developing countries into submission.

References

Agricultural Statistics at a Glance- August, 2004. Indian Farmers Fertilizer Co-operative Limited, New Delhi.

Agricultural Statistics at a Glance. New Delhi: Ministry of Agriculture, Government of India. 2005.

Anderson, K. 2003. How can agricultural Trade reform reduce Poverty? Discussion paper No. 0321, centre for International Economic Studies, University of Adelaide, Adelaide.

Anonymous, 2005. The damage done: Aid, death and dogma'. May 2005, Christian Aid.

Anonymous, 2009. Mid term Appraisal of Xth Five Year Plan, New Delhi, Planning Commission, Govt. of India.

Berkeley and Kumar V. 2003. Globalization unanswered questions. *http://www.commondreams.org/views* (Accessed on June 12, 2007).

Chakraborty, D. 2004. 'Recent negotiation Trends on Agriculture under WTO' Working paper, International Conference on Globalization and WTO, September, 2004.

Gulati, A. and Kelley, T. 1999. Trade liberalization and Indian Agriculture. Oxford University Press, New Delhi.

Hardikar, J. 2006. 'The rising import of suicides', India together, viewed June 10, 2006, *http://www.indiatogether.org/2006/jun/opi-cotton.htm*

Isaac, G.E. and Kerr, W.A. 2003. Genetically Modified Organisms and Trade Rules: Identifying Important Challenges for the WTO. *The World Economy*, 26 (1): 29-42.

Isaac, G.E. and Kerr, W.A. 2004. Bioprospecting or Biopiracy ? Intellectual Property and Traditional Knowledge in Biotechnology Innovation. *The Journal of World Intellectual Property*, 7 (1): 35-52.

Issac, R. 2005. 'The globalization gap', Pearson Education Inc., New Jersey.

Menon, M. and Jayaraman, N. 2002. India at Doha: Retrospect and Prospect. *Economic and Political Weekly*, 26:3.

Mohankumar, S. and Sharma, R.K. 2006. 'Analysis of farmer suicides in Kerala'. *Economic and Political Weekly*, April 22, 2006.

Sainath, P. 2004. 'Dreaming of water, drowning in debt'. *The Hindu*, July 18, 2004.

Sainath, P. 2004. 'Seeds of suicide'. *The Hindu*, July 20, 2004.

Sainath, P. 2005. 'As you sow, so shall you weep'. *The Hindu*, May 30, 2005.

Sainath, P. 2005. 'No lessons from past mistakes'. *The Hindustan Times*, viewed 4 June, 2006.

Sainath, P. 2005. 'Spice of life carries whiff of death'. *The Hindu*, February 13, 2005.

Sidhu, R.S. and Bhullar, A.S. 2004. Changing structure of the livestock economy in Punjab: Impact of livestock on income and employment patterns. *Indian Journal of Agricultural Economics*, 59(3): 579-587.

Whitwell, G. 1998. 'What is economic rationalism'. *ABC Viewed*, June 10, 2006.

❑❑❑

Ecologically Based Integrated Pest Management, 2012
© D.P. Abrol and Uma Shankar (eds.), pp. 925-954
New India Publishing Agency, New Delhi (India)
e-mail : info@nipabooks.com; website : www.nipabooks.com

47

Implications of Statistical Tools in Pest Management

S.E.H. RIZVI AND MANISH SHARMA

Variation is inherent in almost all biological populations. In any science, research work, that is, the study and explanation of development changes in a phenomenon can be theoretical as well as experimental. Thus, in order to have solution to a problem the need arises to collect relevant information (called data) their analysis and interpretation through appropriate statistical tools. Statistical analyses help the biologists to draw appropriate conclusions. For instance, in Integrated Pest Management (IPM) the entomologists are primarily concerned with the estimation of insect-pest population, its distribution pattern over time and space, its relationship/fluctuation in population with the prevailing weather parameters and ultimately its control measure through the judicious use of chemicals, biological control agents and resistant varieties etc. There are two broad modes of collection of data:

- Through surveys or
- Through experiments

Survey versus experiment: Without exercising any control on factors that may influence it, we conduct what may be called a survey of the phenomenon of the interest e.g. epidemiological studies, diseases infestation, estimation of insect-pest population etc. On the other hand, when the object is to study the variation produced in response to changes purposively made on influencing factors, we conduct a controlled experiment. For example, we may conduct a controlled experiment to study the effect of various pesticides on the yield of a crop etc.

Methods of sampling: In survey, we think of statistical population which is defined as an aggregate of objects whether animate or inamimate having some common characteristics. Since it is difficult to handle the population as such through complete enumeration (census survey), the statistician usually has to select a subset (called sample) by using an appropriate sampling techniques which is based on some probability mechanism. For instance, in pest management the observation on incidence of insect-pest are time consuming and often require the removal of the plants from the field, it is obvious that such estimates of incidence should be based on sampling observations only. Through random selection of units from the study population samples are drawn which in turn results in the reduced cost of the survey, greater speed of getting results, greater accuracy of results. Sampling procedures can be broadly classified as:

(i) Probability or random sampling methods

(ii) Non-probability sampling methods

In an efficient sampling programme, it is necessary to determine the optimum size of sample as well as proper method of sample selection. The basic sampling techniques which are commonly employed, as per the situations, are simple random sampling, systematic sampling, probability proportional to size sampling, stratified sampling, cluster sampling and multi-stage sampling, etc. There are some other techniques like sequential sampling, which are used for the estimation of mobile population such as flying insects or pests etc.

The number of units in the population is called its population size represented by N. The total number of units in the sample is termed as sample size and is denoted by n. Statistical constants like population mean (μ), population variance (σ^2), population correlation co-efficient (ρ) etc. are called parameters. It is a function of population values. On the other hand a statistic is defined as a function of sample values only. Statistical measures like sample mean ($\bar{y}$), sample variance (s^2) sample correlation coefficient (r) are the statistics corresponding to population parameters, μ, σ^2 and ρ, respectively. A statistic gives an estimate to the corresponding parameter. The difference between the value of a statistic and its parameter is called sampling error. Standard error (S.E.) is the standard deviation of the sampling distribution of a statistic.

Simple random sampling is a method of selecting n units out of the N units of a homogeneous population such that every one of the $^{N}C_{n}$ distinct samples has got an equal change of being drawn. A simple random sample is drawn unit by unit. The units in the population are numbered from 1 to N. If the unit once drawn is replaced back in the population before next draw is made; it is called simple random sampling with replacement (SRSWR). Further, if the unit once drawn is not replaced back in the population before next draw is made, it is called simple random sampling without replacement (SRSWOR). Obviously, the difference between with replacement and without replacement sampling becomes less important when the population size is large and the sample size is noticeably smaller than it.

When the population under study is heterogeneous an improved sampling tool known as stratified random sampling is used. In stratified sampling, heterogeneous population consisting of N units is first divided into L homogeneous non-overlapping sub-populations (known as strata) of N_1, N_2, ---, N_L units, respectively. Then samples are drawn independently from these strata. Generally, the stratification is done according to the administrative groupings, geographical areas and on the basis of auxiliary characters correlated with the character under study.

Estimation of population parameters : Let Y be the character of interest and $Y_1, Y_2, \cdots, Y_i, \cdots, Y_N$ be the values of the character on N units of the population. Further, let $y_1, y_2, \cdots, y_i, \cdots, y_n$ be the units in the sample of size n selected by simple random sampling. The procedure of calculation of various parameter and statistic as well as their S.E. is given below with the help of numerical example:

Numerical illustration: Suppose a hypothetical population consists of 5 units(plants) with their values (number of insects per plant) as 9, 7, 5, 8, 11.A random sample of 3 units has been drawn from it giving their values as 5, 8, 9 by using with replacement as well as without replacement sampling procedures. The procedure of computation of various descriptive statistics has been given below:

$$\text{Population mean} = \bar{Y} = \frac{1}{N}\sum_{i=1}^{N} Y_i = \frac{1}{5}\sum_{i=1}^{5} Y_i = \frac{1}{5}(9+7+5+8+11) = \frac{40}{5} = 8$$

Sample mean = $\bar{y}=\frac{1}{n}\sum_{i=1}^{n}x_i=\frac{1}{3}\sum_{i=1}^{3}x_i=\frac{1}{3}(5+8+9)=\frac{22}{3}=7.33$ i.e. approx. 7 insects/plant.

Estimate of population total= $\hat{Y}=N\bar{y}$ =5x7.33=36.65 i.e. approx. 37 insects.

Sampling error = $\bar{y}-\bar{Y}$ = 7.33 8 = -0.67

$$\text{Population variance } (\sigma^2)=\frac{1}{N}\sum_{i=1}^{n}(Y_i-\bar{Y})^2=\frac{1}{5}(20)=4$$

$$\text{Sample variance} = s^2=\frac{1}{n-1}\sum_{i=1}^{n_1}(y_i-\bar{y})^2=\frac{1}{3-1}(8.666)=4.33$$

Now, assuming sampling with replacement, standard error (S.E.) of sample mean is given by

$$\text{S.E.}(\bar{y})=\sqrt{\text{var}(\bar{y})}=\sqrt{\frac{\sigma^2}{n}}=\sqrt{\frac{4}{3}}=\frac{2}{\sqrt{3}}=1.15$$

When σ^2 is unknown, it is estimated by sampling variance (s^2)

$\text{Var}(\bar{y})=\frac{s^2}{n}$, therefore,

$$\text{Estimated S.E.}(\bar{y})=\sqrt{\frac{s^2}{n}}=\sqrt{\frac{4.33}{3}}=1.20$$

Confidence Interval: Here we give certain limits known as confidence limits. In point estimation, we give single value as an estimate for population parameters under study on the basis of sample values. In interval estimation some interval is provided which has two values- known as upper and lower limits within which the value of the parameter is expected to lie. Taking α = 0.05 or 0.01, (1-α) 100% confidence limits are obtained as below:

$$\hat{\bar{Y}}_u=\bar{y}+t_{\alpha/2,N-1}\text{S.E.}(\bar{y})$$
$$\hat{\bar{Y}}_L=\bar{y}-t_{\alpha/2,N-1}\text{S.E.}(\bar{y})$$

Where N is large then

$$\hat{\bar{Y}}_u=\bar{y}+Z_{\alpha/2}*\text{S.E.}(\bar{y})$$
$$\hat{\bar{Y}}_L=\bar{y}-Z_{\alpha/2}*\text{S.E.}(\bar{y})$$

Now, for instance if we take=0.05 then 95% confidence interval for population mean can be put, for large sample, as follows:

$$[\bar{y}-1.96\hat{\text{S}}.\text{E}.(\bar{y}),\bar{y}+1.96\text{S.E.}(\bar{y})]$$

Thus, for the above example, 95% confidence interval for population mean, presuming that σ^2 is unknown, can be easily obtained as [4.98, 9.68] or is approximately put as [5, 10].

Determination of sample size: In any sample survey it is desirable to determine the sample size in advance so as to plan the survey in a proper way by specifying the degree of risk (or precision) in terms of permissible loss and level of confidence. Sample size is calculated by using the following formulae:

$n = \frac{NS^2}{ND + S^2}$ if we are using SRSWOR, and $n = \frac{S^2}{D}$ if we are using SRSWR.

Where D = $\left(\frac{\varepsilon}{t}\right)^2$, $t_{\alpha, N-1}$ is the tabulated value of Student's t-distribution and ε is permissible error.

In case S^2 is unknown its estimate i.e. sampling variance (s^2) is used which can be obtained through pilot survey.

Example: A study of sampling method was conducted in a population having 500 units, by total count it was obtained $\bar{y}$ = 49, S2 = 44.6.In a simple random sample, how many sampling units be selected to estimate Y with a permissible margin error of 5% of population & 95% confidence coefficient when sampling is done by using (i) WR (ii) WOR.

(i) If WR sampling is to be used

ε = 55 of 49 = 5/100*49 = 2.45

$t_{0.025,\ 499} = 1.96$

$$D = \left(\frac{\varepsilon}{t}\right)^2 = \left(\frac{2.45}{1.96}\right)^2 = 1.5625$$

$$n = \frac{NS^2}{ND + S^2} = \frac{500 \times 44.6}{500 \times 1.5625 + 44.6} = 27.002 \cong 27$$

(ii) If WOR sampling is to be used

$$n = \frac{S^2}{D} = \frac{44.6}{1.5625} = 28.544 \cong 29$$

Estimation of population size through capture-recapture method: For the estimation of biological populations, which are generally mobile in nature, sample size can not be fixed in advance. For such a situation, another sampling approach known as sequential sampling or inverse sampling, which uses capture-recapture technique, is used.

Example: A large number of field rats inhabit a state sugarcane farm and are causing severe damage to the sugarcane crop. The farm officials are concerned about it. Before applying the suitable rodenticide, the officials wish to estimate the total population of the rats in the farm. They trapped 125 rats at random locations. The rats were marked and then released at scattered points at the farm. After three days 256 rats had to be trapped from the farm until 25 were found to be marked ones. Treating the 256 rats as (i) WR (ii) WOR random sample, estimate the rat population.

N = the size of the population being studied
n = the size of the sample being studied which is random = 256
t = the number of marked individuals in the population =125
s = the number of marked individuals recovered in the sample =25

Now, $\hat{N}$ the estimated population size is calculated as below:

In case of WR sampling $\hat{N} = \frac{nt}{s} = \frac{256 * 125}{25} = 1280$

In case of WOR sampling $\hat{N} = \frac{nt}{s} - 1 = \frac{256 * 125}{25} - 1 = 1279$

Testing of Hypothesis

Before conducting survey or any experiment, a researcher generally frame certain hypothesis and wishes to obtain valid inferences from the collected information regarding parameter to be studied. The statistical statement regarding the form of a parameter of the distribution is called hypothesis. A hypothesis of no difference or of no effect is known as null hypothesis and is denoted by H_0. Any statement which is alternative to H0 is called alternative hypothesis and is denoted by H_1.

Types of error in decision making: During the course of investigation, one may commit two types of errors which are explained below:

Decision	Actually in the population H_0 True	H_1 True
Reject H_0	Wrong decision (Type I error)	Correct decision
Accept H_0	Correct decision	Wrong decision (Type II error)

The size of type I error denoted by Greek letter (alpha) is the probability α = [Reject H0/ H0]. It is also known as level of significance and abbreviated as l.o.s. Generally we take = 0.5 or 0.01. Here = 0.05 indicates that there is possibility of taking wrong decision in 5% cases.

The size of type II error is obtained as $\beta = P$ [Accept H_0/H_1]. Power of the test is judged by the quantity 1-β. Therefore, in decision making minimization of both the types of errors are important in order to increase the confidence level as well as power of the test. The value of test statistic which separates the rejection region and the acceptance region is called critical value or significant value.

Procedure of Decision Making: The steps involved in testing a statistical hypothesis have been given schematically as follows:

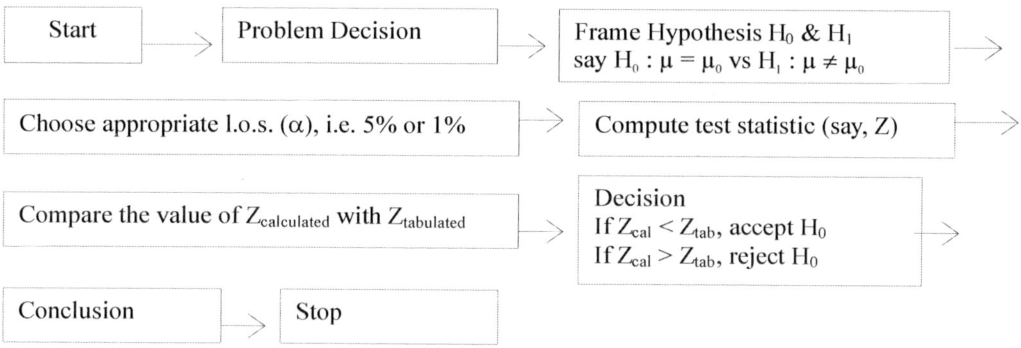

Several types of tests can be found in literature for various situations depending on the nature of the problem under study. These tests are broadly categorized as parametric tests (used for quantitative type of data) and non-parametric tests (used for qualitative type of data or when assumptions of parametric test is not fulfilled). Below are given some tests which are frequently used for entomological studies:

Testing hypothesis about one population mean ($\mu=\mu_0$) & difference of two population means ($\mu_1=\mu_2$)

Following table provides the situations under which Z-test or Student's t-test is used:

Actually in the population	Sample size large ($n \geq 30$)	Sample size small ($n<30$)
variance (σ^2) known	Z-test	Z-test
variance (σ^2) unknown	Z-test	Student's t-test

One Sample Test

Example: A random sample of 100 plants is found to have mean plant height of 66.5 cm. can it be reasonably regarded as a sample from a large population with mean height of 65.5 cm and standard deviation 3.4 cm? How would you proceed if sample size is considered as 25 with same mean values?

Solution: Here we have to test the null hypothesis

H_0: μ = 65.5 cm. against H_1: $\mu \neq$ 65.5 cm.

Case I: We are given that n = 100, $\bar{x}$ = 66.5, μ_0 = 65.5 cm, σ = 3.4 cm. Since sample is of large size and population variance is known, Z-test is to be applied. The value of test statistic is to be calculated as follows:

$$Z = \frac{\bar{x} - \mu_0}{\sigma / \sqrt{n}}$$

$$Z = \frac{66.5\text{-}65.5}{3.4/\sqrt{100}} = \frac{1.0}{0.34} = 2.94$$

Let us take α = 0.05 for which Z_{tab} = 1.96. Since Z_{cal} = 2.94 > 1.96, null hypothesis is being rejected. Thus, we conclude that the given sample is not a true representative of the population with mean height 65.5 cm. and standard deviation 3.4 cm.

Case II: If we take n = 25 i.e. sample is of small size. But the population variance is known, therefore, we still use z-test and the value of test statistic is calculated as follows:

$$Z = \frac{66.5\text{-}65.5}{3.4/\sqrt{25}} = \frac{1.0}{0.68} = 1.47 < 1.96$$

Hence, for this case H_0 is accepted and the sample is said to be true representative of the population.

Student's t- test is used when sample is of small size (n<30) and population variance is unknown, for which the test statistic is given as $t = \frac{\bar{x}\text{-}\mu_0}{s/\sqrt{n}}$

Example: A random sample of 10 framers selected from a large group of farmers regarding per ha N (in kg) applied gave the values as 105, 117, 135, 109, 124, 121, 117, 108, 126, and 128 respectively. Do you agree with the claim that the farmers adopted the recommended dose of nitrogen for wheat crop as 120 kg/ha?

Solution: In this problem we have to test the null hypothesis

H_0: μ = 120 kg. N/ha

against H_1: μ ? 120 kg. N/ha (two tailed test)

Since sample is of small size and variance is unknown t-test is to be used. The test statistic is given by

Now, sample mean ($\bar{x}$) and sample S.D. (s) is to be computed from sample values.

Here n = 10, $\bar{x}$ = 1190/10 = 119 and the sample variance will be computed as

$$s^2 = \frac{1}{n-1}\sum(x_i - \bar{x})^2 = \frac{1}{9} \times 840 = 93.33$$

Which gives s = 9.66.

Therefore,

$$t = \frac{119\text{-}120}{9.66/\sqrt{10}} = \frac{-1}{3.055} = -0.327$$

Tabulated value of t-statistic with (10-1) i.e. 9 d.f. at 5% l.o.s. for two-tailed test is 2.262.

Since calculated value of $|t| = 0.327$ is less than $t_{9,\,0.05}$, H_0 may be accepted at 5% level of significance and it may be concluded that he farmers adopt the recommended dose of nitrogen i.e. 120 kg N/ha.

Two Samples Test

Testing H_0: $\mu_1 = \mu_2$, against the possible alternatives hypotheses viz. H_1: $\mu_1 \neq \mu_2$ or H_0: $\mu_1 > \mu_2$ or $H_0 : \mu_1 < \mu_2$

Z-test statistics for the two situations viz. (a) two population variances are equal, and (b) unequal, are given below :

(a) $\sigma_1^2 = \sigma_2^2 = \sigma^2$ (common variance)

$$Z = \frac{\bar{x} - \bar{y}}{\sqrt{\sigma^2\left(\frac{1}{n_1} + \frac{1}{n_2}\right)}} = Z_c \text{ (say)}$$

(b) $\sigma_1^2 \neq \sigma_2^2$ (Two known population variances are unequal)

$$Z = \frac{\bar{x} - \bar{y}}{\sqrt{\left(\frac{\sigma_1^{\,2}}{n_1} + \frac{\sigma_2^{\,2}}{n_2}\right)}}$$

Remark : In the above situations if population variances (σ_1^2 and σ_2^2) unknown then they are to be estimated through sample values.

Student's t- test is used when samples' sizes are small (n<30) and population variances are unknown and are assumed equal, for which the test statistic is given as follows:

$$t = \frac{\bar{x} - \bar{y}}{\sqrt{s^2\left(\frac{1}{n_1} + \frac{1}{n_2}\right)}}$$

where $s^2 = \frac{(n_1 - 1)s_1^{\,2} + (n_2 - 1)s_2^{\,2}}{n_1 + n_2\text{-}2} = \frac{[\sum(x_i\text{-}\bar{x})^2 + \sum(y_i\text{-}\bar{y})^2]}{n_1 + n_2\text{-}2}$ is the pooled sample variance.

Test Criterion

H_1: $\mu_1 \neq \mu_2$	(two -tailed test)
If $\lvert t_c \rvert < t_{n_1+n_2-2,\ \alpha/2}$	Accept H_0
If $\lvert t_c \rvert \geq t_{n_1+n_2-2,\ \alpha/2}$	Reject H_0 at α% l.o.s.

Example

Paired t-test : Sometimes we may be interested to test the significance of difference between two sample means for the case where samples are dependent to each other. For instance, we may wish to see whether there is any difference between number of insects per plant before and after the application of a particular insecticide. Here, it is worth mentioning that this test is applicable to the cases where object for which observation is to be recorded is same. The null hypothesis in this case is formulated as H_0: x-y = 0 or d = 0

This is to be tested against the alternative hypothesis

H_1 : x-y ≠ 0 or d ≠ 0

The test statistic under null hypothesis is given by:

$$t = \frac{\bar{d}}{sd/\sqrt{n}} \sim t_{n-1,\alpha} \quad \text{where} \quad \bar{d} = \frac{1}{n}\sum_{i=1}^{n} d_i$$

The following table shows the mean number of bacterial colonies per plate from soil samples taken at 4 P.M. and 8 P.M. respectively.

Time	**1**	**2**	**3**	**4**
4 P.M.	29.75	27.50	30.25	27.80
8 P.M.	39.20	40.60	36.20	42.40

Is there any significant difference in the bacteria in the soil samples collected at 4 P.M. & 8 P.M.?

Here the observations are in pairs hence paired t-test is to be used and the hypotheses are framed as follows:

H_0: d = 0 against H_1: d ≠ 0

The sample variance of difference is calculated as

$$sd^2 = \frac{1}{n-1}\left[\sum d_i^2 - \frac{\left(\sum d_i\right)^2}{n}\right] = \frac{1}{3}\left[509.475 - \frac{(43.10)^2}{4}\right]$$

$$= \frac{1}{3}[509.475 - 464.4025] = 15.024$$

So, $sd = \sqrt{15.024} = 3.876$ and $\bar{d} = \frac{1}{4} \times 43.10 = 10775$

Thus,

$$t = \frac{10.775}{3.876/\sqrt{4}} = \frac{21.55}{3.876} = 5.56$$

The tabulated value of $t_{3,\ 0.05}$ = 3.18. Since t_{cal} > t3, 0.05 therefore null hypothesis is rejected and we infer that the difference in number of bacterial colonies at two different timings is statistically significant.

Test for proportions: Sometimes we may be interested in the testing of qualitative type of data such as proportion or percentage infestation. Let 'p' be the population proportion (the ratio of members possessing certain attributes 'A' to the entire population) then the hypotheses for the test of single proportion are framed as:

H_0: p = p_0

H_1: p ≠ p_0 or H_1: p > p_0 or H_1: p < p_0

Example : A researcher has developed a new insecticide which is believed to kill more than 95% of cotton borers upon contact. In a separate laboratory test, the insecticide was applied to 200 cotton borers. Although all 200 eventually died, only 180 died immediately upon contact. Do the data contradict at = 0.01 the claim that the insecticide can kill 95% of cotton borers on contact ?

In this example we are given population proportion p=0.95. The sample proportion is obtained as:

$$\hat{p} = \frac{\text{no. of members possessing the given attribute}}{\text{Total no. of individual in the sample}} = \frac{180}{200} = 0.90$$

Under null hypothesis the value of test statistic is obtained as:

$$Z = \frac{\hat{p} - p_0}{\sqrt{\frac{p_0 q_0}{n}}} = \frac{0.90 - 0.95}{\sqrt{\frac{0.95*0.05}{200}}} = \frac{0.05}{0.0154} = 3.246$$

Test criterion: Here $|Z_{cal}|$ > 2.58,therefore null hypothesis is rejected indicating thereby that the claim is true.In this case confidence interval will be

$$\hat{p} \pm Z_{\alpha/2}\sqrt{\frac{pq}{n}}$$

Test for difference of two proportions: Let there are two population proportions p1 & p2, respectively as regards to certain attributes say 'A'. Now we draw independent random samples from the given population & compute the sample potions as

$$\hat{p}_1 = \frac{x_1}{n_1} \text{ and } \hat{p}_2 = \frac{x_2}{n_2}$$

For testing the null hypothesis $H_0 : p_1 = p_2$ against $H_1 : p_1 \neq p_2$, the test procedure is explained by the following example:

Example : There were two insecticides A and B. insecticide A when applied killed 25 out of 35 insects and B when applied killed 32 out of 40 insects. It is desired to test that two insecticides are equally efficient. Here

$$\hat{p}_1 = \frac{x_1}{n_1} = \frac{25}{35} = 0.71 \text{ and } \hat{p}_2 = \frac{x_2}{n_2} = \frac{32}{40} = 0.80$$

Therefore, $\hat{p} = \frac{n_1\hat{p}_1 + n_2\hat{p}_2}{n_1 + n_2} = \frac{25+32}{35+40} = 0.76$ and $\hat{q} = 1 - 0.76 = 0.24$

Now, $$Z = \frac{0.71\text{-}0.80}{\sqrt{0.47 * 0.24\left(\frac{1}{35} + \frac{1}{40}\right)}} = -0.09$$

We get $Z_{cal} = |-0.09| = 0.09 < 1.96$, therefore null hypothesis is to be accepted indicating thereby that two insecticides under comparison are equally effective.

χ^2 test for homogeneity of several proportions : If there are more than two populations on which study is to be made to observe whether proportion that an individual is having certain common attribute say A differ from population to population.

H_0: $p_1 = p_2 = \ldots\ldots\ldots\ldots\ldots = p_k = p$

H_1: not all p_i are same.

In this case χ^2 test as given below is used to test the homogeneity of several proportions.

$$\chi^2 = \frac{\sum p_i x_i - \bar{p}A}{\bar{p}\bar{q}} \sim \chi^2_{k-1} \quad \textit{Where } \bar{q} = 1 - \bar{p}$$

If $\chi^2_{cal} < \chi^2_{\alpha, k-1}$, we may conclude that all the population are homogenous regarding the proportion of given attribute.

Correlation and Regression

Correlation: If we record the yield of a crop (y) from n number of plots of same area along with fertilizer applied (x) to them, then the set representing pairs of such observations is known as bivariate distribution. A distribution involving more than two variables is termed as multivariate distribution. And, for studying the relationship between two or more variables we use statistical tool what is known as correlation. Correlation refers to degree and direction of association ship between related variables.

Simple Correlation : If we confine to two variables only, it is termed as *simple correlation*. It is measured with the help of *coefficient of correlation* denoted by r_{xy} or singly by r. It is also known as Karl Pearson's coefficient of correlation, and is given, under usual notations, by the following formula:

$$r_{xy} = \frac{\text{Cov}(X,Y)}{\sqrt{\text{Var}(X)\text{Var}(Y)}} = \frac{n\sum x_i y_i - \sum x_i \sum y_i}{\sqrt{[n\sum x_i^2 - (\sum x_i)^2][n\sum y_i^2 - (\sum y_i)^2]}}$$

On the basis of degree and direction of association ship between variables, simple correlation can be categorized as follows :

(i) *Positive correlation*: When increase (decrease) in one variable result in an increase (decrease) in other variable, correlation is called positive correlation ($0 < r_{xy} < 1$).

(ii) *Negative correlation*: When increase (decrease) in one variable result in a decrease (increase) in other variable, relationship is termed as negative correlation ($1 < r_{xy} < 0$).

(iii) *Perfect correlation*: When increase (decrease) in one variable is associated with increase (decrease) in other variable in same proportion then correlation is said to be perfect (positively or negatively) ($r_{xy} = \pm 1$).

Note

1. The simple correlation coefficient r_{xy} lies between – 1 and + 1 and these are known as limits of correlation coefficient. If X and Y are independent or uncorrected then $r_{xy} = 0$, but not vice- versa.

2. To have a rough idea about above mentioned categories scatter diagram is also helpful. The n points plotted with co-ordinates (x_i, y_i) (i =1, 2,.., n) is called *scatter diagram.*

Multiple correlation : When there are more than two variables then the association ship between a variable & the joint effect of rest variables is termed as multiple correlation which is measured by *multiple correlation coefficient* denoted by R. For instance, let us take 3 variables X_1, X_2 & X_3 , then $R_{1.23}$ is the measure to find out the relationship between X_1 & joint effect of two other variables X_2 & X_3 on it and is obtained by the following formula:

$$R_{1.23} = \sqrt{\frac{r_{12}^2 + r_{13}^2 - 2r_{12}r_{13}r_{23}}{1 - r_{23}^2}}$$

Where r12, r13 & r23 are the simple correlation coefficients between respective variables. Value of R lies between 0 and 1.Square of multiple correlation coefficient (R2) is called coefficient of multiple determination or coefficient of determination. When expressed in terms of percentage gives an idea about the percent variation in the dependent variable due to the joint effect of all the independent variables. This serves as an indicator for the validity of regression model.

Partial Correlation: While studying the relationship between more than two variables sometimes one may be interested to have the relationship between two variables only by eliminating the effect of rest of the variables. This sort of relationship is called partial correlation can be measured with the help of partial correlation coefficients denoted by r12.3 etc. The partial correlation coefficient also lies between 1 and + 1. It is calculated by the formula (taking three variables only) as

$$r_{1.23} = \frac{r_{12} - r_{13}r_{23}}{\sqrt{1-r_{13}^2}\sqrt{1-r_{23}^2}}$$

Correlation matrix : When all possible simple correlations are expressed in the form of matrix, it is known as Correlation matrix. For example with four variables X1, X2, X3 & X4 it can be expressed as:

$$R = \begin{bmatrix} 1 & r_{12} & r_{13} & r_{14} \\ r_{21} & 1 & r_{23} & r_{24} \\ r_{31} & r_{32} & 1 & r_{34} \\ r_{41} & r_{42} & r_{43} & 1 \end{bmatrix}$$ which is obviously a symmetric matrix.

Testing significance of simple correlation coefficients:

H_0: $\rho_{12} = 0$

Vs H_1: $\rho_{12} \neq 0$

$$t = \frac{r}{\sqrt{1-r^2}}\sqrt{(n-2)} \sim t_{\alpha,n-2}$$

Testing significance for multiple correlation coefficients:

H_0: $R_2 = 0$

Vs H_1: $R_2 \neq 0$

$$F = \frac{R^2}{1-R^2} \cdot \frac{n-k-1}{k} \sim F_{\alpha;k,n-k-1}$$

n = total no. of observations.

k = no. of independent variables.

Testing significance for partial correlation coefficients:

$$H_0: \rho_{12.3} = 0$$

$$\text{Vs} \quad H_1: \rho_{12.3} \neq 0$$

$$t = \frac{r_{12.3}}{\sqrt{1 - r_{12.3}^{\ 2}}} \sqrt{(n-2)-k} \sim t_{\alpha, n-3}$$

Regression : In the preceding part the problems related with correlation have been dealt with in which there was no consideration of dependent or independent variable. But, the term regression refers to functional relationship between two or more variables in which one variable is taken as dependent and the other variables as independent ones. Let us restrict ourselves to the case of two variables. The functional form can be looked at graphically by plotting pair of points (xi, yi) (i = 1, 2,..., n) on xy-plane and joining them through free hand curve. The curve so obtained is known as regression curve. In particular if we get a straight line, it is called linear regression. If the functional form is non-linear, it is known as non-linear regression. Further, with one dependent and only one independent variable, the functional relationship is called simple linear/ non-linear regression whereas with more than two independent variables, it is termed as multiple linear/ non-linear regressions. Let Y be dependent variable X_1, X_2, ..., X_k be k independent variables. Then the population multiple linear regression equation will be

$$Y = \beta_0 + \beta_1 X_1 + \beta_2 X_2 + \ldots\ldots\ldots + \beta_k X_k + C$$

In the above regression equation β_i's are called the parameters which are unknown and are to estimated And, the fitted regression line obtained is

$$y = a + b_1 X_1 + b_2 X_2 + \ldots\ldots\ldots\ldots + b_k X_k$$

Example: Following are the observation showing mean population of insect /plant (averaged over five plants) of a vegetable crop alongwith weather data during the study period:

Mean population	Temp.Max.	Temp.Min.	RH(Mor.)	RH.(Evn.)	RainFall(mm.)
0.15	30.80	17.40	76.00	48.00	2.80
0.40	33.40	14.60	73.00	24.00	1.60
0.65	37.80	17.70	54.00	17.00	0.00
0.95	39.60	20.40	50.00	16.00	0.00
0.75	36.20	20.40	59.00	29.00	5.40
1.15	38.40	23.40	51.00	25.00	0.20
0.95	34.90	22.00	69.00	33.00	13.40
1.70	38.90	22.10	45.00	24.00	0.00

2.30	36.30	24.40	66.00	46.00	50.10
3.20	33.90	24.90	84.00	62.00	78.40
4.80	33.90	22.80	80.00	54.00	76.20
6.10	33.40	23.70	84.00	63.00	72.80
2.00	32.60	25.00	89.00	72.00	208.80
6.55	32.80	25.30	80.00	70.00	38.40
2.00	33.30	24.50	91.00	70.00	129.40
6.70	35.70	26.80	78.00	61.00	0.00
6.00	31.90	25.00	90.00	74.00	119.20
7.70	33.20	25.30	89.00	71.00	16.40
8.20	32.10	23.30	90.00	69.00	43.50
7.20	34.10	25.30	87.00	65.00	3.00
7.60	33.00	23.60	86.00	64.00	8.40
8.15	32.40	21.50	83.00	60.00	12.80
7.40	34.80	23.00	85.00	53.00	0.00
6.80	31.60	20.50	88.00	62.00	5.00
4.50	33.70	20.70	89.00	50.00	0.00
3.30	33.60	22.40	89.00	53.00	17.00

Correlation and regression analyses were performed through statistical software using SPSS 14.0 version and the out put of the result along with interpretation of results are presented below:

Model Summary

Model	R	R Square	Adjusted R Square	Std. Error of the Estimate
1	.903(a)	.815	.769	1.38425

a Predictors: (Constant), Rainfall, Temp Max, Temp Min, Rel Hum Mrn, Rel Hum Even

ANOVA (b)

Model		Sum of Squares	df	Mean Square	F	Sig.
1	Regression	169.133	5	33.827	17.653	.000(a)
	Residual	38.323	20	1.916		
	Total	207.456	25			

a Predictors: (Constant), Rainfall, Temp Max, Temp Min, Rel Hum Mrn, Rel Hum Even

b Dependent Variable: Mean Pop

Coefficients (a)

Model		Unstandardized Coefficients		Standardized Coefficients	t	Sig.
		B	Std. Error	Beta		
1	(Constant)	-33.511	14.301		-2.343	.030
	Temp Max	.948	.459	.762	2.064	.052
	Temp Min	-.412	.328	-.406	-1.256	.224
	Rel Hum Mrn	.014	.057	.068	.240	.813
	Rel Hum Even	.282	.087	1.827	3.233	.004
	Rainfall	-.033	.006	-.586	-5.116	.000

a Dependent Variable: Mean Pop

Remark : here standardized coefficients are also called *path coefficients.*

Regression model can now be written as:

$Y = -33.511 + 0.948X_1 - 0.412X_2 + 0.014X_3 + 0.282X_4 - 0.033X_5$ ($R^2 = 0.815^{**}$)

From the analysis it can be inferred that the model is of best fit. Further, the abiotic factors contribute 81.5% towards population build up.the rest variation i.e. 18.5% is due to some other factors not included in the model. The values of the coefficients of independent variables indicate the amount of change in the value of dependent variable with a unit change in the corresponding independent variable. Further improvement in the model can be done through stepwise, a procedure which is done through:

Correlation Matrix

	Mean Pop	Temp Max	Temp Min	Rel Hum Mrn	Rel Hum Even	Rainfall
Mean Pop	1	-.470(*)	.525(**)	.647(**)	.724(**)	-.033
Temp Max		1	-.075	-.868(**)	-.769(**)	-.346
Temp Min			1	.373	.664(**)	.428(*)
Rel Hum Mrn				1	.893(**)	.409(*)
Rel Hum Even					1	.527(**)
Rainfall						1

* Correlation is significant at the 0.05 level (2-tailed).

** Correlation is significant at the 0.01 level (2-tailed).

Forward selection: Add one variable at a time to the model, on the basis of its *F* statistic

Backward elimination: Remove one variable at a time, on the basis of its *F* statistic

Stepwise regression: Adds variables to the model and subtracts variables from the model, on the basis of the *F* statistic.

Analysis of Variance

In agricultural field trials experience shows that even if the same variety was to be sown on all plots there would still be substantial variation in yield from plot to plot. Similar is the case with manurial trials, management trails and other agricultural experiments. In such experiments it may be realized that neighbouring plots are generally more alike than distant plots in term of yield of a crop. There may be locally periodic variations across the field or even if treatment is repeated over place or time, there may be substantial change in mean yield. The aim of such experiments is the comparison of treatments and is called '*Comparative experiments*'. When comparison is to be made among more than two treatments, analysis of variance is used.

The concept of analysis of variance (ANOVA) was originally introduced by Prof. Ronald Arthur Fisher in 1920s to deal with the problem in the analysis of agronomical data. According to him "ANOVA is the technique of separation of variance ascribable to one group of causes from the variance ascribable to other group".

Thus, Total variation = Controlled variation + Uncontrolled variation

Assumptions in ANOVA

ANOVA is applied for testing the homogeneity of several means for which F-test is applied. And, for the validity of F-test following assumptions must be satisfied.

i) *Normality* : Population from which the observations are taken is normal.

ii) *Independence* : Observations are independent to each other.

iii) *Homoscedastcity* : Variances of experimental error are constant

iv) *Additivity* : Various effects are additive.

Common Violations in Agricultural Experiments & their Remedial Measures

For most experimental data in agricultural research the assumptions underlying the analysis of variance are, in general, satisfied, but one may face certain types of experiments where one or more assumptions are violated.

One of the most important assumptions made in analysis of variance is that the observations follow normal distribution. In general, in agricultural field experiments characteristics like yield measurements, height and girth measurements of the plants etc. follow normal distribution. However, certain types of measurements like percentage infestation, germination percentage, insect population, weeds per plot etc. do not follow normal distribution. It is, therefore, necessary to transform the data into a different scale so that the transformed data may follow normal distribution. Several types of transformations have been evolved for different types of data. Some guidelines are given below which indicate the appropriate transformation for a given set of data.

Firstly, mean and variance for each treatment may be worked out and a graph between mean and variance may be drawn. In case we get a straight line, we may conclude that the variance is varying with the mean.

If variance is proportional to mean, $y = \sqrt{x}$ transformation (square root transformation),where x is the observed value, may be used. This is useful especially when the data consists of integral values such as number of bacterial columns in a plate count, or the number of insects per trap etc. The use of $y = \sqrt{x + 1/2}$ instead of $y = \sqrt{x}$ is recommended when zeros occur in the data.

When mean is proportional to standard deviation, the transformation used is y = log x (*logarithmic transformation*). This transformation is particularly useful when the observed data consists of big integers such as index numbers or biological populations like nematodes etc.

When data relate to binomial populations, an angular transformation (also termed as *sine inverse transformation*) as given by the equation $\theta = \text{Sin}^{-1}\sqrt{p}$ may be used, where θ is in degrees and p is the percentage value expressed as a fraction (or proportion).

However, when all the percentage values lie between 30 and 70, the data are expected to follow approximately a normal distribution. Hence, no transformation is required. When there are zeros also, we put 1/4 for 0 and n-1/4 for n and obtain the percentage. Further, if all the values lie either between 0% to 30% or between 70% to 100%, but not both, square root transformation is applied.

Here, it should be noted that once the transformation has been made the analysis is carried out in usual way with the transformed data and the conclusions are drawn on the basis of transformed variate.

Design of experiment: Design of experiment may be defined as the logical construction of the experiment in which the degree of uncertainty with which the inference is drawn may be well defined.

In other words, the plan in which an experiment has to be conducted so that all situations, expect that of treatments or factors are kept under control; as much as possible, are known as experimental design. These designs may be for crop experiments including experiments on IPM, animal experiments, engineering, chemical industry etc.

Principles of Designing an Experiment

According to Prof. R.A. Fisher, the basic principles of the design of experiments are replication randomization, and local control. The details of these principles can be found in any standard book on 'Design of Experiments'. The importance of these principles can be depicted with the help of Fisher's Diagram (fig.1) as follows:

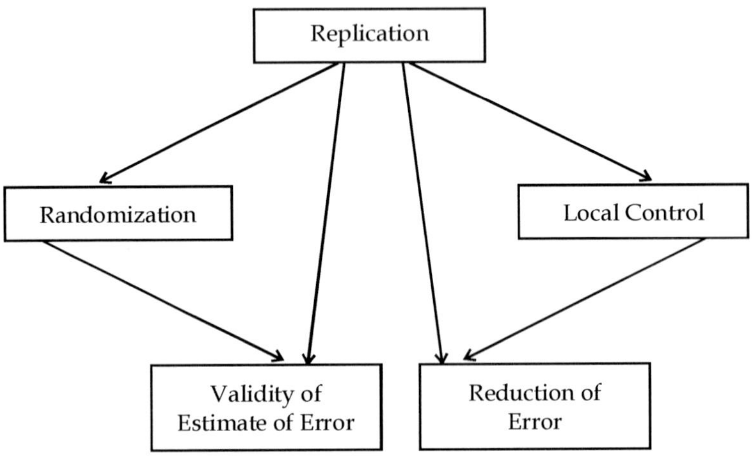

Now in the forthcoming discussions the concept of some statistical designs commonly used in agricultural experiments for data analysis have been given as follows:

Simple Experimental Designs

Completely Randomized Design: It is simplest of all the designs, based on principles of randomization and replication only Missing plot or unequal replications do not create any difficulty in the analysis of *Completely Randomized Design* (CRD).This design is most useful in laboratory techniques and methodological studies e.g. in physics, chemistry or cookery, in chemical and biological experiments, in some green house studies etc. where either experimental material is homogeneous or the intrinsic variability between units can be reduced. It is also recommended in situations where an appreciable fraction of units is likely to be destroyed or fail to respond.

Model: The linear model, assuming various effects to be additive, is given by

$$y_{ij} = ì + á_i + e_{ij} \quad \begin{pmatrix} i = 1, 2, ..., t \\ j = 1, 2, .., r_i \end{pmatrix}$$

Where y_{ij} is the yield or response of ith treatment from the jth unit.
μ is the general mean effect
α_i is the effect due to i^{th} treatment
e_{ij} is the error effect due to chance such that $e_{ij} \sim N(0, \sigma_e^2)$

ANOVA table for CRD

Source of variation	Degree of freedom (d.f.)	Sum of squares (SS)	Mean sum of squares (MSS)	Variance Ratio $F_{calculated}$
Treatment (between classes)	t-1	$SST = \sum_i \left(\frac{T_{i.}^2}{n_i} \right) - C.F.$	s_R^2	$s_T^2 / s_E^2 \sim F_{\alpha ;(t-1),(n-t)}$
Error (within the classes)	n-t	SSE = TSS – SST	s_E^2	
Total	n-1	$TSS = \sum_i \sum_j y_{ij}^2 - C.F.$		

- Correction factor (C.F.) = $\frac{G^2}{n}$ and grand total $G = \sum_i T_{i.} = \sum_i \sum_j y_{ij}$
- The SS divided by its degrees of freedom gives the corresponding mean sum of squares.

Test Procedures : We compare the calculated value of F with its tabulated value and if the F_{cal} value exceeds $F_{k-1, n-k, \alpha}$ then we reject the null hypothesis H_0: $\alpha_1 = \alpha_2 = \ldots = \alpha_t$ and conclude that effects of all the levels are not equal at á% level of significance otherwise not. In case the treatment effects come out to be significant we go for further pair comparison by using proper tests viz. *Least Significance Difference (lsd) test or Critical Difference (C.D.), Duncan's Multiple Range Test (DMRT), Tukey's hsd* test etc. However, the simplest and frequently used ones are *lsd* or *C.D.* and DMRT which are discussed as follows:

The estimated standard error of the difference between i^{th} and j^{th} treatment means is

$$SE_d = \sqrt{s_E^2\left(\frac{1}{r_i}+\frac{1}{r_j}\right)} \text{ if } r_i \neq r_j \text{ and } SE_d = \sqrt{\frac{2s_E^2}{r}} \text{ if } r_i = r_j = r.$$

Critical Difference: The critical difference (C.D.) can be calculated as

C.D. = $SE_d \times t_{\acute{a},\ error\ d.f.}$

Note: Any two treatment means are said to differ significantly if their difference is larger than the C.D. value. Here it may be noted that value of C.D. is generally three times the standard error of treatment means.

Duncan's Multiple Range Test: In a set of t treatment if comparison of all possible pairs of treatment means is required we can use Duncan's multiple range test (DMRT). It involves number of steps, which are as follows:

Step 1: Arrange the treatments in decreasing order, that is, according to their ranks.

Step 2: Calculate the standard error of mean as $SE_d = \sqrt{\frac{2s_E^2}{r}}$

Step 3: From statistical tables write the significant studentized ranges (r_p) for p=2, 3... t treatments and error degrees of freedom.

Step 4: Calculate the shortest significance ranges (R_p), where $R_p = r_p \times SE_m$

Step 5: From the largest mean subtract the R_p for largest p. Declare all the means less than this value as significantly different from the largest mean.

For the remaining treatments whose values are larger than the difference (Largest mean-larges R_p), compare the differences with appropriate R_p value. If two treatments are remaining compare with R_2; if three are remaining compare with R_3; and so on.

Step 6: From the second largest mean subtract the second largest R_p and compare the treatments as in step 5.

Continue this process till all treatments are covered.

Step 7: Present the results by using either the line notation or the alphabet notation to indicate which treatments are at par and which are significantly differ.

Randomized Block Design: If experimental units show the variability in one direction, the effect of this kind of heterogeneity is taken care of by dividing the experimental units into homogeneous groups of units known a blocks.

The treatments are randomly allocated separately to each of these blocks such that each treatment occurs precisely once in each block. This procedure gives rise to a design known as *Randomized Block Design* (RBD). It is the most frequently used design in agricultural experiments. The principal advantage of RBD is that it increases the precision of the experiment. This is due to the reduction of experimental error by adoption of local control.

When the number of treatments is increased, the block size is large and it may be difficult to maintain homogeneity within blocks. Consequently, the experimental error will be increased. Hence, RBD may not be suitable for large number of treatments. When the number of treatments tested is more than 10, the complete block designs are to be abandoned for experimentation and incomplete block designs are needed in such situations.

Model : The model of an RBD is expressed in linear additive form as $y_{ij} = \mu + \alpha_i + \beta_j + e_{ij}$

Where y_{ij} is the value of the variate for the ith treatment in the jth block (i = 1, 2, .., t; j = 1, 2, ..., r) and the rest symbols have their usual meanings.

ANOVA table for RBD

Sources	**d.f.**	**SS**	**MSS**	$\mathbf{F_{calculated}}$
Replicates or Blocks	(r-1)	$SSR = \sum_j \frac{R_j^2}{t} - C.F.$	s_R^2	s_R^2 / s_E^2
Treatments	(t-1)	$SST = \sum_i \frac{T_i^2}{r} - C.F.$	s_T^2	s_T^2 / s_E^2
Error	(r-1)(t-1)	By subtraction	s_E^2	
Total	rt-1	$TSS = \sum_i \sum_j y_{ij}^2 - C.F.$		

Test criterion is similar as discussed in case of CRD. For the calculation of C.D. we use $SE_d = \sqrt{\frac{2s_E^2}{r}}$

Factorial Experiments: When different factors influence a character under study, it is always desirable to test different combinations of the factors at various levels. Such experiments are called *factorial experiments*. Consider a factorial experiment with n factors, the jth factor being experimented with s_j levels (j = 1, 2, ..., n) Clearly there are $S_1 \times S_2 \times \ldots \times S_n$ different treatment combinations. The experiment is said to be *symmetrical* if $s_j = s$ for all j = 1, 2,

..., n otherwise it is called *asymmetrical factorial experiments*. For example, 2 x 2 (2^2) ; 2 x 2 x 2 (2^3);3 x 3 (3^2); 3 x 3 x3 (3^3) etc. are symmetrical whereas 2 x 3; 3 x 4; 3 x 5 ; 4 x 6 etc. are asymmetrical factorial experiments It is important to note that a factorial experiment is not a factorial design. It can be conducted in any of the designs.

Advantages

- In factorial experiments, we have an opportunity to evaluate the combined effect of two or more factors when they are used simultaneously. Hence, we study the individual effects of each factor as well as their interactions.
- In factorial experiments, a wide range of factor combinations is used. This helps us to predict what will happen when two or more factors are used in combinations. For example, we can evaluate the responses of crop varieties when they are subjected to different fertilizer treatments and crop protection measures. This will provide us a reliable basis for making practical recommendation.
- The factorial experiments will result in a considerable saving of the experimental resources. For example, consider an experiment in which we want to evaluate the responses of five varieties of paddy under three levels of nitrogen. For a satisfactory precision, we require at least 12 degrees of freedom for error variance. Hence, we require four replications for the experiment on varieties alone. The total number of plots required for the experiment on varieties is 20. Similarly, the total number of plots required for the experiment on nitrogen is 21. We need 41 plots for two experiments. But for conducting a factorial experiment with 15 combinations of the two factors we require only 30 plots for similar precision.
- Further the time required for the combined experiment is less than that required for the separate experiments.

Disadvantages

- When the number of factors or levels of factors or both are increased, the number of treatment combinations will increase. Consequently, the block size will increase. If the block size increases it may be difficult to ensure homogeneity of experimental material. This will lead to increased experimental error and loss of precision in the experiment.
- In some cases, we may not be interested in certain treatment combinations. However, we are forced to incorporate them in the experiment. This leads to wastage of the experimental material as well as time.

Three types of effects are measured in factorial experiments:

Simple effects: The simple effect of a factor is the difference between its responses for a fixed level of other factors.

Main effects: The main effect of a factor may be defined as a measure of change in response due to change in the level of the factor averaged over all levels of all other factors in a factorial experiment.

Interaction effect(s): The interaction between two factors may be defined as the failure of a factor to give the same response at various levels of the other factor(s).

In general, an interaction is an additional effect due to the combined effect of two or more factors.

For illustrations, consider the following data related to wheat yield (average of replicates):

Variety(V)	Nitrogen(N)		Mean	Simple effect of N
	n_0	n_1		
v_1	32	40	36.0	40-32=8
v_2	37	47	42.0	47-37=10
Mean	34.5	43.5	39.0	
Simple effect of V	37-32=5	47-40=7		

Main effect of variety = (5+7)/2 = 12/2 = 6.0
Main effect of nitrogen = (8+10)/2 = 18/2 = 9.0
Interaction effect of N vs. V = (7-5)/2 = 2/2 =1.0

Factorial experiment with two factors at more than two levels

In this case the linear mathematical model is given by

$$y_{ij} = \mu + \alpha_i + \beta_j + (\alpha\beta)_{ij} + e_{ij} \ (i = 1, 2, .., p\ ; j = 1, 2, \ldots, q).$$

Example: A factorial experiment was conducted in RBD to study the effect of seed treatments in controlling white tip nematodes in paddy seeds. There were two factors, (1) seed treatment, and (2) duration of treatment. The four levels of the seed treatments were m0: water; m_1: 0.036% monocrotophos; m_2: 0.054% monocrotophos; m_3: 0.072% monocrotophos. The two levels of duration of treatment were d0: 30 minutes, and d_1: 1 hour. The numbers of nematodes per 100 seeds were observed after the treatment.

Here, the total treatment combinations are 8 (D_0M_0, D_0M_1, D_0M_2, D_0M_3, D_1M_0, D_1M_1, D_1M_2 and D_1M_3). Therefore, the number of mutually orthogonal contrasts is 8-1 = 7, which is equal to the number of degrees of freedom for treatments. ANOVA table for this problem will take following form:

Source of variation	d.f.
Replication	(r-1)=3
Duration (D)	(p-1)=1
Seed treatment (M)	(q-1)=3
D x M	(p-1) (q-1)=3
Error	(r-1) (pq-1)=21
Total	(rpq-1)=31

Split-Plot Design: In field experiments, sometimes some factors have to be applied to large experimental units in addition to some requiring plots of smaller size. In such situations split-plot designs are used. A split-plot design using an RBD for the first set of treatments (called the main-plot treatments), is obtained by allotting the main-plot treatments at random to the whole-plots of a block and then randomizing the second set of treatments (called the sub-plot treatments) within each whole-plot. This enables to test for the main-effects of the sub-plot treatments and the interaction of the whole-plot treatments and the sub-plot treatments more efficiently than the main effects of the main-plot treatments. This is because of the fact that main effects of the whole-plot treatments get confounded with block effects.

The model used is $y_{ijk} = \mu + \alpha_i + \beta_j + e_{ij} + \gamma_k + (\alpha\gamma)_{jk} + e_{ijk}$

Where y_{ijk} is the value of Observation in the k-th sub-plot, j-th main plot and i-th replication, (i = 1, 2, ..., r; j = 1, 2,..., p; k = 1, 2, .., q) the other symbols have their usual meanings.

The layout of the ANOVA will be as follows:

Sources	d.f.	SS	MSS	Fcalculated
Replications	(r-1)	SS_R	sR^2	$sR^2/ s EI^2$
Whole plot treatments (A)	(p-1)	SS_A	sA^2	sA^2/ sEI^2
Whole plot error (EI)	(r-1) (p-1)	SS_{EI}	$s EI^2$	
Sub-plot treatments (B)	(q-1)	SS_B	sB^2	$sB^2/ sEII^2$
Interaction (AB)	(p-1) (q-1)	SS_{AB}	sAB^2	$sAB^2/ sEII^2$
Sub-plot error (EII)	p(q-1) (r-1)	SS_{EII}	$sEII^2$	
Total	pqr-1			

Standard Errors and Critical Differences

i) Estimate of Standard Error of difference between two main plot treatment means is $SE_d = \sqrt{\frac{2s^2_{EI}}{rq}}$

ii) Estimate of Standard Error of difference between two sub-plot treatment means is $SE_d = \sqrt{\frac{2s^2_{EII}}{rp}}$

iii) Estimate of Standard Error of difference between two main plot treatment means at the same or different levels of the sub-plot treatments is $SE_d = \sqrt{\frac{[2(q-1)s^2_{EII} + s^2_{EI}]}{rq}}$

iv) Estimate of Standard Error of difference between two sub-plot treatment means at the same level of main plot treatments is $SE_d = \sqrt{\frac{2s^2_{EII}}{r}}$

Critical difference (C.D.) = SE_d x t_α; error d.f.

In case of (iii) the above formula for calculating C.D. can not be used because the ratio of the treatment difference to its standard error does not follow the t-distribution. For practical purpose, an approximation to exact test involves calculation of 5 per cent level of t by the following formula:

$$t = \frac{(q-1)s^2_{EII}t_\beta + s^2_{EI}t_\alpha}{(q-1)s^2_{EII} + s^2_{EI}}$$

Where t_α and t_β are t-values at error (EI) and error (EII) d.f.

C.D. for (iii) = SE_d x t (as obtained above)

Bioassays

A chemical test to establish what a substance contains is called an *assay*. In some experiments the researchers are interested to compare the efficacy of two or more substances in respect of some of their common effects. Bioassays may be defined as the technique through which the efficacy of two or more substances, or preparations like drug, insecticides, pesticides, etc. may be compared by using the responses produced by them on suitable living organism.

The objective in bioassay work is to estimate what level of stimulus is necessary to bring about the responses in a given percentage of individuals in the population. The parameter usually estimated in LD_{50} (median lethal dose) or ED_{50} (median effective dose) which is an indicator of response in 50 per cent of the individual in the population.

Normally, two preparations having a common effect are taken for assaying. One of the preparations is known strength and is called standard preparation and the other is of unknown strength and is called test preparation.

Let D_s and D_t denote the doses of standard and the test preparation, respectively, such that each of them produces a pre-assigned response in same living organism. Then the ratio r = D_s/D_t is called the relative potency of test preparation w.r.t. standard preparation.

If $\rho > 1$, smaller dose of test preparation produces comparatively much response

$\rho < 1$, potency of test preparation is smaller than standard preparation

$\rho = 1$, both the preparations are equally effective

Estimation of LD_{50} values

The probit transformation leads to a method of estimating LD_{50} values. A reliable estimate of this parameter would be available if an indefinitely large number of individuals were exposed to each dose of a given drug. However, for practical reasons, the number of animals given each dose can not be idenfinitely large.

Assume that a researcher is studying the toxicity of an insecticide using insects as the test organism. He uses insecticide concentrations varying from 0.3 units to 2.4 units. Assume further that the distribution of tolerance doses is normal. In this case the LD_{50} value is 1.2, which means that a dose of 1.2 will kill 50 per cent of the insects in the population. The standard deviation of the distribution is 0.3.

Dose	Percent killed	c	Probit values(c+5)
0.3	0.13	-3	2
0.6	2.27	-2	3
0.9	15.87	-1	4
1.2	50.00	0	5
1.5	84.13	1	6
1.8	97.73	2	7
2.1	99.87	3	8

If dose is plotted against per cent killed, the result is a sigmoid curve (Fig. 2). This curve can be drawn by using the table giving the areas under the normal curve. A dose of 0.3 units is three standard deviations left of the median; therefore, c = -3. For this dose 0.13 per cent of the insects will be killed. For a dose of 0.9 units, 15.87per cent of the insects will die etc. From such a sigmoid curve the LD_{50} value is read off by determining the value on the X axis for a value of 50% on the Y axis.

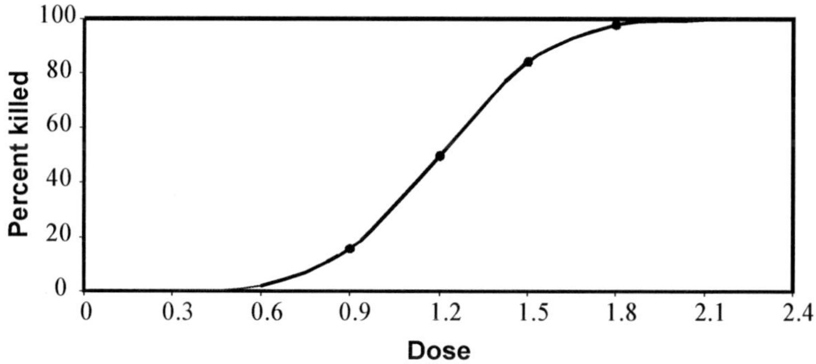

Fig. 1 : Relationship between dose and percent killed

Since a curvilinear relationship is cumbersome to work with, a transformation can be made to another scale so that a linear relationship exists between dose and response. One transformation is to convert the percentage values (proportion values) to c values. However, negative numbers can be a nuisance; therefore, a value of 5 is added to each c value. This is known as the *probit transformation*.

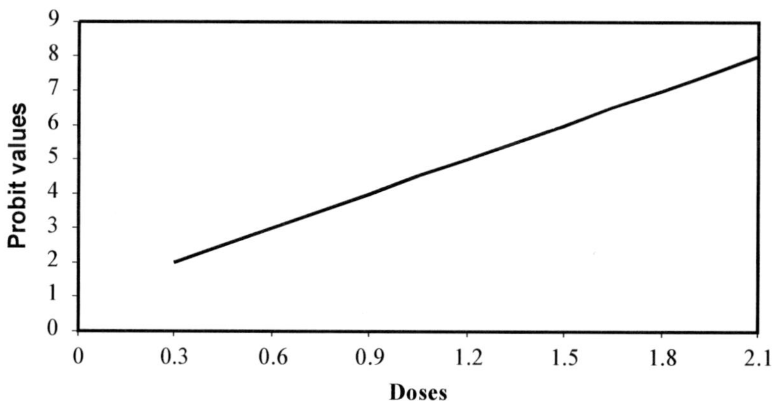

Fig. 2 : Relationship between dose and Probit values

The linear relationship between doses and probit values is illustrated in Fig. 3. The LD_{50} value on the X axis now corresponds to a probit value of 5.0 on the Y axis.

Reliability is also increased if the investigator is able to bracket the estimate well between the highest and the lowest dose, and at the same time use a small interval between doses.

References

Cochran, W.G. and Cox, D.R. 1987. *Experimental Designs.* John Wiley and Sons, New York.

Cox, D.R. 1958. *Planning of Experiments.* John Wiley and Sons, New York.

Das, M.N. and Giri, N.C. 1986. *Design and Analysis of Experiments.* Wiley Eastern, New Delhi.

Federer, W.T. 1963. *Experimental Designs.* Oxford and IBH Publishing Co., New Delhi.

Gomez. K.A. and Gomez. A.A. 1984. *Statistical Procedures for Agricultural Research.* The International Rice Research Institute, Philippines.

Pause, V.G. and Sukhatma, P.V. 1985. *Statistical Methods for Agricultural Workers.* ICAR, New Delhi.

Raghavarao, D. 1987. *Statistical Techniques in Agricultural and Biological Research.* Oxford & IBH Publishing Co., New Delhi.

Snedecor, G.W. and Cochran, W.G. 1989. *Statistical Methods.* Iowa State University Press, Ames, Iowa.

❐❐❐

Ecologically Based Integrated Pest Management, 2012
© D.P. Abrol and Uma Shankar (eds.), pp. 955-963
New India Publishing Agency, New Delhi (India)
e-mail : info@nipabooks.com; website : www.nipabooks.com

48

Future of Integrated Pest Management

D.P. ABROL

Introduction

The contribution of agrochemicals in improving food security and human health cannot be undermined. They are, however, like a double-edged weapon; their indiscriminate use could result in a serious threat to the sustainability of the agricultural production system and human health. Though their long-term effects on environment and human health are yet to be fully understood, the short-term adverse effects of their indiscriminate use became apparent soon after their invention during the World War II. Many insect pests have developed resistance to chemical pesticides, and a number of beneficial insects that are natural enemies of the pests have disappeared. Realizing these threats, the scientific community has been proactive and developed safer alternatives using flora and fauna as substitutes for chemical pesticides. These alternatives are claimed to be as effective as chemical pesticides. Experimental evidences indicate that these provide effective protection against pests when used in conjunction with other methods of pest control, including chemical pesticides. The strategy is often referred to as Integrated Pest Management (IPM). Since the adoption of IPM as a cardinal principle of plant protection in 1985, India has devised and implemented many IPM programmes encompassing research, extension and education with the objective to reduce the use of chemical pesticides, improve farm profitability, conserve environment and reduce adverse effect of pesticides on human health. Their effect is revealed in considerable reduction in pesticide-use, particularly during 1990s. The purpose of this paper is to provide a perspective on the IPM with emphasis on food security and safety.

IPM Research

The effectiveness of chemical pesticides in reducing the pest-induced losses has diminished in recent years. A number of insect pests have developed manifold resistance to the pesticides intended to control them. Further, with the destruction of natural enemies of insect pests, a number of new pests have emerged. These imply that intensive use of chemical pesticides is leading to increased cost of pest control and reduced farm profitability. Under such a situation, alternative technologies such as bio-pesticides could provide some solutions. Research has generated a number of technologies using plants and pathogens. Many of these have, however, not been commercialized perhaps due to lack of their proven economic feasibility, short shelf-life, slow effect and incompatibility with chemical pesticides. Technologies such as, *Trichogramma chilonis* and *Crysoperla carnea* despite their proven effectiveness, do not find favour with industry as well as farmers because of their short shelf-life, sensitivity to chemical pesticides and higher cost of application. Plant-based pesticides are often slow in action. This suggests that the research should target overcoming these technological problems. Genetic manipulation of seed varieties for pest resistance is an important constituent of plant protection strategy. Genetically modified varieties of some crops, such as cotton and rapeseed-mustard, have been developed but these are surrounded by controversies regarding their long-term effect on the environment and human beings. Nevertheless, genetic resistance could be an effective tool in pest management.

Public-Private Sector Interface

While most of the technologies have been developed by the public sector, private sector does not find investment in commercial production mainly because of their short shelf-life and stochastic pest behaviour. Most of the bio-pesticides are produced by the public sector firms. These hardly comprise 2 per cent of the agrochemical market. Further, the pesticide has been biased towards chemicals, and views bio-pesticides as a threat to the existing chemical industry. Moreover, the firms engaged in production and promotion of bio-pesticides face stiff competition from the pesticide industry.

There is no denying the fact that transition from chemicals to bio-pesticides would be less remunerative in the short run. But, in view of global concerns of environmental conservation and rising consumer awareness about food quality, the industry has to switch over to bio-pesticides to harness the emerging opportunities. Nevertheless there is considerable scope to promote bio-pesticide industry as a small-scale industry with use of local resources, but with strict quality control.

Economic Feasibility

Scientists claim IPM to be an effective way of protecting the crops against insect pests. The claims are based on controlled experimental evidences and its wide scale testing under field conditions is yet to prove its economic feasibility. It's environmental and health benefits are well recognized. But farmers in the developing countries have a myopic view, and heavily discount the environmental and health benefits. They adopt a new technology only if it generates as much economic returns as the current technology. Evidences on economic feasibility are limited and scattered. Nevertheless, these indicate IPM as profitable as chemical control. Thus, in order to make IPM acceptable under field conditions, it is necessary to demonstrate its economic worth through large scale on-farm trials. In other words, there is a need for greater integration of biological and social science research.

Area-wide Adoption

There is hardly any information available on area protected with IPM. Estimates based on production statistics of bio-pesticides indicate that only about 1 per cent of the gross cropped area receives application of IPM inputs. One of the major impediments is the lack of availability of bio-pesticides and information thereon to the farmers. IPM is akin to a new technology and farmers often resist its adoption because of risk aversion. Further, as indicated above, many of the bio-pesticides are slow in action and are sensitive to chemicals. Since pest is a detrimental common property resource, it requires common action for its effective management. Application of chemicals in the neighbourhood of IPM farms reduces the effectiveness of IPM. But, the technological characteristics of bio-pesticides are such that demand greater involvement of community for realizing their full potential. The current efforts are largely individual-centered. The future of IPM would largely be determined by the community participation. There is a need to devise an 'incentive system' for the farmers who participate in community pest management. Involvement of local administrative units (*Panchayats*) and Non-Governmental Organizations (NGOs) could be of great help in pushing IPM forward.

Agricultural Extension

India has a well-developed agricultural extension system. It has, however, not been tuned to the emerging technological requirements of the farmers. Extension personnel often lack awareness on the IPM inputs in terms of their technological characteristics, application rates and method of application. In recent years though considerable efforts have been made to train extension personnel in IPM, the required skills have not percolated down to the farmers. A system of reward and punishment for extension personnel should be devised.

Funding

The current efforts to promote IPM are largely on account of the initiatives of the Government of India through its Central Integrated Pest Management Centres. Under the national programme for promotion of IPM, the state governments are required to allocate at least 50 per cent of the plant protection funds for promotion of IPM.

Regulations

The production of bio-pesticides is controlled by the same regulations as applicable to that of chemical pesticides. The process of registration is often cumbersome and costly. This encourages small entrepreneurs to undertake production of bio-pesticides. Moreover, there are more than 150 pesticides registered for use in agriculture. There are many pesticides that have been banned in the developed countries, but these are freely available in India. Bio-pesticides require entirely a different set of registration norms. It is, therefore advisable to relax registration norms for bio-pesticides considering heir environmental and health benefits. Banning hazardous pesticides would help emergence of bio-pesticide industry.

Food Security and Quality

Until recently, food security has been an overriding policy concern. Now with sufficient stocks of food grains, this has dissipated. A few years back it was apprehended that reduction in pesticide-use would adversely affect the production of food as well as non-food crops. And this might endanger the food security. Recent evidences, however, have indicated that gradual reduction in pesticide-use may not have much adverse effects on agricultural productivity. Further, there is a rising awareness about the food safety, particularly among the rich consumers. These concerns are going to be stronger in the future. Promotion IPM besides ensuring environmental protection would also ensure production of quality food. At present, IPM is not very popular among the farmers. A number of technological, socio-economic, institutional and infrastructure related factors are responsible for this. The success of IPM would be determined by the extent to which these constraints are alleviated.

The Future of Integrated Pest Management

As the pest management industry and academic researchers continue to study pests and discover better ways to combat them, IPM will evolve into an even better pest management methodology. From hand held data-collection technology that enables real-time tracking and analysis of pest-activity data, to the isolation of a cockroach sex pheromone that may lead to better lures and traps, the future of IPM is one of the smarter and more targeted treatments. Just as medical innovations have allowed surgeons to perform more complex procedures with fewer after-effects, so IPM enables pest management professionals to isolate and control pest problems with minimal unintended impact. In health care facilities, where tolerance for both pests and pesticides is extremely low, such precision is a must.

IPM in the future will emphasize biological and ecological knowledge in managing pests. Pest control methods are evolving and diversifying in response to public awareness of environmental and health impacts of synthetic chemical pesticides and resulting legislation. The strong growth of the organic foods market— 20% annual expansion for the past several years— may also be a factor in the accelerated development of organic pest management methods. Agricultural pests are developing resistance to many synthetic agrichemicals, and new synthetic chemicals are being registered at a slower rate than in the past. This situation has helped open the market for a new generation of bio-pesticides.

Most important developments have been anticipated in novel IPM tactics, with both plant and insect (natural enemy) products of genetic engineering heading the list. Exploitation of hitherto underused natural enemies (such as entomopathogenic nematodes), development of novel biorational pesticides, management of resistance to extend the useful life of essential pesticides and of transgenic plants, and clever use of semiochemicals all bode a vastly expanded repertoire of tactical options to enrich extant IPM strategies. The strategic innovations are fewer, but some show much promise. One that provides new opportunities for ecosystem level IPM is the strategic approach of area wide IPM.

The ease and speed of information dissemination through the Internet will certainly impact IPM in the next century (IPMNet, 1997). The promise of reliable predictive models based on real-time weather data may finally become reality, as programs and weather information are accessible through the Web. Extension information and educational programs via the Web are already available and, when cleverly used, have provided positive results.

The excitement about genetic engineering, however, dominates the futurist literature in IPM. If there is a lesson to learn from the past 35 years, it is that a silver bullet is unlikely to come out of any of the new technologies, and nothing would have been learned from the past if genetic engineering is emphasized over all other technologies that are also blossoming. IPM has proven to be a robust construct. It does not need qualifiers such as biointensive or ecologically based to convey the gamut of opportunities for innovators in the laboratories, the experimental plots, or on the farms. Opportunities to be creative within the confines of the original IPM formulation are limitless. Review of the vast literature on IPM confirms that success has come from a fundamental understanding of the ecology of crop/pest interactions, rarely from a revolutionarily new control tactic. The advance of IPM to higher levels of integration will hinge on the depth of understanding of agro-ecosystem structure and dynamics. As the agro-ecosystem is the result of the overlap of the ecological and socioeconomic scales, no technological innovation will be adopted unless it contributes to producers' economic goals and meets the requisites for acceptance by society.

The future of Integrated Pest Management is closely allied to the future of systems devoted to production and distribution of plant and animal commodities. As a sustainable approach combining biological, cultural, physical, and chemical tools to regulate pest populations while minimizing economic, environmental, and human health risks, IPM combines essential aspects of efficacy and safety to meet expectations of those who produce and market commodities and of the general public. It is beyond doubt IPM as an essential aspect of all production and distribution systems for plant and animal commodities implies much broader awareness and acceptance by those working with these commodities than currently exists. Certainly, education of those working in production and distribution systems regarding alternatives to pesticides must be a priority. Integrated control programs with employment of cultural measures, host resistance, and biological agents along with attention to appropriate decision-making processes for application of pesticides must be emphasized.

IPM needs to be emphasized as the means for effective regulation of pest species that stresses safety for the human population and for the environment in which we live. Our basic contention that the future of IPM is closely allied to the future of plant and animal production systems must be appreciated by the general public. The public must view development and implementation of IPM as a sound investment of resources.

In a trend that is often considered to have gained impetus with publications such as "Silent Spring" (Carson, 1962) and "Only One Earth" (Ward and Dubos, 1972), concerns for food safety, environmental contamination, endangered species, and safety of farm workers continue to escalate. Another area of concern regarding continued use of chemical pesticides is the growing problem of contamination of ground and surface water supplies (Christensen and Rea, 1993). Economic considerations often rank foremost among factors that influence decisions relating to pest management. These considerations have been basic to IPM since its inception in the earliest definitions for the economic threshold and economic injury level (Stern *et al.*, 1959). These parameters have not only served as criteria for decisions regarding the use of chemical pesticides in many plant and animal production systems, but have also helped to define the goal of effective pest management. Rather than an insistence on complete eradication of pests, an understanding is developing among practitioners of IPM that maintaining pest populations below the economic injury level by use of all available control methods is indeed successful control. There remains, however, a perception that using economic thresholds as criteria for management decisions increases the risk of economic losses. While progress has been made in defining simple economic thresholds and economic injury levels, especially for key insect pests, much research is needed for development of comprehensive thresholds for insect pests, pathogens, and weeds. Also, there has been relatively little accomplished in terms of integrating economic considerations relating to pest management activities into comprehensive benefit/cost analyses for agricultural enterprises. Also, the balance among social, environmental, and economic considerations has been a central theme of IPM since its inception. This theme is consistent with expectations for the entire system of agricultural production. The extent to which balance among these considerations can be maintained depends greatly on understanding and acceptance of IPM methods by the public and the degree to which principles of IPM are implemented in both rural and urban settings.

The scope of IPM implementation will continue to change greatly in the future. What began as limited approaches to promote integration of control methods for key pests and provide guidelines for decisions regarding pesticide applications is evolving to an ecologically-based systems approach (Higley and Wintersteen, 1992; Cate and Hinckle, 1994). Where there was once limited cooperation among disciplines, now there are teams comprised of scientists from several disciplines that work with consultants and producers to incorporate effective pest management into production systems. The complexity of these systems requires a broad base of expertise to address issues such as timing and efficacy of pest control recommendations; dealing with the ever-growing problem of pesticide resistance; addressing issues of worker safety; avoiding pollution of soils and water by agrichemicals; and assuring that foods contain no harmful residues of pesticides. While these issues must be dealt with for the forseeable future, it is important to assure an effective transition from pest control programs with heavy reliance on chemical pesticides to ecologically-based biointensive IPM. A recent report by the Consumers Union states that the way we control pest problems is at a crossroads with the necessity of decreasing reliance on chemical-dependent approaches becoming steadily more evident (Benbrook *et al.*, 1996). An organizational strategy must be adopted to increase acceptance of integrated approaches by; 1) providing intensive educational outreach involving coordinated efforts of extension specialists, industry representatives, and private consultants, and 2) emphasizing regional and national coordination for preparation and dissemination of educational materials.

Those who develop and implement IPM technology must be proactive in design of information delivery systems. This challenge has several aspects including effective communication of research information to extension specialists to promote the formulation and delivery of integrated programs; providing timely advisories of the status of pest populations so essential to adoption of ecologically-based programs by IPM practitioners; and feedback to scientists from practitioners regarding effectiveness of programs.

Modest programs that have begun in schools to promote environmental awareness in a more global sense may be expanded to teach young people how to preserve the quality of urban environments where they live. The grower meetings, field days, and fact sheets which have been the mainstay of agricultural extension work will never reach the volume of people necessary for educational outreach in towns and cities. Much work remains to prepare the organizational framework and enlist the personnel required for this educational effort.

In summary, it is concluded that the principles of IPM are essential to sustainability of agricultural production systems. Implementation of IPM programs is essential to profitable production of a safe and wholesome food supply while minimizing hazards of environmental pollution that could threaten entire ecosystems. To make the transition from chemically-based to ecologically-based pest management strategies, production and processing systems must be aggressive in developing the human and information resources required. Only through enhancement of research and educational resources will safety and profitability of food and fiber production and processing be assured. A unique opportunity exists now for IPM to bring great benefits to both agricultural and urban environments.

References

Benbrook, C.M., Groth, E. III, Halloran, J.M., Hansen, M.K. and Marquardt, S. 1996. Pest management at the crossroads. Consumers Union. Yonkers, NY.

Carson, R. 1962. Silent Spring. Fawcett Crest, New York, NY.

Cate, J.R. and Hinkle, M.K. 1994. Integrated pest management: the path of a paradigm. National Audubon Society. Washington, D.C.

Christensen, S. and Rea, A. 1993. Ground water quality in the Oklahoma City area. In W.A. Alley (Ed.). Regional ground-water quality. Van Nostrand Reinhold, New York, NY.

Higley, L.G. and Wintersteen, W. 1992. A new approach to environmental risk assessment of pesticides as a basis for incorporating environmental costs into economic injury levels. *Am. Entomol.* 38:34-39.

IPM Net, 1997. Directory of IPM Resources. Geneva, NY : Consort.Int. Crop Prot. *http://www.ipmnet.org/DIR*

Stern, V.M., Smith, R.F., Van den Bosch, R. and Hagen, K.S. 1959. The integrated control concept. *Hilgardia.* 29:81-101.

Ward, B. and Dubos, R. 1972. Only one earth: the care and maintenance of a small planet. George J. McLeod Ltd., Toronto, Ontario, Canada.

❑❑❑

Index

A

B

D

E

F

G

H

I

J

K

L

M

N

O

P

Q

R

S

T

V

W

X

Y

□□□